Lecture Notes in Computer Science

Lecture Notes in Computer Science

Vol. 23: Programming Methodology. 4th Informatik Symposium, IBM Germany Wildbad, September 25–27, 1974. Edited by C. E. Hackl. VI, 501 pages. 1975.

Vol. 24: Parallel Processing. Proceedings 1974. Edited by T. Feng. VI, 433 pages. 1975.

Vol. 25: Category Theory Applied to Computation and Control. Proceedings 1974. Edited by E. G. Manes. X, 245 pages. 1975.

Vol. 26: GI-4. Jahrestagung, Berlin, 9.–12. Oktober 1974. Herausgegeben im Auftrag der GI von D. Siefkes. IX, 748 Seiten. 1975.

Vol. 27: Optimization Techniques. IFIP Technical Conference. Novosibirsk, July 1–7, 1974. (Series: I.F.I.P. TC7 Optimization Conferences.) Edited by G. I. Marchuk. VIII, 507 pages. 1975.

Vol. 28: Mathematical Foundations of Computer Science. 3rd Symposium at Jadwisin near Warsaw, June 17–22, 1974. Edited by A. Blikle. VII, 484 pages. 1975.

Vol. 29: Interval Mathematics. Procedings 1975. Edited by K. Nickel. VI, 331 pages. 1975.

Vol. 30: Software Engineering. An Advanced Course. Edited by F. L. Bauer. (Formerly published 1973 as Lecture Notes in Economics and Mathematical Systems, Vol. 81) XII, 545 pages. 1975.

Vol. 31: S. H. Fuller, Analysis of Drum and Disk Storage Units. IX, 283 pages. 1975.

Vol. 32: Mathematical Foundations of Computer Science 1975. Proceedings 1975. Edited by J. Bečvář. X, 476 pages. 1975.

Vol. 33: Automata Theory and Formal Languages, Kaiserslautern, May 20–23, 1975. Edited by H. Brakhage on behalf of GI. VIII, 292 Seiten. 1975.

Vol. 34: GI – 5. Jahrestagung, Dortmund 8.–10. Oktober 1975. Herausgegeben im Auftrag der GI von J. Mühlbacher. X, 755 Seiten. 1975.

Vol. 35: W. Everling, Exercises in Computer Systems Analysis. (Formerly published 1972 as Lecture Notes in Economics and Mathematical Systems, Vol. 65) VIII, 184 pages. 1975.

Vol. 36: S. A. Greibach, Theory of Program Structures: Schemes, Semantics, Verification. XV, 364 pages. 1975.

Vol. 37: C. Böhm, λ-Calculus and Computer Science Theory. Proceedings 1975. XII, 370 pages. 1975.

Vol. 38: P. Branquart, J.-P. Cardinael, J. Lewi, J.-P. Delescaille, M. Vanbegin. An Optimized Translation Process and Its Application to ALGOL 68. IX, 334 pages. 1976.

Vol. 39: Data Base Systems. Proceedings 1975. Edited by H. Hasselmeier and W. Spruth. VI, 386 pages. 1976.

Vol. 40: Optimization Techniques. Modeling and Optimization in the Service of Man. Part 1. Proceedings 1975. Edited by J. Cea. XIV, 854 pages. 1976.

Vol. 41: Optimization Techniques. Modeling and Optimization in the Service of Man. Part 2. Proceedings 1975. Edited by J. Cea. XIII, 852 pages. 1976.

Vol. 42: James E. Donahue, Complementary Definitions of Programming Language Semantics. VII, 172 pages. 1976.

Vol. 43: E. Specker und V. Strassen, Komplexität von Entscheidungsproblemen. Ein Seminar. V, 217 Seiten. 1976.

Vol. 44: ECI Conference 1976. Proceedings 1976. Edited by K. Samelson. VIII, 322 pages. 1976.

Vol. 45: Mathematical Foundations of Computer Science 1976. Proceedings 1976. Edited by A. Mazurkiewicz. XI, 601 pages. 1976.

Vol. 46: Language Hierarchies and Interfaces. Edited by F. L. Bauer and K. Samelson. X, 428 pages. 1976.

Vol. 47: Methods of Algorithmic Language Implementation. Edited by A. Ershov and C. H. A. Koster. VIII, 351 pages. 1977.

Vol. 48: Theoretical Computer Science, Darmstadt, March 1977. Edited by H. Tzschach, H. Waldschmidt and H. K.-G. Walter on behalf of GI. VIII, 418 pages. 1977.

Vol. 49: Interactive Systems. Proceedings 1976. Edited by A. Blaser and C. Hackl. VI, 380 pages. 1976.

Vol. 50: A. C. Hartmann, A Concurrent Pascal Compiler for Minicomputers. VI, 119 pages. 1977.

Vol. 51: B. S. Garbow, Matrix Eigensystem Routines – Eispack Guide Extension. VIII, 343 pages. 1977.

Vol. 52: Automata, Languages and Programming. Fourth Colloquium, University of Turku, July 1977. Edited by A. Salomaa and M. Steinby. X, 569 pages. 1977.

Vol. 53: Mathematical Foundations of Computer Science. Proceedings 1977. Edited by J. Gruska. XII, 608 pages. 1977.

Vol. 54: Design and Implementation of Programming Languages. Proceedings 1976. Edited by J. H. Williams and D. A. Fisher. X, 496 pages. 1977.

Vol. 55: A. Gerbier, Mes premières constructions de programmes. XII, 256 pages. 1977.

Vol. 56: Fundamentals of Computation Theory. Proceedings 1977. Edited by M. Karpiński. XII, 542 pages. 1977.

Vol. 57: Portability of Numerical Software. Proceedings 1976. Edited by W. Cowell. VIII, 539 pages. 1977.

Vol. 58: M. J. O'Donnell, Computing in Systems Described by Equations. XIV, 111 pages. 1977.

Vol. 59: E. Hill, Jr., A Comparative Study of Very Large Data Bases. X, 140 pages. 1978.

Vol. 60: Operating Systems, An Advanced Course. Edited by R. Bayer, R. M. Graham, and G. Seegmüller. X, 593 pages. 1978.

Vol. 61: The Vienna Development Method: The Meta-Language. Edited by D. Bjørner and C. B. Jones. XVIII, 382 pages. 1978.

Vol. 62: Automata, Languages and Programming. Proceedings 1978. Edited by G. Ausiello and C. Böhm. VIII, 508 pages. 1978.

Vol. 63: Natural Language Communication with Computers. Edited by Leonard Bolc. VI, 292 pages. 1978.

Vol. 64: Mathematical Foundations of Computer Science. Proceedings 1978. Edited by J. Winkowski. X, 551 pages. 1978.

Vol. 65: Information Systems Methodology, Proceedings, 1978. Edited by G. Bracchi and P. C. Lockemann. XII, 696 pages. 1978.

Vol. 66: N. D. Jones and S. S. Muchnick, TEMPO: A Unified Treatment of Binding Time and Parameter Passing Concepts in Programming Languages. IX, 118 pages. 1978.

Vol. 67: Theoretical Computer Science, 4th GI Conference, Aachen, March 1979. Edited by K. Weihrauch. VII, 324 pages. 1979.

Vol. 68: D. Harel, First-Order Dynamic Logic. X, 133 pages. 1979.

Vol. 69: Program Construction. International Summer School. Edited by F. L. Bauer and M. Broy. VII, 651 pages. 1979.

Vol. 70: Semantics of Concurrent Computation. Proceedings 1979. Edited by G. Kahn. VI, 368 pages. 1979.

Vol. 71: Automata, Languages and Programming. Proceedings 1979. Edited by H. A. Maurer. IX, 684 pages. 1979.

Vol. 72: Symbolic and Algebraic Computation. Proceedings 1979. Edited by E. W. Ng. XV, 557 pages. 1979.

Vol. 73: Graph-Grammars and Their Application to Computer Science and Biology. Proceedings 1978. Edited by V. Claus, H. Ehrig and G. Rozenberg. VII, 477 pages. 1979.

Vol. 74: Mathematical Foundations of Computer Science. Proceedings 1979. Edited by J. Bečvář. IX, 580 pages. 1979.

Vol. 75: Mathematical Studies of Information Processing. Proceedings 1978. Edited by E. K. Blum, M. Paul and S. Takasu. VIII, 629 pages. 1979.

Vol. 76: Codes for Boundary-Value Problems in Ordinary Differential Equations. Proceedings 1978. Edited by B. Childs et al. VIII, 388 pages. 1979.

Lecture Notes in Computer Science

Edited by G. Goos and J. Hartmanis

112

CAAP '81

Trees in Algebra and Programming
6th Colloquium
Genoa, March 5–7, 1981
Proceedings

Edited by E. Astesiano and C. Böhm

Springer-Verlag

Editors
Egidio Astesiano
Istituto di Matematica
Via L.B. Alberti 4, 16132 Genova, Italy

Corrado Böhm
Istituto Matematico
"Guido Castelnuovo", Piazzale Aldo Moro 5, 00185 Roma, Italy

AMS Subject Classifications (1979): 68E10, 68B15
CR Subject Classifications (1981): 5.32, 4.34

ISBN 3-540-10828-9 Springer-Verlag Berlin Heidelberg New York
ISBN 0-387-10828-9 Springer-Verlag New York Heidelberg Berlin

Printing and binding: Beltz Offsetdruck, Hemsbach/Bergstr.
2145/3140-543210

FOREWORD

The preceding five Colloquia on Trees in Algebra and Programming were held in Lille (France), under the name of "Colloque de Lille sur les Arbres en Algèbre et en Programmation", starting in 1976.

Though the title looks restrictive, trees and related algebraic structures enter in almost every conceptual structure of Computer Science, so that "Trees in Algebra and Programming" includes a wide range of topics. A prominent feature of the presented papers is their mathematical character and so contributions related to algebra, mathematical logic and arithmetical complexity are also included.

The 6th Colloquium took place in Genoa (Italy), 5-7 March 1981, organized by E.Astesiano and G.Costa, under the sponsorship of the European Association for Theoretical Computer Science (EATCS).

The Program Committee consisted of C.Böhm (Rome, chairman), A.Arnold (Poitiers), E.Astesiano (Genova), A.Bertoni (Milano), St.L.Bloom (Hoboken), G.Cousineau (Paris), M.Dauchet (Lille), M.Dezani Ciancaglini (Torino), Ph.Flajolet (Paris), A.Maggiolo-Schettini (Pisa), K.Mehlhorn (Saarbrücken), J.Winkowski (Warszawa).

Of a total of 61 submitted papers, 21 were selected by the Program Committee, helped by the following referees: L.Aiello, K.R.Apt, G.Ausiello, A.Blilke, E.Börger, D.P.Bovet, R.Castanet, M.Coppo, R.Cori, G.Costa, B.Courcelle, S.Crespi Reghizzi, W.Damm, P.Della Vigna, J.Engelfriet, H.Ehrig, M.Fontet, I.Guessarian, J.R.Hindley, K.Indermark, M.Karpinski, L.Kott, J.J.Lévy, F.Luccio, A.Machì, D.Mandrioli, G.Mauri, P.Miglioli, J.C.Raoult, J.D.Rutledge, A.Salvicki, M.Soria, J.Thatcher, R.Tindell, J.Tiuryn, P.Torrigiani, G.Uccella, M.Vanneschi, B.Vauquelin, M.Venturini Zilli, H.Wedde.

This volume, together with the accepted papers, contains also four invited lectures by H.Ehrig, R.Milner, J.Nievergelt and M.Nivat.

We gratefully acknowledge the financial support provided by the following institutions and firms:

- Consiglio Nazionale delle Ricerche (Comitato per la Matematica, Gruppo Nazionale di Sistemistica e di Informatica dell'Ingegneria, Gruppo Nazionale per l'Informatica Matematica, Progetto Finalizzato Informatica P1, CNET, METOD)
- Istituto di Matematica, Università di Genova
- Burroughs Italiana Spa.
- Cassa di Risparmio di Genova ed Imperia.

Finally we wish to express our gratitude to the other members of the Institute of Mathematics of the University of Genoa, who helped in the organization.

Egidio Astesiano

Istituto di Matematica
Università di Genova

Corrado Böhm

Istituto Matematico "G.Castelnuovo"
Università di Roma

March 1981

CONTENTS

ALGEBRAIC THEORY OF PARAMETERIZED SPECIFICATIONS WITH REQUIREMENTS

Hartmut Ehrig

Technical University Berlin, FB 20
Institute for Software and Theoretical Computer Science
D-1000 Berlin 10, Germany (West)

ABSTRACT

Parameterized specifications of abstract data type are studied within the theory of algebraic specifications. In the algebraic theory as introduced by the ADJ-group a parameterized specification, like set(data), consists of a parameter declaration data and a target specification set(data). This basic algebraic approach is combined with a very general notion of requirements which have to be satisfied for the parameters of the specification. Especially we can use fixed basic types like bool or nat in the parameter part, a feature which is already included in the algebraic specification language CLEAR. This allows to specify bounded types like bounded natural numbers nat(bound) with variable bound or bounded arrays. Moreover the requirement feature allows to use arbitrary predicate formulas which are also used in logical requirement specifications for software systems. In spite of this generality the theory developed for the basic algebraic approach can be fully extended to the case with requirements. The basic result is an extension lemma which allows to show correctness of parameter passing and associativity of nested parameterized specifications like set(stack(nat)). Correctness of such composite specifications is automatically induced by correctness of the parts.
This theory with requirements is still based on initial algebra semantics but with slight modifications it can also be used for final algebraic semantics.

PRELIMINARY REMARK

Within the past five years algebraic specifications have been proved to be a powerful method for the specification of abstract data types in programming languages and software systems.
Trees - the main subject of this colloquium - are playing a fundamental role in syntax and semantics of algebraic specifications because terms of operations are trees. It was one of the great merits of the ADJ-group to point out that the initiality of termalgebras is the key to study abstract data types independent of their particular tree representation. The universal properties of initial algebras are still reflecting most of the fundamental properties of termalgebras.
Having this in mind the theory of algebraic specifications is mainly a theory on universal properties of trees.

1. INTRODUCTION

One of the most important issues within the theory of algebraic specifications is the specification of parameterized data types. Most common data types like stacks, queues, arrays and sets are in fact parameterized types stack(attr), queue(par), array(item) and set(data) respectively. The key idea is to consider the parameter parts attr, par, item and data as formal algebraic specifications which can be actualized by other predefined algebraic specifications like nat, int or bool. Similar to procedures in programming languages the process to replace formal by actual parameters is called parameter passing. Hence parameter passing allows to obtain from one parameterized specification like set(data) the three value specifications set(nat), set(int) and set(bool) corresponding to sets of natural numbers, sets of integers and sets of boolean values. Moreover we can also pass another parameterized specification like array(item) as actual parameter leading to the parameterized value specification set*array(item). The benefit of this process is not only economy in presentation but we also have automatically correctness of all the value specifications provided that the parameterized specification set(data) and all the actual specifications nat, int, bool and array(item) are correct. This is a most important property in order to build up larger data types and software systems from small pieces in a correct way. Similar to procedures in programming languages parameterized specifications promise to become one of the most important structuring principle for the design of software systems.

A closer look at these problems shows that we need precise notions of syntax, semantics, correctness and parameter passing for parameterized specifications to be able to show induced correctness of the value specifications and some other desirable properties. One of these properties is certainly associativity of composite parameterized specifications: Since parameterized specifications include unparameterized ones (taking the parameter part to be empty) we would like to be sure that the value specification set*array(nat), where nat is passed as actual parameter for item in set*array(item), is the same as set(array(nat)), where array(nat) is passed as parameter for data in set(data).

Precise notions for syntax, semantics and correctness of parameterized specifications were first given in /ADJ 78/. While syntax of parameter passing was first studied in /Ehr 78/ and /EL 79/ the semantics of parameter passing including all the results mentioned above was presented in /ADJ 80a+b/. In the following we will refer to this approach as "the basic algebraic case" which is reviewed in Section 2. In this approach a parameterized specification PSPEC=⟨SPEC,SPEC1⟩ consists of a pair of specifications where the parameter declaration SPEC is included in the target specification SPEC1. Parameter passing from the formal parameter SPEC to an actual parameter SPEC' is given by a "specification morphism" f:SPEC⟶ SPEC'. The value specification SPEC1' is more or less a "renaming" of the target specification SPEC1 where the SPEC-parts of SPEC1 are renamed by the corresponding SPEC'-parts of the actual parameter. Mathematically SPEC1' is the pushout object of SPEC1 and SPEC' via f in the category CATSPEC of algebraic specifications and specification morphisms. Although the theory in the basic algebraic case is very smooth and elegant it turns out that the applicability to common data types in software practice is somewhat limited. In several applications we need an equality predicate on the formal parameter, like EQ:data data⟶ bool in the parameterized specification set(data). In order to show correctness of such specifications we need requirements for the operation EQ making sure that EQ is really an equality predicate on all admissable parameter algebras A, i.e. EQ_A(d,d')=if d=d' then TRUE else FALSE for all

$d,d' \in A_{data}$. Unfortunately there seem to be no equations but only negative conditional axioms to assure this property, e.g. EQ(X,X)=TRUE and $X \neq Y \Rightarrow EQ(X,Y)=FALSE$ (see /ADJ 78/). Moreover we have to make sure that the bool-part of A consists exactly of two distinct elements TRUE and FALSE. The most convenient way to obtain these properties is the requirement "initial(bool)" which makes sure that the bool-part of A is isomorphic to the initial boolean algebra T_{bool}.
Such requirements are called "initial restrictions" in /Rei 80/ and "constraints" in /BG 80/ which become special cases of our more general notion of requirements. In this paper a set R is called "set of requirements" on a specification SPEC if for all $r \in R$ there is a well-defined subclass VALID(r) of all SPEC-algebras. This very general definition is easy to handle and allows also to state all kinds of predicate formulas as requirements, especially the negative conditional axioms for EQ as given above. Hence a parameterized specification of set(data) with requirements can be given as follows where bool is some correct specification of boolean values including TRUE, FALSE and an IF-THEN-ELSE-operation.

1.1 EXAMPLE (set(data))

```
PARAMETER DECLARATION: data =
    bool +
    sorts:  data
    opns:   EQ: data data ⟶ bool

REQUIREMENTS:
    initial(bool)
    EQ(X,X)=TRUE
    X≠Y ⇒ EQ(X,Y)=FALSE

TARGET SPECIFICATION: set(data) =
    data +
      sorts: set
      opns:  CREATE: ⟶ set
             INSERT: data set ⟶ set
             DELETE: data set ⟶ set
             MEMBER: data set ⟶ bool
             EMPTY: set ⟶ bool
             IF-THEN-ELSE: bool set set ⟶ set
      eqns:  INSERT(d,INSERT(d's))=IF EQ(d,d')THEN
                      INSERT(d,s)ELSE INSERT(d',INSERT(d,s))
             DELETE(d,CREATE)=CREATE
             DELETE(d,INSERT(d',s))=IF EQ(d,d')THEN
                      DELETE(d,s)ELSE INSERT(d',DELETE(d,s))
             MEMBER(d,CREATE)=FALSE
             MEMBER(d,INSERT(d',s))=IF EQ(d,d')THEN TRUE
                                    ELSE MEMBER(d,s)
             EMPTY(CREATE)=TRUE
             EMPTY(INSERT(d,s))=FALSE
             IF TRUE THEN s1 ELSE s2=s1
             IF FALSE THEN s1 ELSE s2=s2
```

This parameterized specification set(data) with requirements is similar to that in /ADJ 78/ except of different requirement handling which has significant consequences for the semantics (see 3.9.2).

Another important feature of parameterized specifications with requirements is the possibility to specify bounded data types such as arrays of fixed bounded length B. The bound B, however, is supposed to be a parameter which may take different values for different actual parameters. In most cases it seems to be convenient to construct the specification of bounded types as an extension of bounded natural numbers nat(bound) where nat is some correct specification of natural numbers with operations O, SUCC and ADD.

1.2 EXAMPLE (nat(bound))

```
PARAMETER DECLARATION: bound =
    nat +
      opns: BOUND:⟶ nat

REQUIREMENTS:
    initial(nat)

TARGET SPECIFICATION: nat(bound) =
    bound +
      sorts: bnat
      opns:  MOD: nat ⟶ bnat
             MODO:⟶ bnat
             MODSUCC: bnat⟶ bnat
             MODADD: bnat bnat⟶ bnat
      eqns:  MOD(ADD(BOUND,n))=MOD(n)
             MODO=MOD(O)
             MODSUCC(MOD(n))=MOD(SUCC(n))
             MODADD(MOD(n1),MOD(n2))=MOD(ADD(n1,n2))
```

Remark: Note that the requirement initial(nat) implies that BOUND picks out some well-defined natural number B. Without the initiality requirement BOUND may define a new value which does not correspond to any natural number. The semantics of our specification is $\mathbb{N}$ mod B (natural numbers modulo B). See 3.9.1 for more details.

Finally let us give an overview of Sections 3-5. In Section 3 we introduce requirements and parameterized specifications with requirements and we study problems of correctness. The main results of our theory are given in Section 4 where parameter passing with requirements is studied. The key result is an R-EXTENSION LEMMA which generalizes the EXTENSION LEMMA of /ADJ 80b/ to the case with requirements. As direct consequences of this lemma we are able to show correctness of parameter passing and associativity as well as induced correctness of composite parameterized specifications provided that the given parameterized specifications are"persistent". Actually we are able to extend all the results known in the basic algebraic case to the case with requirements. In Section 5 we sketch possibilities for further development. Especially it is worthwhile to mention that with slight modifications in the semantics our approach with requirements can also be used for parameterized specifications with final algebra semantics as studied in /Gan 80b/ without requirements. Moreover parameter passing with requirements allows to handle identification of common subtypes in an easy way, an issue which has caused considerable difficulties in the semantics of CLEAR.

2. THE BASIC ALGEBRAIC CASE

We shall assume the algebraic background of ADJ /76-78/, /EKP 78/ or /Kre 78/ but we will review the most important notions in connection with this paper. For basic terminology of category theory like categories and functors we refer to /AM 75/. We shall introduce the basic algebraic case of parameterized data types and specifications as given in /ADJ 78/. Moreover we study standard parameter passing as in /ADJ 80a+b/ but with simplified parameter passing morphisms.

An abstract data type is regarded as (the isomorphism class of) a many-sorted (heterogeneous) algebra which is minimal, meaning that all data elements are "accessible" using constants and operations of the algebra. A many-sorted algebra consists of an indexed family of sets (called carriers) with an indexed family of operations between those carriers. The indexing system is called a signature and consists of a set S of sorts which indexes the carriers and a family $\langle \Sigma_{w,s} | w \in S^* \text{ and } s \in S \rangle$ of operation names (Σ is called the operator domain); a symbol $\sigma \in \Sigma_{w,s}$ with w=s1...sn names an operation $\sigma_A : A_{s1} \times \ldots \times A_{sn} \longrightarrow A_s$ in an algebra A with signature Σ. The pair $\langle S, \Sigma \rangle$ determines the category $\underline{Alg}_{\langle S,\Sigma \rangle}$ of all S-sorted Σ-algebras with Σ-homomorphisms between them.

A specification, SPEC=$\langle S,\Sigma,E \rangle$, is a triple where $\langle S,\Sigma \rangle$ is a signature and E is a set of equations. $\underline{Alg}_{SPEC}$ is the category of all SPEC-algebras, i.e., all S-sorted Σ-algebras satisfying the equations E. When we write the combination SPEC'=SPEC+$\langle S',\Sigma',E' \rangle$ we mean that S and S' are disjoint, that Σ' is an operator domain over S+S' which is disjoint from Σ, and that E' is a set of axioms over the signature $\langle S+S', \Sigma+\Sigma' \rangle$.

Although some authors see the equations as "semantics", we follow /ADJ 76-78/ in saying that the semantics of a specification SPEC is the (isomorphism class of the) algebra T_{SPEC} which is initial in $\underline{Alg}_{SPEC}$. T_{SPEC} can be constructed as a quotient $T_{SPEC} = T_{\langle S,\Sigma \rangle / E}$ of the term algebra $T_{\langle S,\Sigma \rangle}$ (corresponding to the signature $\langle S,\Sigma \rangle$) by the congruence generated from the equations E.

A specification SPEC=$\langle S,\Sigma,E \rangle$ is called correct with respect to a model algebra A in $\underline{Alg}_{MSPEC}$ if the model specification MSPEC=$\langle MS,M\Sigma,ME \rangle$ is included in SPEC, i.e. $MS \subseteq S$, $M\Sigma \subseteq \Sigma$ and $ME \subseteq E$, and the MSPEC-reduct of T_{SPEC} is isomorphic to A. (The MSPEC-reduct of T_{SPEC} consists of those carriers and operations belonging to sorts of MS and operation symbols of $M\Sigma$ respectively.) Note that this definition allows to use "hidden functions" which are included in the specification but not in the model specification. Moreover in most cases ME will be the empty set of equations.

Now let us consider parameterized data types and specifications:

2.1 DEFINITION

A parameterized data type PDAT=$\langle SPEC, SPEC1, T \rangle$ consists of the following data:

PARAMETER DECLARATION	SPEC=$\langle S,\Sigma,E \rangle$
TARGET SPECIFICATION	SPEC1=SPEC+$\langle S1,\Sigma1,E1 \rangle$

and a functor $T:\underline{Alg}_{SPEC} \longrightarrow \underline{Alg}_{SPEC1}$. PDAT is called persistent (strongly persistent) if T is, i.e. for every SPEC-algebra A, we have $V \circ T(A) \cong A$ (resp. $V\,T(A)=A$) where V is the forgetful functor from SPEC1 to SPEC-algebras (see 2.5.3).

Remark: If T is equipped with a (natural) family of homomorphisms $\langle I_A : A \to V(T(A)) \rangle$ such that for each SPEC-algebra A, the set $\{I_A(a) \mid a \in A\}$ generates T(A) then PDAT is called abstract parameterized data type. (This generalizes the condition that an algebra has to be minimal to be considered as abstract data type.) As discussed in /ADJ 78/, the family I tells how to find each parameter algebra A in the result of the construction T(A). Each abstract parameterized data type must be equipped with such a natural transformation but not the model data types. The motivation for persistence is given in /ADJ 78/; the idea is that the parameter algebra "persists" (up to isomorphism) in the result of the construction T. In contrast to /ADJ 78/ and /ADJ 80a+b/ we only allow equations in the specifications but negative conditional and universal Horn axioms may be included in the requirements.

To illustrate our definitions we will construct the model functor SETO corresponding to a simplified version set0(data0) of our set specification in Example 1.1 where only the operations CREATE and INSERT are considered. In this simplified version we do not need the equality predicate and hence also not bool in the parameter declaration.

2.2 EXAMPLE

PARAMETER DECLARATION (SPEC=$\langle S,\emptyset,\emptyset\rangle$):data0 =
sorts(S): data

TARGET SPECIFICATION (MSPEC1=SPEC+$\langle S1,\Sigma 1,\emptyset\rangle$):Mset0 =
data0 +
sorts(S1): set
opns(Σ1): CREATE: $\to$ set
INSERT: data set $\to$ set

The model functor SETO:$\underline{Alg}_{data0} \to \underline{Alg}_{Mset0}$ takes each data0-algebra E, which is simply a set of parameter elements, to the Mset0-algebra $A=\langle A_{data}, A_{set}, CREATE_A, INSERT_A\rangle$ with $A_{data}=E$, $A_{set}=\mathcal{P}_{fin}(E)$ the set of all finite subsets of E, $CREATE_A=\emptyset$ and $INSERT_A(e,s)=\{e\}\cup s$ for all $e\in E$ and $s\in\mathcal{P}_{fin}(E)$. The model functor is strongly persistent because we have $V(SETO(E))=V(A)=A_{data}=E$. In the following we shall show that the simplified version set0(data0) of our set specification is correct with respect to the parameterized model data type PMDAT=$\langle$data0,Mset0,SETO$\rangle$ defined above.

2.3 DEFINITION

1. A parameterized specification PSPEC=$\langle$SPEC,SPEC1$\rangle$ consists of the following data:

PARAMETER DECLARATION SPEC=$\langle S,\Sigma,E\rangle$

TARGET SPECIFICATION SPEC1=SPEC+$\langle S1,\Sigma 1,E1\rangle$

The semantics of the specification is the free construction (see /ADJ 78/), $F:\underline{Alg}_{SPEC} \to \underline{Alg}_{SPEC1}$, i.e., the (abstract) parameterized type PDAT=$\langle$SPEC,SPEC1,F$\rangle$.

Remark: We will talk about the "parameterized type$\langle$SPEC,SPEC1$\rangle$" and mean the type whose (model) functor is the free construction from SPEC-algebras to SPEC1-algebras.

2. Let PDAT=$\langle$MSPEC,MSPEC1,T$\rangle$ be a parameterized data type and PSPEC=$\langle$SPEC,SPEC1$\rangle$ a parameterized specification. Then PSPEC is called correct with respect to PDAT if we have MSPEC$\subseteq$ SPEC, MSPEC1$\subseteq$ SPEC1 and (up to isomorphism) $T\circ U=U1\circ F$ with surjective forgetful functor $U:\underline{Alg}_{SPEC} \to \underline{Alg}_{MSPEC}$, forgetful functor $U1:\underline{Alg}_{SPEC1} \to \underline{Alg}_{MSPEC1}$ and $F:\underline{Alg}_{SPEC} \to \underline{Alg}_{SPEC1}$ the semantics (free construction) of PSPEC.

Remark: If U and U1 are identity functors correctness means that the free construction F is equal to the given model functor T. Otherwise they have to be equal up to renaming and forgetting of those sorts and operations which are in SPEC1 but not in MSPEC1. Surjectivity of U (which is not assumed in /ADJ 78/ and /ADJ 80a+b/ makes sure that for each model parameter algebra in $\underline{Alg}_{MSPEC}$ there is also a corresponding parameter algebra in $\underline{Alg}_{SPEC}$.

2.4 FACT

The parameterized specification setO(dataO)- given by dataO and setO=MsetO+E1 as in Example 2.2 and E1 consisting of the equations INSERT(d,INSERT(d,s))=INSERT(d,s) and INSERT(d,INSERT(d',s))=INSERT(d',INSERT(d,s)) - is correct with respect to the parameterized type PMDAT=⟨dataO,MsetO,SETO⟩ (see 2.2).

Proof: In our case U and U1 are identity functors and it remains to show that the functor SETO considered as functor from $\underline{Alg}_{dataO}$ to $\underline{Alg}_{setO}$ is the free construction with respect to the forgetful functor $V:\underline{Alg}_{setO} \longrightarrow \underline{Alg}_{dataO}$. This means that we have to show for each $B \in \underline{Alg}_{setO}$ and each function $f:E \longrightarrow V(B)$ that there is a unique setO-morphism $g:SETO(E) \longrightarrow B$ with $g_{data}=f$. Since g must become a setO-morphism we have for the setO-component of $g:g(\emptyset)=g(CREATE_A)=CREATE_B$ and $g(\{e\} \cup s)=g(INSERT_A(e,s))= INSERT_B(f(e),g(s))$. Using the INSERT-equations for $INSERT_B$ it is easy to show that the equations above define a well-defined setO-morphism $g:SETO(E) \longrightarrow B$ with $g_{data}=f$.
□

We consider now the problem of paramter passing which was sketched already in the introduction. We need a mechanism which allows us to replace the formal parameters, given by the parameter declaration of a parameterized specification, by actual parameters, given by actual specifications. This mechanism will be called "standard parameter passing". The problem of"parameterized parameter passing" where the actual parameters are parameterized specifications will be studied in Section 4.
The main problem for parameter passing is to develop suitable morphisms, called "parameter passing morphisms", from the formal to the actual parameters taking into account possible renamings and/or identifications of sorts and operations. Assume that we have the actual parameter nat (natural numbers with O,SUCC and ADD). Then there is an obvious "specification morphism" h:dataO⟶ nat which identifies the sort data with the sort nat in nat. The result of standard parameter passing is the value specification setO(nat)=nat+⟨S1,h(Σ1),h(E1)⟩ with S1,Σ1 and E1 as given in 2.2 and 2.4 respectively. h(Σ1) and h(E1) means a translation of operations Σ1 and equations E1 where the sort data is translated into nat. Mathematically this translation is given as pushout object in the category CATSPEC of specifications and specification morphisms:

2.5 DEFINITION

1. A specification morphism $h:\langle S,\Sigma,E\rangle \longrightarrow \langle S',\Sigma',E'\rangle$ consists of a mapping $h_S:S \longrightarrow S'$ and an $(S^* \times S)$-indexed family of mappings, $h_\Sigma:\Sigma \longrightarrow \Sigma'$ (where $h_{\Sigma(w,s)}:\Sigma_{w,s} \rightarrow \Sigma_{h_S(w),h_S(s)}$). This data is subject to the condition that every equation of E, when translated by h, belongs to E', short $h(E) \subseteq E'$. The morphism h is called simple if $\langle S,\Sigma,E\rangle \subseteq \langle S',\Sigma',E'\rangle$

and h_S, h_Σ are inclusions.

2. The category of all specifications and specificatin morphisms is called CATSPEC. If s:SPEC⟶ SPEC1 is a simple and h:SPEC⟶ SPEC' an arbitrary specification morphism the following diagram is called <u>parameter passing diagram</u> or <u>pushout</u> in CATSPEC.

$$\begin{array}{ccc} SPEC & \xrightarrow{\ s\ } & SPEC1 \\ {\scriptstyle h}\downarrow & & \downarrow{\scriptstyle h'} \\ SPEC' & \xrightarrow[\ s'\ ]{} & SPEC1' \end{array}$$

where h is given as above, s and s' are simple specification morphisms and SPEC1', called <u>value specification</u>, is defined by

$$SPEC1' = SPEC' + \langle S1', \Sigma 1', E' \rangle,$$

with $S1' = S1, \Sigma 1' = h'(\Sigma 1)$, and $E1' = h'(E1)$ where

$$h' : SPEC1 \longrightarrow SPEC1'$$

is a specification morphism defined by

$$h'_S(x) = \text{if } x \in S1 \text{ then } x \text{ else } h_S(x) \text{ and}$$
$$h'_\Sigma(y) = \text{if } y \in \Sigma 1 \text{ then } y \text{ else } h_\Sigma(y).$$

<u>Remark</u>: In the EXTENSION-LEMMA of /ADJ 80b/ it is shown that the parameter passing diagram defined above satisfies the universal pushout properties in the category CATSPEC.

3. For each specification morphism h:SPEC⟶ SPEC' there is a functor $V_h : \underline{Alg}_{SPEC'} \longrightarrow \underline{Alg}_{SPEC}$, called <u>forgetful functor</u> with respect to h, defined for all $A' \in \underline{Alg}_{SPEC'}$ by $V_h(A') = A$ with $A_s = A'_{h(s)}$ for all $s \in S$ and $\sigma_A = h_\Sigma(\sigma)_{A'}$ for all $\sigma \in \Sigma$. $V_h(f')$ is defined by $V_h(f')_s = f'_{h(s)}$ for all $s \in S$.

In the following we define standard parameter passing as in /ADJ 80a+b/ where, however, unlike /ADJ 80a+b/ the parameter passing morphisms are specification morphisms. This simplifies the theory of /ADJ 80a+b/ considerably but restricts the applicability of the basic algebraic case.

2.6 <u>DEFINITION (Standard Parameter Passing)</u>

Given a parameterized specification PSPEC=⟨SPEC,SPEC1⟩, a specification SPEC', called <u>actual parameter</u>, and a specification morphism h:SPEC⟶ SPEC', called <u>parameter passing morphism</u>, then the <u>value specification</u> SPEC1' is given as pushout object (see 2.5.2) in the following parameter passing diagram

$$\begin{array}{ccc} SPEC & \xrightarrow{\ s\ } & SPEC1 \\ {\scriptstyle h}\downarrow & & \downarrow{\scriptstyle h'} \\ SPEC' & \xrightarrow[\ s'\ ]{} & SPEC1' \end{array}$$

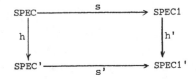

The mechanism of <u>standard parameter passing</u> is called <u>correct</u> if the following two conditions are satisfied:

1. actual parameter protection, i.e. $V_{s'}(T_{SPEC1'}) = T_{SPEC'}$
2. passing compatibility, i.e. $F \circ V_h(T_{SPEC'}) = V_{h'}(T_{SPEC1'})$

where $T_{SPEC'}$ and $T_{SPEC1'}$ are initial algebras and F the semantics of PSPEC (see 2.3).

Interpretation: The value specification SPEC1', also written SPEC1(SPEC') like set0(nat), is the result of replacing the formal parameter SPEC in SPEC1, also written SPEC1(SPEC) like set0(data0), by the actual parameter SPEC'.
Actual parameter protection means that the actual parameter SPEC' is protected in the value specification SPEC1'. This means in our example that the nat-part of the initial semantics of set0(nat) is equal to the initial nat-algebra $\mathbb{N}$ (natural numbers).
Passing compatibility means that the semantics of parameter passing, especially the transformation from $T_{SPEC'}$ to $T_{SPEC1'}$ is compatible with the semantics F of PSPEC.
This means in our example that we have $SETO(\mathbb{N}) = V_{h'}(T_{set0(nat)})$.

The main result for standard parameter passing is the following:

2.7 THEOREM (Correctness of Standard Parameter Passing)

Standard parameter passing is correct (with respect to all actual parameters SPEC' and all parameter passing morphisms h:SPEC⟶ SPEC') if and only if the given parameterized specification PSPEC=⟨SPEC,SPEC1⟩ is (strongly) persistent.

Remark: The proof of this theorem is given in Theorem 6.2 of /ADJ 80b/ and the if-part follows also from Corollary 4.6 in this paper.

The concept of parameterized parameter passing, where actual parameters are parameterized specifications, associativity and induced correctness of composite parameterized specifications will be studied in Section 4 immediately for the case with requirements.

3. PARAMETERIZED SPECIFICATIONS WITH REQUIREMENTS

In this section we shall introduce a notion of requirements such that specifications with fixed types like bool in set(data) (see 1.1) and bounded types like nat(bound) can be formulated in a comprehensive way. We define requirements on a fully abstract level. We only assume to have an abstract set R, called set of requirements on SPEC, such that for each r∈R there is a well-defined subset VALID(r) of all SPEC-algebras. Since we do not want to include the definition of VALID(r) in the requirement part of our specifications we consider three special types of requirements which seem to be sufficient for most of our applications: predicate formulas, initial restrictions and functor image restrictions. The latter ones turn out to be special cases of "functor restriction"-requirements which are generalizing the initial restrictions in /Rei 80/ and the constraints in /BG 80/. Using the fully abstract notion of requirements we are going to define parameterized specifications and data types with requirements and study problems of semantics and correctness. Parameter passing will be considered in Section 4.

3.1 DEFINITION (Requirements)

1. A set R is called set of requirements on a specification SPEC if for each r∈R

there is an assignment to a subset VALID(r) of all SPEC-algebras. For each subset R' of R we define

$$\text{VALID}(R') = \bigcap_{r \in R'} \text{VALID}(r)$$

The full subcategory of $\underline{Alg}_{SPEC}$ with objects VALID(R) is denoted $\underline{Alg}_{RSPEC}$ where RSPEC=⟨SPEC,R⟩ (see also 3.6).

2. For each specification morphism f:SPEC1⟶ SPEC2 and all requirements r on SPEC1 the translation r_f is a requirement on SPEC2 with

$$\text{VALID}(r_f) = \{A2 \in \underline{Alg}_{SPEC2} \mid V_f(A2) \in \text{VALID}(r)\}$$

For a set R of SPEC1-requirements we have in the obvious way $R_f = \{r_f / r \in R\}$. If f is simple (inclusion) we also write R instead of R_f provided that it is clear whether R has to be considered as a set of requirements on SPEC1 or on SPEC2.

Remark: If R1 and R2 are sets of requirements on SPEC then also the disjoint union R1+R2 is a set of requirements on SPEC with VALID(R1+R2)=VALID(R1)∩VALID(R2).

3.2 EXAMPLE (Predicate Formulas)

Each set R of predicate formulas on the signature ⟨S,Σ⟩ is a set of requirements on SPEC=⟨S,Σ,E⟩ where for each predicate formula r∈R we define

$$\text{VALID}(r) = \{A \in \underline{Alg}_{SPEC} \mid A \text{ satisfies } r\}$$

For most applications it seems to be sufficient to consider first order predicate formulas but such a restriction is not significant for our theory. Especially this type of requirements includes negative conditional axioms as used in Example 1.1.

Other important types of requirements are initial restrictions and functor image restrictions which will be used in 3.10-3.12.

3.3 DEFINITION (Initial and Functor Image Restrictions)

1. Given a simple specification morphism (inclusion) s:SPEC0⟶ SPEC1 then the requirement initial(SPEC0) is called initial restriction on SPEC1 and we define

$$\text{VALID}(\text{initial}(SPEC0)) = \{A1 \in \underline{Alg}_{SPEC1} \mid V_s(A1) = T_{SPEC0}\}$$

2. Given a specification morphism f:SPEC⟶ SPEC1, sets R and R1 of requirements on SPEC and SPEC1 respectively, and a persistent functor $RT:\underline{Alg}_{RSPEC} \to \underline{Alg}_{RSPEC1}$ then the requirement image(RT) is called functor image restriction on SPEC1 and we define

$$\text{VALID}(\text{image}(RT)) = \{A1 \in \underline{Alg}_{SPEC1} \mid \exists A \in \text{VALID}(R) \text{ with } RT(A) \cong A1\}$$
$$= \{A1 \in \underline{Alg}_{SPEC1} \mid V_f(A1) \in \text{VALID}(R) \text{ and } A1 \cong RT(V_f(A1))\}$$

Remark: Note that the last equation holds because persistency of RT means $V_f(RT(A)) \cong A$ such that we have $A \cong V_f(A1)$. The notion RSPEC1 means ⟨SPEC1,R1⟩ (see 3.1.1).

Initial restrictions and functor image restrictions turn out to be special cases of functor restrictions in the following sense:

3.4 DEFINITION AND REMARK (Functor Restrictions)

Let f:SPEC⟶ SPEC0 and f1:SPEC0⟶ SPEC1 be specification morphisms, R and R0 sets of requirements on SPEC and SPEC0 respectively, $RT:\underline{Alg}_{RSPEC} \to \underline{Alg}_{RSPEC0}$ a functor with RSPEC0=⟨SPEC0,R0⟩ ,then (RT,f,f1) is called functor restriction on SPEC1.

(RT,f,f1) becomes a requirement on SPEC1 if we define

$$\mathrm{VALID}(RT,f,f1)=\{A1\in \underline{Alg}_{SPEC1} / V_f(V_{f1}(A1))\in \mathrm{VALID}(R) \text{ and } V_{f1}(A1)=RT(V_f \circ V_{f1}(A1))\}$$

An initial restriction can be regarded as a functor restriction in the case SPEC=∅, R=RO=∅, f and f simple and $F:\underline{Alg}_{\emptyset}\longrightarrow \underline{Alg}_{SPECO}$ the free functor assigning to the empty (initial) algebra A∅ in $\underline{Alg}_{\emptyset}$ the initial SPECO-algebra T_{SPECO} in $\underline{Alg}_{SPECO}$. In this case we have VALID(initial(SPECO))=VALID(F,f,f1). Moreover a functor image restriction can be regarded as a functor restriction in the case SPECO=SPEC1, f simple, f1 identity and RT persistent because we have VALID(image(RT))=VALID(RT,f,f1).

Now we are able to define parameterized specifications with requirements in the abstract sense of Definition 3.1. But in all our applications we will only use those explicitly defined in 3.2 and 3.3.

3.5 DEFINITION (Parameterized R-Specifications)

A parameterized specification with requirements, short parameterized R-specification, PRSPEC=⟨SPEC,SPEC1,R⟩ consists of a parameterized specification PSPEC=⟨SPEC,SPEC1⟩ (in the sense of 2.3) and a set R of requirements on SPEC. The semantics of PRSPEC is the functor $RF_O=F\circ I$, where

$I:\underline{Alg}_{RSPEC}\longrightarrow \underline{Alg}_{SPEC}$	is the inclusion functor of the full subcategory $\underline{Alg}_{RSPEC}$ (see 3.1.1) of $\underline{Alg}_{SPEC}$
$F:\underline{Alg}_{SPEC}\longrightarrow \underline{Alg}_{SPEC1}$	is the free construction (see 2.3)

Remarks: The semantics of a parameterized R-specification is a restriction of the semantics F of the underlying parameterized specification PSPEC=⟨SPEC,SPEC1⟩ . Hence it is always well-defined. In the worst case $\underline{Alg}_{RSPEC}$ is empty and RF_O becomes the empty functor. The requirements R are only defined on SPEC. Even the translated requirements R on SPEC1 will not be valid for $RF_O(A)$ with $A\in\underline{Alg}_{RSPEC}$ in general. But they will be valid if RF_O is persistent. In this case RF_O can be restricted to a functor $RF:\underline{Alg}_{RSPEC}\longrightarrow \underline{Alg}_{RSPEC1}$ which will be constructed in 3.10. Theorem 3.12 will show that we can also use RF as semantics of PRSPEC which will be done in Section 4.

Examples of parameterized R-specifications were given already in the introduction. Bounded natural numbers nat(bound) using the requirement initial(nat) are given in 1.2, and sets of data set(data) using the requirements initial(bool) and negative conditional axioms for EQ are given in 1.1.

There are two interesting special cases of parameterized R-specifications. If R is empty then we have a parameterized specification (without requirements) as in Section 2. If in addition also SPEC is empty then we have usual algebraic specifications with initial algebra semantics. In 3.6 we shall consider another special case, namely SPEC=SPEC1.

3.6 DEFINITION (Requirement Specifications)

A requirement specification, short R-specification, RSPEC=⟨SPEC,R⟩ consists of a specification SPEC together with a set R of requirements on SPEC.

Remark: In contrast to an algebraic specification SPEC with initial algebra semantics T_{SPEC} the semantics of a requirement specification would be the full subcategory $\underline{Alg}_{RSPEC}$ of $\underline{Alg}_{SPEC}$, because F in 3.5 becomes the identity functor. This means that we do not have a specific semantical algebra in mind but only a class of algebras satisfying the given requirements R. This corresponds to requirement or functional specifications in the sense of software engineering (see /Flo 81/). Algebraic specifications with initial algebra semantics on the other hand correspond to design specifications in the sense of software engineering (see /Flo 81/). Hence parameter passing, as going to be defined in Section 4, in this special case becomes a formal technique to come from a requirement specification to another requirement specification or (in the case of standard parameter passing) to a design specification. Moreover correctness of parameter passing would imply the correctness of this specification technique. This problem will be studied in more detail in a subsequent paper.

In order to be able to define correctness of parameterized specifications with requirements we also have to introduce parameterized data types with requirements.

3.7 DEFINITION (Parameterized R-Data Types)

A parameterized data type with requirements, short parameterized R-data type, PRDAT= $\langle SPEC, SPEC1, R, RT_O\rangle$ consists of a parameterized R-specification PRSPEC= $\langle SPEC, SPEC1, R\rangle$ (see 3.5) and a functor $RT_O:\underline{Alg}_{RSPEC} \longrightarrow \underline{Alg}_{SPEC1}$. PRDAT and RT_O are called (strongly) persistent if we have $V_s(RT_O(A)) \cong A$ (resp. equal to A) for all $A \in \underline{Alg}_{RSPEC}$ where V_s is the forgetful functor associated with the inclusion $s:SPEC \longrightarrow SPEC1$.

Remarks: 1. Note, that in general the range of V_s is $\underline{Alg}_{SPEC}$ rather than $\underline{Alg}_{RSPEC}$
2. The semantics RF_O of a parameterized R-specification PRSPEC= $\langle SPEC, SPEC1, R\rangle$ becomes the following parameterized R-data type PRDAT= $\langle SPEC, SPEC1, R, RF_O\rangle$.

3.8 DEFINITION (Correctness)

Let PRDAT= $\langle MSPEC, MSPEC1, RM, RT_O\rangle$ be a parameterized R-data type and PRSPEC= $\langle SPEC, SPEC1, R\rangle$ a parameterized R-specification with semantics RF_O. Then PRSPEC is called correct with respect to PRDAT if $MSPEC \subseteq SPEC$ with surjective forgetful functor $RU:\underline{Alg}_{RSPEC} \longrightarrow \underline{Alg}_{RMSPEC}$ and $MSPEC1 \subseteq SPEC1$ with forgetful functor U1 such that the following diagram commutes (up to isomorphism).

$$\begin{array}{ccc} \underline{Alg}_{RMSPEC} & \xrightarrow{RT_O} & \underline{Alg}_{MSPEC1} \\ \uparrow RU & & \uparrow U1 \\ \underline{Alg}_{RSPEC} & \xrightarrow[RF_O]{} & \underline{Alg}_{SPEC1} \end{array}$$

where RMSPEC= $\langle MSPEC, RM\rangle$ and RSPEC= $\langle SPEC, R\rangle$.

3.9 EXAMPLE (Correctness of nat(bound) and set(data))

1. The bounded type nat(bound) given in 1.2 is correct with respect to the following parameterized R-data type

$\langle$ bound, Mnat(bound), initial(nat), NATMOD $\rangle$

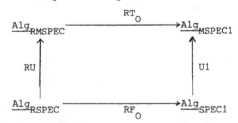

where Mnat(bound) is nat(bound) without equations for MOD, MODO, MODSUCC and MODADD and NATMOD is the following persistent functor NATMOD: $\underline{Alg}_{\text{Rbound}} \longrightarrow \underline{Alg}_{\text{Mnat(bound)}}$

$$NATMOD(\mathbb{N},B)_{\underline{bnat}} = \begin{cases} \mathbb{N} & \text{for } B=0 \\ \{0,\ldots,B-1\} & \text{for } B>0 \end{cases}$$

The Rbound-part of NATMOD($\mathbb{N}$,B) is ($\mathbb{N}$,B). For B=0 the operations MOD_B, $MODO_B$, $MODSUCC_B$ and $MODADD_B$ in NATMOD($\mathbb{N}$,B) are the identity resp. those of ($\mathbb{N}$,B). For B>0 we have $MOD_B(n)=n$ mod(B), $MODO_B=0$ mod(B) while $MODSUCC_B$ and $MODADD_B$ are the usual successor and addition functions mod(B). Hence $NATMOD(\mathbb{N},B)_{\underline{bnat}}=\mathbb{N}$ mod(B) for B>0.

Since we have MSPEC=SPEC and RM=M the forgetful functor RU is the identity and hence surjective. In order to show the correctness it remains to show that NATMOD($\mathbb{N}$,B) $\in \underline{Alg}_{\text{nat(bound)}}$ and that it is free over ($\mathbb{N}$,B) with respect to all nat(bound)-algebras A. But $A_{\underline{nat}}$ is not necessary isomorphic to $\mathbb{N}$ because nat(bound) does not include the requirement initial(nat). By construction we have NATMOD($\mathbb{N}$,B) $\in \underline{Alg}_{\text{nat(bound)}}$ and it remains to show that for each bound-morphism f:($\mathbb{N}$,B) $\longrightarrow$ V(A) there is a unique nat(bound)-morphism g:NATMOD($\mathbb{N}$,B) $\longrightarrow$ A s.t. the restriction of g to bound is equal to f. This means we have $g(0)=f(0)=0_A$ g(n)=f(n) for all $n \in \mathbb{N}$ and $g(B)=f(B)=BOUND_A$. Since g must be a nat(bound)-morphism we have for the bnat-component of g

$$g(n \bmod(B))=g(MOD_B(n))=MOD_A(g(n))=MOD_A(f(n))$$

But this is already a well-defined definition for $g_{\underline{nat}}$. Actually n mod(B)=n' mod(B) implies w.l.o.g. $n=n'+m\cdot B$. Hence f bound-morphism and the MOD-equations for A imply $MOD_A(f(n))=MOD_A(f(n'+m\cdot B))=MOD_A(f(n')+_A m\cdot f(B))=$

$$=MOD_A(f(n')+_A m\cdot BOUND_A)=MOD_A(f(n'))$$

Now it is easy to check that the equations for MODO, MODSUCC and MODADD in nat(bound) imply that g is a nat(bound)-morphism.

2. The parameterized R-specification set(data) in Example 1.1 is correct with respect to the following parameterized R-data type:

PMDAT=⟨Mdata, Mset, SET, R⟩ with Mdata=bool+{data}, R={initial(bool)}, Mset=Mdata+ ⟨{set}, {CREATE, INSERT, DELETE, MEMBER, EMPTY}⟩ and SET: $\underline{Alg}_{\text{RMdata}} \longrightarrow \underline{Alg}_{\text{Mset}}$ is the functor which takes each RMdata-algebra A (which has $A_{\underline{bool}}=T_{\underline{bool}}$) to the Mset-algebra (also denoted A) with $A_{\underline{set}}=\mathcal{P}_{fin}(A_{\underline{data}})$ (finite subsets of $A_{\underline{data}}$) and with the expected operations $CREATE_A$ and $INSERT_A$ as in 2.2, $DELETE_A(a,s)=s-\{a\}$, MEMBER(a,s)=(if $a \in s$ then TRUE else FALSE) and EMPTY(s)=(if $s=\emptyset$ then TRUE else FALSE). In contrast to /ADJ 78/ our parameter algebras A have a fixed initial bool-part which allows to use Prop 14 in /ADJ 78/ as stated. (Note that it does not work if $A_{\underline{bool}}$ has more than two elements which should be forbidden in /ADJ 78/.) For similar reasons the semantics of our set(data) is persistent but not the semantics of the corresponding type in /ADJ 78/ because it does not include the requirement initial(bool).

Now we want to consider the special case of persistent parameterized R-data types and R-specifications. In this case the translated requirements R on SPEC1 are valid for all $RF_O(A)$ with $A \in \underline{Alg}_{RSPEC}$. Hence we can add the translated requirements R to the target specification SPEC1. For technical reasons (see 4.5) it is convenient to add also the functor image restriction image(RF_O) to SPEC1 such that $\underline{Alg}_{RSPEC1}$ consists exactly of all parameter algebras in the sense of /ADJ 80b/. Using functor image restrictions we can avoid the technical burden of generalized parameter passing which was necessary in /ADJ 80a+b/ to prove associativity of composite parameterized specifications.

Obviously we can restrict RF_O to a functor $RF:\underline{Alg}_{RSPEC} \longrightarrow \underline{Alg}_{RSPEC1}$ (see 3.10). This leads to a reformulation of persistent parameterized R-data tpyes and to the notion of persistent semantics which is convenient to use for the main results concerning parameter passing in Section 4.

3.10 <u>FACT(Persistent Restrictions)</u>

Let PRDAT=⟨SPEC,SPEC1,R,RT_O⟩ be a (strongly) persistent parameterized R-data type and RSPEC1=⟨SPEC1,R+image(RT_O)⟩ where R is the translated set of requirements on SPEC1 (see 3.1.2) and image(RT_O) is the functor image restriction (see 3.4.2). Then we have

1. VALID(image(RT_O)) $\subseteq$ VALID(R), s.t. $\underline{Alg}_{RSPEC1} = \underline{Alg}_{\langle SPEC1, image(RT_O)\rangle}$
2. $RT_O:\underline{Alg}_{RSPEC} \longrightarrow \underline{Alg}_{SPEC1}$ can be restricted to $RT:\underline{Alg}_{RSPEC} \longrightarrow \underline{Alg}_{RSPEC1}$ and the forgetful functor $V:\underline{Alg}_{SPEC1} \longrightarrow \underline{Alg}_{SPEC}$ can be restricted to $RV:\underline{Alg}_{RSPEC1} \longrightarrow \underline{Alg}_{RSPEC}$ such that $RV \bullet RT(A)=A$ for all $A \in \underline{Alg}_{RSPEC}$.

If in addition RT_O is the semantics RF_O of the (strongly) persistent R-specification PRSPEC=⟨SPEC,SPEC1,R⟩ ,i.e. $RT_O=RF_O$ is restriction of the free functor F with respect to V, then we have:

3. RF becomes a (strongly) persistent free functor with respect to RV.

Vice versa given PRSPEC and RSPEC1 as above such that the restriction RV of V exists but RF_O is not assumed to be (strongly) persistent, then we have:

4. A (strongly) persistent free functor RT with respect to RV is not necessary the restriction of the free functor F with respect to V.

<u>Proof:</u> 1. $A1 \in$ VALID(image(RT_O)) implies $V(A1) \in$ VALID(R) on SPEC by 3.3.2 and hence $A1 \in$ VALID(R) on SPEC1 by 3.1.2.

2. By 3.3.2 we have $RT_O(A) \in$ VALID(image(RT_O)) and hence $RT_O(A) \in$ VALID(R) by part 1 which implies $RT_O(A) \in \underline{Alg}_{RSPEC1}$ for all $A \in \underline{Alg}_{RSPEC}$. As in part 1 we have for each $B \in \underline{Alg}_{RSPEC1}$ $V(B) \in$ VALID(R) and hence $V(B) \in \underline{Alg}_{RSPEC}$. This means that V can be restricted to RV and $RV \circ RT(A)=A$ holds by persistency of RT_O.

3. For each $A \in \underline{Alg}_{RSPEC}$ the universal properties of RF(A) with respect to an arbitrary $B \in \underline{Alg}_{RSPEC1}$ are a direct consequence of the universal properties of F(A)=RF(A) with respect to the same $B \in \underline{Alg}_{SPEC1}$.

4. The proof in part 3 cannot be reversed because an object $B \in \underline{Alg}_{SPEC1}$ does not belong to $\underline{Alg}_{RSPEC1}$ in general. For a counter example let SPEC=$\langle \{s\}, \{0:\to s\}, \emptyset \rangle$,SPEC1=SPEC+$\langle \emptyset, \{SUCC:s \to s\}, \emptyset \rangle$ and R=$\{x=y\}$. Obviously each of

the categories $\underline{Alg}_{RSPEC}$ and $\underline{Alg}_{RSPEC1}$ has (up to isomorphism) only one algebra, say A and A1, which has cardinality 1. Hence $RT:\underline{Alg}_{RSPEC} \longrightarrow \underline{Alg}_{RSPEC1}$ with RT(A)=A1 is a persistent free functor with respect to RV satisfying RV(A1)=A. But the free functor $F:\underline{Alg}_{SPEC} \longrightarrow \underline{Alg}_{SPEC1}$ has $F(A) \cong \mathbb{N}$ such that RT is not the restriction of F.

□

3.11 DEFINITION (Persistent R-Data Types and Persistent Semantics)

1. If $PRDAT=\langle SPEC, SPEC1, R, RF_0 \rangle$ is a persistent parameterized R-data type we will also write $PRDAT=\langle SPEC, SPEC1, R, RF \rangle$ where $RF:\underline{Alg}_{RSPEC} \longrightarrow \underline{Alg}_{RSPEC1}$ is defined as in 3.10.

2. If RF_0 is persistent and the semantics of $PRSPEC=\langle SPEC, SPEC1, R \rangle$ then RF will be called (strongly) persistent semantics of PRSPEC.

Finally let us show that for persistent parameterized R-specifications it is equivalent whether we use the original semantics RF_0 or the persistent semantics RF to show the correctness with respect to a parameterized R-data type.

3.12 THEOREM (Implications of Persistency)

If $PRSPEC=\langle SPEC, SPEC1, R \rangle$ is a persistent parameterized R-specification and $PRDAT=\langle MSPEC, MSPEC1, RM, RT_0 \rangle$ a parameterized R-data type then we have:

1. If PRSPEC is correct with respect to PRDAT then also PRDAT and RT_0 are persistent such that RT_0 can be restricted to $\underline{Alg}_{RMSPEC1}$ with $RMSPEC1=\langle MSPEC1, RM+image(RT_0) \rangle$, i.e.

$$RT:\underline{Alg}_{RMSPEC} \longrightarrow \underline{Alg}_{RMSPEC1}$$

2. PRSPEC is correct with respect to PRDAT if and only if we have $MSPEC \subseteq SPEC$ with surjective RU and $MSPEC1 \subseteq SPEC1$ with restricted forgetful functor $RU1:\underline{Alg}_{RSPEC1} \longrightarrow \underline{Alg}_{RMSPEC1}$ such that the following diagram commutes (up to isomorphism)

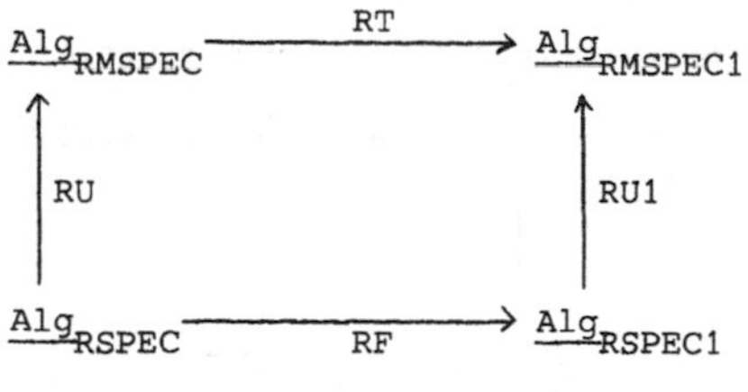

Proof: 1. RF_0 persistent implies RT_0 persistent, because RU is surjective and we have

$$V_t \circ RT_0 \circ RU(A) = V_t \circ U1 \circ RF_0(A) = U \circ U_s \circ RF_0(A) = U(A) = RU(A)$$

where V_t, V_s and U are the forgetful functors associated with the inclusions $t:MSPEC \longrightarrow MSPEC1$, $s:SPEC \longrightarrow SPEC1$, and $MSPEC \subseteq SPEC$.

2. RU1 is well-defined because for $A1 \in \underline{Alg}_{RSPEC1}$ we have $U1(A1)=U1 \circ RF_0(A)=RT_0 \circ RU(A)$. Hence the diagram in 3.8 is commutative if and only if that in 3.12.2 commutes.

4. PARAMETER PASSING WITH REQUIREMENTS

In this section we study parameter passing for parameterized R-specifications (specifications with requirements). We will show that constructions and results in the basic algebraic case can be extended to the case with requirements. We only need an additional semantical property, called "passing consistency". It means that the actual parameter satisfies the requirements of the given parameterized specification. In contrast to Section 2 we immediately allow parameterized specifications as actual parameters which corresponds to "parameterized parameter passing" in /ADJ 80b/. The main idea is to use the persistent semantics including functor image restrictions on the target specifications (see 3.10) and to consider only passing consistent parameter passing morphisms $h:SPEC \longrightarrow SPEC1'$ which means that the corresponding forgetful functor V_h is compatible with the requirements on SPEC1' and SPEC. The key lemma is a generalization of the EXTENSION LEMMA in /ADJ 80b/ to an R-EXTENSION LEMMA including requirements. This allows to show correctness of parameter passing with requirements including parameter protection and passing compatibility. Moreover we obtain associativity of iterated parameter passing and induced correctness of the value specifications.

As in the basic algebraic case we assume for all results persistency of the corresponding parameterized specifications. The basic definitions for parameter passing, however, are independent of persistency.

4.1 <u>DEFINITION (Parameter Passing with Requirements)</u>

Given parameterized R-specifications $PRSPEC=\langle SPEC,SPEC1,R\rangle$ and $PRSPEC'=\langle SPEC',SPEC1',R'\rangle$ and a specification morphism $h:SPEC \longrightarrow SPEC1'$, called <u>parameter passing morphism</u>, then the mechanism of <u>parameter passing with requirements</u> is given by the following syntax, semantics and correctness conditions:

<u>SYNTAX:</u> Let $s:SPEC \longrightarrow SPEC1$ be the simple specification morphism associated with PRSPEC and

$$\begin{array}{ccc} SPEC & \xrightarrow{\quad s \quad} & SPEC1 \\ {\scriptstyle h}\downarrow & & \downarrow{\scriptstyle h'} \\ SPEC1' & \xrightarrow[\quad t \quad]{} & SPEC2' \end{array}$$

the <u>parameter passing</u> diagram in CATSPEC (see 2.5.2) where h' is called <u>induced parameter passing morphism</u>. Then the <u>value specification</u> or <u>composite parameterized specification with requirements</u> is given by

$$PRSPEC *_h PRSPEC' = \langle SPEC', SPEC2', R'\rangle$$

<u>SEMANTICS:</u> The semantics is given by $(RF_O, RF'_O, RF_O *_h RF'_O)$, or short $RF_O *_h RF'_O : \underline{Alg}_{RSPEC'} \longrightarrow \underline{Alg}_{SPEC2'}$, where the three functors are the semantics of PRSPEC, PRSPEC' and $PRSPEC *_h PRSPEC'$ respectively (see 3.5).

The parameter passing morphism h is called <u>passing consistent</u> if we have $V_h(RF'_O(A')) \in \underline{Alg}_{RSPEC}$ for all $A' \in \underline{Alg}_{RSPEC'}$.

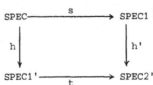

<u>CORRECTNESS:</u> <u>Parameter passing with requirements</u> is called <u>correct</u> if for all passing consistent parameter passing morphisms h:SPEC⟶ SPEC1' the following two conditions are satisfied:

1. <u>parameter protection (with requirements)</u>, i.e.

   $V_{t}\circ(RF_{O}*_{h}RF'_{O})=RF'_{O}$

2. <u>passing compatibility (with requirements)</u>, i.e.

   $V_{h'}\circ(RF_{O}*_{h}RF'_{O})=RF_{O}\circ(V_{h}\bullet RF'_{O})$

where the right hand side is well-defined because of passing consistency.

<u>Remark:</u> In the special case of standard parameter passing we have PRSPEC'=⟨∅,SPEC1',∅⟩. In this case the value specification is SPEC2' and passing consistency means $V_h(T_{SPEC1'})\in \underline{Alg}_{RSPEC}$, i.e. the initial algebra of the actual parameter SPEC1' satisfies the requirements R. For interpretation of the conditions see 2.6.

The main correctness result we want to obtain is the following:

4.2 <u>THEOREM (Correctness of Parameter Passing with Requirements)</u>

Parameter passing with requirements is correct for persistent parameterized R-specifications. In more detail: Given (strongly) persistent parameterized R-specifications PRSPEC=⟨SPEC,SPEC1,R⟩ and PRSPEC'=⟨SPEC',SPEC1',R'⟩ and a passing consistent specification morphism h:SPEC⟶ SPEC1' then we have:

1. (strong) persistency of the composition $PRSPEC*_{h}PRSPEC'$,
2. parameter protection (with requirements),
3. passing compatibility (with requirements).

<u>Remark:</u> The proof of this theorem is a corollary of the following R-EXTENSION LEMMA and will be given in 4.5 below.

4.3 <u>R-EXTENSION LEMMA</u>

Given a (strongly) persistent parameterized R-data type PRDAT=⟨SPEC,SPEC1,R,RF⟩ as in 3.11, an R-specification RSPEC'=⟨SPEC',R'⟩ and a <u>passing consistent morphism</u> h:RSPEC⟶ RSPEC', i.e. a specification morphism h:SPEC⟶ SPEC' such that the forgetful functor $V_h:\underline{Alg}_{SPEC'}\longrightarrow \underline{Alg}_{SPEC}$ can be restricted to $RV_h:\underline{Alg}_{RSPEC'}\longrightarrow\underline{Alg}_{RSPEC}$.

Further let SPEC1' be the pushout object of the simple specification morphism s:SPEC⟶ SPEC1 and h in CATSPEC with induced morphisms s':SPEC'⟶ SPEC1' and h':SPEC1⟶ SPEC1' as in 2.5.2. Then we have:

1. There is a (strongly) persistent functor $RF':\underline{Alg}_{RSPEC'}\longrightarrow \underline{Alg}_{RSPEC1'}$ with RSPEC1'=⟨SPEC1',R'+image(RF')⟩, called <u>extension of RF via h</u>, and a restriction $RV_{h'}:\underline{Alg}_{RSPEC1'}\longrightarrow \underline{Alg}_{RSPEC1}$ of the forgetful functor $V_{h'}$ such that

$$RV_{h'}\bullet RF'=RF\circ RV_{h}$$

Moreover RF'(A') is uniquely defined by $A'\in\underline{Alg}_{RSPEC'}$ and $B=RF\bullet RV_h(A')$ in the following sense:

For all $B'\in\underline{Alg}_{SPEC1'}$ satisfying $V_{s'}(B')=A'$ and $V_{h'}(B')=B$ we have already B'=RF'(A').

2. The following diagram is a pushout in the category CATRSPEC of R-specifications

and passing consistent morphisms (see above)

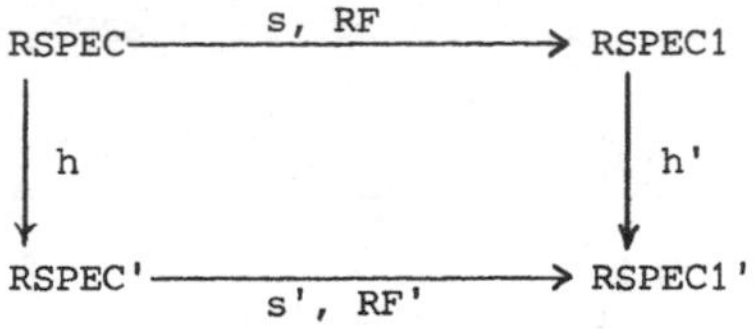

3. If RF is the persistent semantics of PRSPEC=⟨SPEC,SPEC1,R⟩ then also RF' is the persistent semantics of PRSPEC'=⟨SPEC',SPEC1',R'⟩.

Proof: 1. Similar to the proof of the EXTENSION LEMMA in /ADJ 80b/ we define for all $A' \in \underline{Alg}_{RSPEC'}$ $RF'_0(A')$ by $V_{s'} \circ RF'_0(A')=A'$ and $V_{h'} \circ RF'_0(A')=RF_0 \circ RV_h(A')$ such that RF'_0 becomes a well-defined (strongly) persistent functor $RF'_0:\underline{Alg}_{RSPEC'} \longrightarrow \underline{Alg}_{SPEC1'}$. According to the EXTENSION LEMMA $RF'_0(A')$ is uniquely defined by the property given above. By construction 3.10 RF'_0 can be restricted to become a functor $RF':\underline{Alg}_{RSPEC'} \longrightarrow \underline{Alg}_{RSPEC1'}$. It remains to show that $V_{h'}$ can be restricted to $RV_{h'}:\underline{Alg}_{RSPEC1'} \longrightarrow \underline{Alg}_{RSPEC1}$ which means that h' is a passing consistent morphism h':RSPEC1→RSPEC1'. For each $B' \in \underline{Alg}_{RSPEC1'}$ we have to show $V_{h'}(B') \in \underline{Alg}_{RSPEC1}$. $B' \in \underline{Alg}_{RSPEC1'}$ implies $B' \in VALID(image(RF'))$ which means $V_{s'}(B') \in VALID(R')$ and $B'=RF' \circ V_{s'}(B')$. Let $A'=V_{s'}(B') \in \underline{Alg}_{RSPEC'}$ then we have by construction of RF'_0 and RF'
$V_{h'}(B')=V_{h'} \circ RF' \circ V_{s'}(B')=V_{h'} \circ RF'(A')=RF \circ RV_h(A') \in \underline{Alg}_{RSPEC1}$.

2. In part 1 we have shown that h' is passing consistent. Moreover s and s' are passing consistent by definition of translated requirements (see 3.1.2). (This can be independently concluded from the functor image restrictions image(RF) resp. image(RF').) Now let RSPEC2=⟨SPEC2,R2⟩ be an arbitrary specification with requirements and k':RSPEC'⟶ RSPEC2, k1:RSPEC1⟶ RSPEC2 be passing consistent morphisms with $k' \circ h=k1 \circ s$. By the pushout property in CATSPEC we have a unique specification morphism k:SPEC1'⟶ SPEC2 such that $k \circ s'=k'$ and $k \circ h'=k1$. It remains to show that k is passing consistent. It sufficies to show that for each $A2 \in \underline{Alg}_{RSPEC2}$ we have $V_k(A2) \in \underline{Alg}_{RSPEC1'}$. Since $VALID(image(RF')) \subseteq VALID(R')$ we have to show $V_k(A2) =RF' \circ V_{k'}(A2)$. This means $B'=RF'(A')$ for $B'=V_k(A2)$ and $A'=V_{k'}(A2) \in \underline{Alg}_{RSPEC'}$ using passing consistency of k'. By the uniqueness property of part 1 it sufficies to show $V_{s'}(B')=A'$, which follows from $V_{s'} \circ V_k=V_{k'}$, and $V_{h'}(B')=RF \circ RV_h(A')$. This means $V_{h'} \circ V_k(A2)=RF \circ RV_h \circ V_{k'}(A2)$. The last equation is equivalent to $V_{k1}(A2)=RF \circ V_s \circ V_{k1}(A2)$ which means $V_{k1}(A2) \in VALID(image(RF))$ (see 3.3.2). But this follows from passing consistency of k1

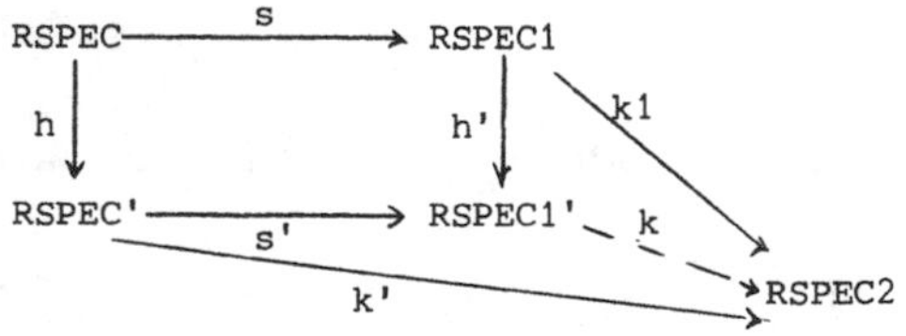

3. Let us assume that RF is the strongly persistent semantics of PRSPEC. This means

by 3.5 and 3.10 that we have for all $A \in \underline{Alg}_{RSPEC}$ $RF(A)=F(A)$ and $V_s \circ F(A)=A$ where F is the free construction $F:\underline{Alg}_{SPEC} \longrightarrow \underline{Alg}_{SPEC1}$. We have to show that RF' is the strongly persistent semantics of PRSPEC'. That means we have to show for all $A' \in \underline{Alg}_{RSPEC'}$ $RF'(A')=F'(A')$ and $V_{s'} \circ F'(A')=A'$ where F' is the free construction $F':\underline{Alg}_{SPEC'} \longrightarrow \underline{Alg}_{SPEC1'}$. Since we have a passing consistent morphism $h:RSPEC \longrightarrow RSPEC'$ we know that for each $A' \in \underline{Alg}_{RSPEC'}$ $V_h(A')$ belongs to $\underline{Alg}_{RSPEC}$. Hence we have $RF \circ RV_h(A')=F \circ V_h(A')$. From part 1 we know that RF'(A') satisfies $V_{s'}(RF'(A'))=A'$ and $V_{h'}(RF'(A'))=RF \circ RV_h(A')=F \circ V_h(A')$ where F is the free functor $F:\underline{Alg}_{SPEC} \longrightarrow \underline{Alg}_{SPEC1}$. Now part 3 and uniqueness in part 2 of the EXTENSION LEMMA implies that we have $RF'(A')=F'(A')$ where F'(A') is free with respect to $V_{s'}$.

□

4.4 NOTE (Freeness of Functor Extensions)

For part 3 of the proof above (or a similar direct proof) it is important that RF is the restriction of the free functor $F:\underline{Alg}_{SPEC} \longrightarrow \underline{Alg}_{SPEC1}$ such that RF(A) has the universal properties of F(A) with respect to V_s. For the proof that RF'(A') is free with respect to $V_{s'}$ in part 3 of the EXTENSION LEMMA we have to take an arbitrary object $B' \in \underline{Alg}_{SPEC1'}$ and have to use freeness of RF(A) with respect to $V_{h'}(B') \in \underline{Alg}_{SPEC1}$. We could not conclude the freeness of RF'(A') with respect to $V_{s'}$ if we would only know freeness of RF with respect to the restriction $RV_s:\underline{Alg}_{RSPEC1} \longrightarrow \underline{Alg}_{RSPEC}$ of V_s. (For a similar reason the pointwise arguments in Theorems 6.2 and 7.3 of /ADJ 80b/ are slightly incorrect, which was first recognized by Eric Wagner.) On the other hand our R-EXTENSION LEMMA would also go through with the assumption that RF is persistent and free with respect to RV_s, provided that RV_s is well-defined. In this case we could conclude that RF' is persistent with respect to a well-defined $RV_{s'}$. But in general we could not conclude $RF'(T_{SPEC'})=T_{SPEC1'}$ as required for standard parameter passing.

Now we return to the proof of Theorem 4.2.

4.5 PROOF OF THEOREM 4.2

Instead of the ordinary semantics RF_O and RF'_O we use the (strongly) persistent semantics $RF:\underline{Alg}_{RSPEC} \longrightarrow \underline{Alg}_{RSPEC1}$ and $RF':\underline{Alg}_{RSPEC'} \longrightarrow \underline{Alg}_{RSPEC1'}$ of PRSPEC and PRSPEC' respectively. Since RSPEC1' includes the functor image restriction image(RF_O) passing consistency of $h:SPEC \longrightarrow SPEC'$ is equivalent to the fact that $h:RSPEC \longrightarrow RSPEC1'$ is a passing consistent morphism (see 4.3). Now we are able to apply the R-EXTENSION LEMMA to PRDAT=⟨SPEC,SPEC1,R,RF⟩ and h. We obtain a (strongly) persistent functor $RG:\underline{Alg}_{RSPEC1'} \longrightarrow \underline{Alg}_{RSPEC2'}$ with $RV_{h'} \circ RG=RF \circ RV_h$ where RSPEC2'=⟨SPEC2',R'+ +image(RF')+image(RG)⟩ because RSPEC1'=⟨SPEC1',R'+image(RF')⟩ (see part1). Moreover part 3 implies that RG is the restriction of the free functor $G:\underline{Alg}_{SPEC1'} \longrightarrow \underline{Alg}_{SPEC2'}$ with respect to V_t.

1. Strong persistency of $PRSPEC *_h PRSPEC'$ means that the restriction $R(G \circ F')$ of the free functor $G \circ F':\underline{Alg}_{SPEC'} \longrightarrow \underline{Alg}_{SPEC2'}$ with respect to $V_{s'} \circ V_t$ is strongly persistent (where we have used the well-known fact that the composition $G \circ F'$ of the free

functors F' and G' is again free). By 3.11 we have $R(G\circ F'):\underline{Alg}_{RSPEC'}\longrightarrow\underline{Alg}_{R2SPEC2'}$ with $R2SPEC2'=\langle SPEC2',R'+image(R(G\circ F'))\rangle$. We will show $R(G\circ F')=RG\circ RF'$ such that (strong) persistency of RG and RF' implies that of $R(G\circ F')$. For all objects $A'\in\underline{Alg}_{RSPEC'}$ we have $RG\circ RF'(A')=G(F'(A'))=(G\circ F')(A')=R(G\circ F')(A')$ and similar for morphisms. It remains to show $\underline{Alg}_{RSPEC2'}=\underline{Alg}_{R2SPEC2'}$ which follows from

$$VALID(image(RF')+image(RG))=VALID(image(RG\circ RF'))=VALID(image(R(G\circ F')))$$

where we have used 3.3.2 repeatedly.

2. Parameter protection in terms of persistent semantics means

$$RV_t\circ(RF*_h RF')=RF'$$

where $RF*_h RF'$ is the persistent semantics of $PRSPEC*_h PRSPEC'$. By part 1 we have $RF*_h RF'=R(G\circ F')=RG\circ RF'$. Hence we have by strong persistency of RG

$$RV_t\circ(RF*_h RF')=RV_t\circ RG\circ RF'=RF'$$

3. Passing compatibility in terms of persistent semantics means

$$RV_{h'}\circ(RF*_h RF')=RF\circ(RV_h\circ RF')$$

Using again $RF*_h RF'=RG\circ RF'$ this follows from $RV_{h'}\circ RG=RF\circ RV_h$ (see above).

□

As a corollary of Theorem 4.2 we want to consider the case of standard parameter passing where the actual parameter is a (nonparameterized) specification SPEC1' with initial algebra semantics $T_{SPEC1'}$. Note that SPEC1' can be considered as persistent parameterized specification $PRSPEC'=\langle\emptyset,SPEC1',\emptyset\rangle$ with semantics RF defined by $RF(A\emptyset)=T_{SPEC1'}$ where $A\emptyset$ is the empty initial algebra in $\underline{Alg}_{\emptyset}$.

4.6 <u>COROLLARY (Standard Parameter Passing)</u>

Given a (strongly) persistent parameterized R-specification $PRSPEC=\langle SPEC,SPEC1,R\rangle$, an actual specification SPEC1' and a specification morphism $h:SPEC\longrightarrow SPEC1'$ satisfying $V_h(T_{SPEC1'})\in\underline{Alg}_{RSPEC}$ (passing consistency). Then the value specification SPEC2' is given as in 4.1 and we have:

1. $V_t(T_{SPEC2'})=T_{SPEC1'}$ (parameter protection)
2. $V_{h'}(T_{SPEC2'})=F(V_h(T_{SPEC1'}))$ (passing compatibility)

where $h':SPEC1'\longrightarrow SPEC2'$ is the induced parameter passing morphism of h and $F:\underline{Alg}_{SPEC}\longrightarrow\underline{Alg}_{SPEC1}$ the free functor.

<u>Proof:</u> Apply Theorem 4.2 with $PRSPEC'=\langle\emptyset,SPEC1',\emptyset\rangle$ where $RF'_0(A\emptyset)=T_{SPEC1'}$, $RF_0*_h RF'_0(A\emptyset)=T_{SPEC2'}$ (because free functors are preserving initial objects) and RF_0 is the restriction of F.

□

Another important consequence of the R-EXTENSION LEMMA is associativity of composite parameterized R-specifications.

4.7 <u>THEOREM (Associativity of Composition)</u>

Given persistent parameterized R-specifications $PRSPECi=\langle SPECi,SPECi',Ri\rangle$ for i=1,2,3 and passing consistent morphisms $h1:RSPEC1\longrightarrow RSPEC2'$ and $h2:RSPEC2\rightarrow RSPEC3'$

with induced morphism h2'. Then also the composition h2'∘h1 is passing consistent and we have

$$(PRSPEC1*_{h1}PRSPEC2)*_{h2}PRSPEC3=PRSPEC1*_{h2'\circ h1}(PRSPEC2*_{h2}PRSPEC3)$$

Remark: The proof is based on part 2 of 4.3 where it is shown that parameter passing diagrams are pushouts in the category CATRSPEC. This includes that passing consistent morphisms are preserved by pushouts and closed under composition. This implies that h2'∘h1 is passing consistent. The associativity of the composition is a corollary of the fact that (horizontal and vertical) composition of pushouts in a triangle of pushout squares is again a pushout. For more detail we refer to the proof of Theorem 9.1 in /ADJ 80b/.

□

A third important consequence of the R-EXTENSION LEMMA is the following result on induced correctness of composite parameterized R-specifications.

4.8 THEOREM (Induced Correctness of Composite Parameterized R-Specifications)

Given persistent parameterized R-specifications PRSPEC=⟨SPEC,SPEC1,R⟩ and PRSPEC'=⟨SPEC',SPEC1',R'⟩ which are correct with respect to PRDAT=⟨MSPEC,MSPEC1,RM, RT⟩ and PRDAT'=⟨MSPEC',MSPEC1',RM',RT'⟩. Further let h:SPEC⟶ SPEC1' be a passing consistent parameter passing morphism which can be restricted to k:MSPEC⟶ MSPEC1', and let RS:$\underline{Alg}_{RMSPEC1'}\longrightarrow \underline{Alg}_{RMSPEC2'}$ be the extension of RT via k. Then we have:

The composite parameterized R-specification $PRSPEC*_{h}PRSPEC'$ is correct with respect to $PRDAT*_{k}PRDAT'$:=⟨MSPEC',MSPEC2',RM',RS∘RT'⟩.

Remark: For the proof we can use the correctness characterization for persistent parameterized R-specifications given in 3.12.2. By assumption we have that the semantics RF and RF' of PRSPEC and PRSPEC' commutes with RT and RT' (and suitable forgetful functors) respectively. The pushout property in 4.3.2 implies that also the extension RG of RF via h commutes with the extension RS of RT via k. Combining the corresponding diagrams we obtain that RG∘RF' commutes with RS∘RT'. This implies the correctness of the composition as stated in the theorem because by 4.5.1 RG∘RF' is the semantics of the composition. For more detail we refer to the proof of Theorem 7.4 in /ADJ 80/.

□

5. FURTHER DEVELOPMENT

In this section we shall sketch some issues for further development of parameterized R-specifications.

First of all the special case of requirement specifications without parameters, as sketched in Remark 3.6, should be studied in more detail. This will allow to consider stepwise refinement of requirement specification (which are connected by passing consistent specification morphisms) and to bridge the gap between requirement and design specification, both of which are still important problems in software engineering. For this purpose it may also be useful to allow translation of requirements (see 3.1.2) along parameter passing morphisms while we only consider translations along simple specification morphisms associated with parameterized specifications (see 3.10-3.12). If we allow arbitrary translations of requirements in the parameter passing diagram of 4.1 then we can also translate requirements from the parameter declaration SPEC to the value specification SPEC2'. This corresponds exactly to what is done for equations in /Gan 80a+b/ where, however, final algebra semantics for parameterized specifications is considered.

Up to now we have only considered the initial algebra approach where the semantics of parameterized types is given by free constructions. In /Kla 78/ it is shown how to make use of inductively specified operations and in /Gan 80a+b/ how to extend the basic algebraic case of /ADJ 80a+b/ from initial to final algebra semantics: The main idea in /Gan 80a+b/ is to replace the free construction $F:\underline{Alg}_{SPEC} \longrightarrow \underline{Alg}_{SPEC1}$ by a quotient functor $CF:\underline{Alg}_{SPEC} \longrightarrow \underline{Alg}_{SPEC1}$ of F defined by $CF(A)=T_{\Sigma+\Sigma 1}(A)/\sim$ where $F(A)=T_{\Sigma+\Sigma 1}(A)/\equiv$. In /Gan 80b/ the relation $\sim$ is defined by $t\sim t'$ for $t,t' \in T_{\Sigma+\Sigma 1}(A)_s$ $(s \in S+S1)$, if for any variable term $\hat{t}(\cdot) \in T_{\Sigma+\Sigma 1}(A+\{\cdot\})_{s'}$, with $s' \in S$ we have $\hat{t}(t) \equiv \hat{t}(t')$. But we suggest to add the condition "and in the case $s \in S$ we have $t \equiv t'$". With this supplement persistency of F is equivalent to persistency of CF. Theorem 1 in /Gan 80b/ shows that if CF is persistent, then it is the right adjoint (cofree functor) of the forgetful functor, provided we take the subcategory of SPEC1-algebras B that are generated by their parameter part as the range of CF. Moreover Theorem 5 in /Gan 80b/ shows that the extension CF' of a persistent cofree functor CF along a passing consistent parameter passing morphism is again a persistent cofree functor. This allows to replace in all those definitions, constructions and results of Section 3 and 4 where persistency is assumed the free construction F by the cofree construction CF. Hence we obtain a final algebraic theory for parameterized R-specifications. Without knowing the contents of the paper we assume that a similar problem is studied in /HR 81/ (this volume) for the special case of inequalities. In /WB 80/ it is suggested to study also other semantics than the initial and final case. This might be extended to parameterized R-specifications provided that part 3 of the R-EXTENSION LEMMA remains valid.

Another important issue is the implementation of parameterized R-data types extending the algebraic implementation concept in /EKP 80/. For the basic algebraic case a first more or less syntactical treatment is given in /Gan 80a/ where, however, a slightly different implementation concept is used. Abstract implementation and parameter substitution based on initially restricting algebraic theories (see /Rei 80/) are studied in /Hup 80/.

A limitation of our theory up to now is that we have to assume persistency for all our parameterized R-specifications. If our semantics RF_0 (see 3.5) is not persistent we might add a persistency restriction persistent(RF_0) to the source specification with $VALID(persistent(RF_0))=\{A \in \underline{Alg}_{SPEC} / V_s(RF_0(A))=A\}$. This would be a new type of requirement in the sense of Definition 3.1.

An interesting application of parameter passing with requirements is the identification of common subtypes in different specifications. Assume that in SPEC1 and

SPEC2 we have the same subtype SPEC, say bool, such that the initial restriction initial(SPEC) is valid for T_{SPEC1} and T_{SPEC2}. Then parameter passing allows to construct a new specification SPEC3 where only one copy of SPEC is included. But we still have that initial(SPEC) is valid for T_{SPEC3} and the reduct of T_{SPEC3} to SPEC1 and SPEC2 is isomorphic to T_{SPEC1} and T_{SPEC2} respectively. In CLEAR (see /BG 80/) identification of common subtypes is handled using general colimit constructions. As shown above this problem should become a special case of parameter passing such that no additional feature in syntax and semantics is needed.

Finally let us note that the algebraic concept of parameterized specifications is already used in the programming language MODLISP (see /Jen 79/) which is used for the implementation of the algebraic manipulation system SRATCHPAD.

ACKNOWLEDGEMENT

This paper is part of a common project together with the ADJ-group (Jim Thatcher, Eric Wagner, Jesse Wright) at IBM Yorktown Heights and the ACT-group (Hartmut Ehrig, Werner Fey, Klaus-Peter Hasler, Reinhold Kimm, Wilfried Koch, Hans-Jörg Kreowski, Michael Löwe, Bernd Mahr, Peter Padawitz, Michaela Reisin) at Technical University of Berlin. Thanks to all of them and also to Rod Burstall for several valuable discussions concerning the topic of this paper which continues our common paper /ADJ 80a+b/ and tries to bridge the gap to /BG 77+80/. Finally I am most grateful to the organizers for my invitation and to H. Barnewitz for excellent typing.

REFERENCES

/ADJ 76-78/ (JAG,JWT,EGW)[+)]: An initial algebra approach to the specification, correctness, and implementation of abstract data types, IBM Research Report RC-6487, Oct 1976. Current Trends in Programming Methodology, IV: Data Structuring (R.T.Yeh, Ed.) Prentice Hall, New Jersey (1978), pp. 80-149

/ADJ 78/ (JWT,EGW,JBW)[+)]: Data Type Specification: parameterization and the power of specification techniques, Proc. SIGACT 10th Annual Symp. on Theory of Computing, May 1978, pp. 119-132, revised version in IBM Research Report RC-7757 (1979)

/ADJ 80a/ (HE,HJK,JWT,EGW,JBW)[+)]: Parameterized data types in algebraic specification languages, Proc. 7th ICALP Nordwijkerhout, July 1980: Lect. Not. in Comp. Sci. (1980), pp. 157-168

/ADJ 80b/ (HE,HJK,JWT,EGW,JBW)[+)]: Parameter passing in algebraic specification languages, Draft Version, TU Berlin, March 1980

/AM 75/ Arbib, M.A., Manes, E.G.: Arrows, Structures and Functors: The categorical imperative, Academic Press, New York, 1975

/BG 77/ Burstall, R.M., Goguen, J.A.: Putting Theories together to make Specifications, Proc.1977 IJCAI,MIT,Cambridge, MA, Aug. 1977

+) ADJ-Authors: J.A. Goguen (JAG), J.W. Thatcher (JWT), E.G. Wagner (EGW), J.B. Wright (JBW),
co-authors: H. Ehrig (HE), H.-J. Kreowski (HJK)

REFERENCES (cont'd)

/BG 80/ Burstall, R.M., Goguen, J.A.: The Semantics of CLEAR, a Specification Language, Proc. 1979 Copenhagen Winter School on Abstract Software Specifications (1980)

/Ehr 78/ Ehrich, H.-D.: On the theory of specification, implementation and parameterization of abstract data types, Research Report, Dortmund 1978

/EL 79/ Ehrich, H.-D., Lohberger, V.G.: Constructing Specifications of Abstract Data Types by Replacements, Proc. Int. Workshop Graph Grammars and Appl. Comp. Sci. and Biology, Bad Honnef 1978, Lect. Not. in Comp. Sci. 73 (1979), pp. 180-191

/EKP 78/ Ehrig, H., Kreowski, H.-J., Padawitz, P.: Stepwise specification and implementation of abstract data types: Technical University of Berlin, Report, Nov. 1977. Proc. 5th ICALP, Udine, July 1978: Lect. Not. in Comp. Sci. 62 (1978), pp. 205-226

/EKP 80/ --: Algebraic Implementation of Abstract Data Types: Concept, Syntax, Semantics, Correctness; Proc. 7th ICALP, Nordwijkerhout, July 1980, Lect. Not. in Comp. Sci. (1980)

/Flo 81/ Floyd, Ch.: Proc. 2nd German Chapter of The ACM-Meeting "Software Engineering - Entwurf und Spezifikation" (editor), Teubner Verlag 1981

/Gan 80a/ Ganzinger, H.: Parameterized Specifications: Parameter Passing and Implementation, version Sept. 1980, to appear in TOPLAS

/Gan 80b/ --: A Final Algebra Semantics for Parameterized Specifications, Draft Version, UC Berkeley, November 1980

/HR 81/ Hornung, G., Raulefs, P.: Initial and Terminal Algebra Semantics of Parameterized Abstract Data Type Specification with Inequalities (this volume)

/Hup 81/ Hupbach, U.L.: Abstract Implementation and Parameter Substitution, submitted to 3rd Hungarian Comp. Sci. Conf., Budapest 1981

/Jen 79/ Jenks, R.D.: MODLISP: An Introduction, Lect. Not. in Comp. Sci. 72 (1979), pp. 466-480, new version in preparation

/Kla 80/ Klaeren, H.A.: On Parameterized Abstract Software Modules using Inductively Specified Operations, Research Report TH Aachen Nr.66, (1980)

/Kre 78/ Kreowski, H.-J.: Algebra für Informatiker; LV-Skript WS 78/79, FB 20, TU Berlin (1978)

/Rei 80/ Reichel, H.: Initially Restricting Algebraic Theories, Proc. MFCS'80, Rydzyna, Sept. 1980, Lect. Not. in Comp. Sci. 88 (1980), pp. 504-514

/WB 80/ Wirsing, M., Broy M.: Abstract Data Types as Lattices of Finitely Generated Models, Proc. MFCS'80, Rydzyna, Sept. 1980, Lect. Not. in Comp. Sci. 88 (1980), pp. 673-685

A modal characterisation of observable machine-behaviour.

R. Milner

Edinburgh University

1. Introduction

This paper is concerned with the interaction of an experimenter with a machine presented as a black box. We are concerned largely with properties of such machines (here called agents) which may be ascertained by a finite amount of experiment. By considering a sufficiently rich class of properties, we seek an equivalence relation over agents (that of having exactly the same properties) which is adequate for practical purposes, so that the meaning of an agent may be taken to be its equivalence class.

We have previously studied a relation, observation equivalence, of agents [HM, Mil]; it was defined intrinsically, in terms of the actions of an agent under experiment, rather than via extrinsic properties. In [HM] this was shown to agree with an extrinsic equivalence relation, but an important feature of agents was ignored - namely, the possibility of divergence. This feature has a strong bearing on what properties of agents may be ascertained effectively, i.e. by finite experiment.

We shall first define equivalence of agents in two ways - first intrinsically, and then extrinsically via properties expressed in a simple modal language. In Section 5 we show that these agree; we then exhibit conditions under which a given property A of an agent p may be effectively ascertained, and (briefly) conditions under which this is not the case.

2. Agents and experiments

Assume given:

(1) A set P, the agents,

(2) A set E, the experiments,

with the following structure:

(3) A family $\{\stackrel{e}{\Longrightarrow} \mid e \in E\}$ of binary relations over P. $p \stackrel{e}{\Longrightarrow} p'$ may be read "p can undergo experiment e to become p' ". If there is such a relation instance for p, we say that p can accept e.

(4) A unary predicate $\uparrow$ over P. $\uparrow(p)$ is written $p\uparrow$ and may be read "p can diverge" or "p can proceed infinitely without accepting experiment". We write $p\not\uparrow$ for the negation of $p\uparrow$.

For the present we assume nothing further about these relations. They may be infinite; even the image of p under some $\stackrel{e}{\Longrightarrow}$ may be infinite.

An agent may be thought of as a black box, equipped with a button for each experiment. It also has a green light, which is lit iff the agent is proceeding without responding to experiment. To attempt an experiment e on agent p we apply continuous pressure to the e-button; if the button goes down (after some time) then p has accepted the experiment, and if the green light goes off without the button moving then p has rejected the experiment. While neither occurs (and if $p\uparrow$ then it is possible that neither will occur) we can conclude nothing.

3. Experimental equivalence of agents

We aim to characterise the behaviour of agents by experiment only. In one method, we define a pre-order, $\sqsubseteq$, over P . Intuitively, $p \sqsubseteq q$ means that p and q have the same experimental behaviour except that p may diverge where q accepts an experiment. We define $\sqsubseteq$ in terms of a sequence $\sqsubseteq_k$ $(k = 0,1,\ldots)$ of preorders, each concerned with depth k of experiment. For example, to experiment on p to depth 2, we may attempt e on p ; if it succeeds - i.e. $p \overset{e}{\Longrightarrow} p'$ for some p' - we may attempt e' on p' .

Definition $p \sqsubseteq_0 q$ always holds.

$p \sqsubseteq_{k+1} q$ iff, for all $e \in E$,

(i) $p \overset{e}{\Longrightarrow} p'$ implies ($q \overset{e}{\Longrightarrow} q'$ and $p' \sqsubseteq_k q'$, for some q').

(ii) if $p\uparrow$ then

(a) $q\uparrow$

(b) $q \overset{e}{\Longrightarrow} q'$ implies ($p \overset{e}{\Longrightarrow} p'$ and $p' \sqsubseteq_k q'$, for some p').

$p \sqsubseteq q$ iff $p \sqsubseteq_k q$ for all $k \geq 0$.

Proposition 1. $\sqsubseteq_0, \sqsubseteq_1, \ldots$ is a decreasing sequence of pre-orders, and $\sqsubseteq$ is a pre-order.

Remark In the absence of the predicate $\uparrow$, $\sqsubseteq$ reduces to the equivalence relation $\approx$ studied in [HM] and [Mil] , for a certain set E . In [HP] a similar, but not identical, pre-order is defined using divergence.

In general we cannot expect to determine $p \sqsubseteq q$, or its negation, effectively - i.e. by a finite amount of experiment. However, one purpose of this paper is to show that this relation is, under certain assumptions, at least as effective as the inclusion of recursively enumerable sets. In fact, we show that for each agent we can, by experiment, enumerate all its properties, and that $p \sqsubseteq q$ iff q has all the properties of p (and possibly more).

For equivalence of agents, we merely define

Definition $p \approx_k q$ iff $p \sqsubseteq_k q \sqsubseteq_k p$

$p \approx q$ iff $p \sqsubseteq q \sqsubseteq p$

For this paper, we may consider the equivalence class of p , under $\approx$, to be the behaviour of p , since the relation is defined in terms of experimental or observational attributes of agents. It was shown in [HM] and [Mil] that, when certain combinators are allowed over agents, yielding an algebra, the equivalence needed slight refinement to become a congruence relation in the algebra.

4. Properties of agents

A property is any formula A of a simple modal language L . After defining L , we define a relation $\models \subseteq P \times L$ of affirmation. We may read $p \models A$ as "p affirms A" or "p has property A" .

Definition of L

$$\begin{array}{ll} \text{True} & \in L \\ \neg A & \in L \text{ if } A \in L \\ A \wedge B & \in L \text{ if } A,B \in L \\ \langle e \rangle A & \in L \text{ if } e \in E \text{ and } A \in L . \end{array}$$

These are all the formulae of L .

Abbreviations. We write

$$\begin{array}{ll} \text{False} & \text{for } \neg\text{True} \\ A \vee B & \text{for } \neg(\neg A \wedge \neg B) \\ [e] A & \text{for } \neg \langle e \rangle \neg A \end{array}$$

The language L and its relation to observation equivalence (ignoring divergence) was stated in [HM] . Here the attention to divergence requires a more refined treatment.

Definition For $A \in L$, depth(A) is the maximum nesting depth of modal operators $\Diamond$ (and $\Box$) in A .

$L_k = \{A \in L \mid \text{depth}(A) \le k\}$ $(k = 0,1,\ldots)$.

We define A and B to be propositionally equivalent $(A \equiv B)$ iff they are equivalent under the normal Boolean laws. We shall loosely identify propositionally equivalent formulae, since for example our relation $\models$ will respect this equivalence. For $L' \subseteq L$, we abbreviate "L' is finite up to $\equiv$" by "L' is finite". By induction on k we can easily prove

Proposition 2 If E is finite then each L_k is finite.

This will allow us to say that $\wedge L_k$, the conjunction of all members of L_k, "is" a formula (and in L_k itself).

To interpret L , we define a pair $\models$, $\not\models$ of relations $\subseteq P \times L$. At most one of $p \models A$, $p \not\models A$ can hold, but neither may hold. They may be read "p affirms A", "p denies A" .

Definition of $\models$ and $\not\models$, on the structure of formulae.

$p \models \text{True}$ always.

$p \models \neg A$ iff $p \not\models A$.

$p \models A \wedge B$ iff

$p \models A$ and $p \models B$.

$p \models \langle e \rangle A$ iff for some p'

$p \overset{e}{\Longrightarrow} p'$ and $p' \models A$.

$p \not\models \text{True}$ never.

$p \not\models \neg A$ iff $p \models A$.

$p \not\models A \wedge B$ iff

$p \not\models A$ or $p \not\models B$.

$p \not\models \langle e \rangle A$ iff $p \Uparrow$ and for all p'

$p \overset{e}{\Longrightarrow} p'$ implies $p' \not\models A$.

Proposition 3 (1) $\models$ and $\not\models$ are disjoint.

(2) $A \equiv B$ implies $\begin{cases} p \models A \text{ iff } p \models B \\ p \not\models A \text{ iff } p \not\models B \end{cases}$.

The second part justifies our identification of logically equivalent formulae when discussing affirmation.

It is only due to the extra condition $p \Uparrow$, in the last clause of the definition, that both $p \models A$ and $p \not\models A$ may fail to hold. For example if $p\uparrow$ and if $p \overset{e}{\Longrightarrow} p'$ holds for no p', then neither $p \models \langle e \rangle \text{True}$ nor $p \not\models \langle e \rangle \text{True}$ holds.

We require the condition to make $\not\models$, and hence also $\models$, an effective relation.

Since both these relations are defined rather naturally in terms of experiment, we regard this simple formulation as strong evidence that the preorder $\underset{\sim}{\sqsubset}$ over agents is natural and important; this is because of our main result which relates $\underset{\sim}{\sqsubset}$ to inclusion of affirmed properties.

We believe L is the simplest language for this purpose. In [HM] it was indicated that weaker languages, gained by omitting one or both propositional connectives, do not suffice. It is worth remarking that L is a small sublanguage of propositional dynamic logic [Pra] but that it plays a different rôle here. In PDL, non-trivial programs appear within the diamond operator $\langle\rangle$; although we could allow complex (e.g. disjunctive, or iterative) experiments to appear in $\langle\rangle$, we are mainly concerned with the affirmation relation <u>between</u> agents (e.g. programs) and formulae.

Definition

$$AFF_k(p) = \{A \in L_k \mid p \models A\}$$

$$DEN_k(p) = \{A \in L_k \mid p \not\models A\}$$

$$AFF(p) = \bigcup_k AFF_k(p)$$

$$DEN(p) = \bigcup_k DEN_k(p)$$

Proposition 4

$$AFF_k(p) \subseteq AFF_k(q) \text{ iff } DEN_k(p) \subseteq DEN_k(q)$$

$$AFF(p) \subseteq AFF(q) \text{ iff } DEN(p) \subseteq DEN(q)$$

We can now proceed to establish the connection between $\models$ and $\underset{\sim}{\sqsubset}$.

5. Affirmation characterises behaviour

We show in this section that L is sufficiently rich to express behavioural difference, under simple assumptions.

Definition. A binary relation $R \subseteq X \times Y$ is image-finite if, for all $x \in X$, $\{y \mid xRy\}$ is finite.

Characterization Theorem

If either E is finite or each relation $\stackrel{e}{\Longrightarrow}$ is image-finite, then for all $p,q \in P$ and all $k \geq 0$

$$p \underset{\sim}{\sqsubset}_k q \quad \text{iff} \quad AFF_k(p) \subseteq AFF_k(q).$$

Corollary. Under the same conditions

$$p \underset{\sim}{\sqsubset} q \quad \text{iff} \quad AFF(p) \subseteq AFF(q).$$

Before giving the proof, we show the need for one or other of the conditions of the theorem. For the following example we show $AFF(q) \subseteq AFF(q')$ but not $q \underset{\sim}{\sqsubset} q'$.

Let $P = \{q,q',q_0,q_1,\ldots,q_\omega\}$, and

$E = \{0,1,\ldots\ldots\}$. Let $p\uparrow$ for all $p \in P$, and let the experiment relations be given by

$$q \stackrel{0}{\Longrightarrow} q_i \quad \text{and} \quad q' \stackrel{0}{\Longrightarrow} q_i \quad (0 \leq i < \omega).$$

$$q \stackrel{0}{\Longrightarrow} q_\omega$$

$$q_i \stackrel{j}{\Longrightarrow} q_0 \qquad (0 \leq j < i \leq \omega)$$

We show the relations for q in a diagram:

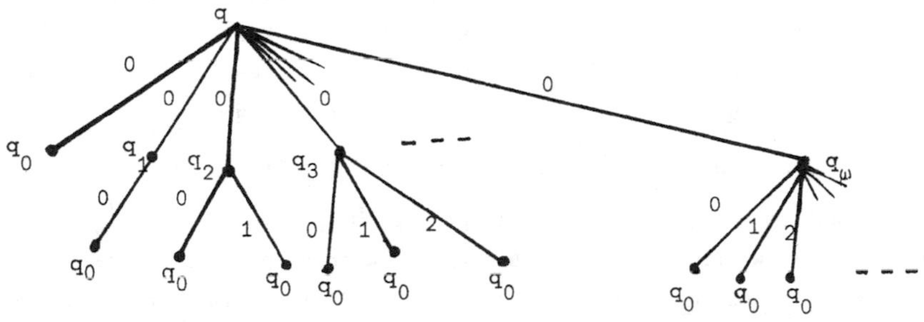

The diagram for q' is the same, but without the right branch to q_ω.

Now $q \not\underset{\sim}{\sqsubset}_2 q'$. To see this, note that $q \stackrel{0}{\Longrightarrow} q_\omega$, while $q' \stackrel{0}{\Longrightarrow} q_i$ only for $i < \omega$; moreover $q_\omega \not\underset{\sim}{\sqsubset}_1 q_i$ $(i < \omega)$, since $q_\omega \stackrel{i}{\Longrightarrow} q_0$ but q_i does not admit experiment i.

On the other hand $AFF(q) \subseteq AFF(q')$. For suppose $A \in AFF(q)$, i.e. $q \models A$, and let k be the maximum for which $\langle k \rangle$ occurs in A. Then – in outline – the truth of $p \models A$ (for any p) depends only upon instances of $\stackrel{i}{\Longrightarrow}$ for $i \le k$. But q and q' are identical with respect to such instances, since $q_{k+1}, \ldots, q_\omega$ are identical with respect to such instances. Hence $q' \models A$ also, i.e. $A \in AFF(q')$.

Proof of Theorem

At $k = 0$ it is trivial; merely note that for all p $AFF_0(p) = \{True\}$ up to logical equivalence (formulae in L_0 are purely propositional). We now proof the inductive step.

($\Longrightarrow$) Assume $p \lesssim_{k+1} q$. It is easy to see that it is enough to consider members of $AFF_{k+1}(p)$ of form $\langle e \rangle A$ or $\neg \langle e \rangle A$, where $A \in L_k$, since any formula is logically equivalent to a disjunction of conjunctions of such.

If $p \models \langle e \rangle A$, then $p \stackrel{e}{\Longrightarrow} p' \models A$. But by assumption, for some q', $q \stackrel{e}{\Longrightarrow} q'$ and $p' \lesssim_k q'$; by induction $q' \models A$, hence $q \models \langle e \rangle A$.

If $p \models \neg \langle e \rangle A$, then $p \not\models \langle e \rangle A$, so $p \not\uparrow$ and, for all p', $p \stackrel{e}{\Longrightarrow} p'$ implies $p' \not\models A$. But by assumption $q \not\uparrow$; hence if not $q \models \neg \langle e \rangle A$, i.e. if not $q \not\models \langle e \rangle A$, then, for some q', $q \stackrel{e}{\Longrightarrow} q'$ and not $q' \not\models A$. By assumption again, for some p', $p \stackrel{e}{\Longrightarrow} p' \lesssim_k q'$, whence by induction and Proposition 4, not $p' \not\models A$. But this is a contradiction, so $q \models \neg \langle e \rangle A$.

($\Longleftarrow$) Assume $p \not\lesssim_{k+1} q$. There are three cases; in each case we find $A \in L_{k+1}$ such that $p \models A$ but not $q \models A$.

(i) $p \stackrel{e}{\Longrightarrow} p'$ but, for all q', $q \stackrel{e}{\Longrightarrow} q'$ implies $p' \not\lesssim_k q'$. Let $\Omega = \{q_1, q_2, \ldots\}$ be the set of q' for which $q \stackrel{e}{\Longrightarrow} q'$. By induction, there are formulae $B_1, B_2, \ldots$ in L_k such that $p' \models B_i$ but not $q_i \models B_i$, for each i.

Now if E is finite then L_k is finite (Proposition 2) while if $\stackrel{e}{\Longrightarrow}$ is image-finite then Ω is finite; in either case $\{B_1, B_2, \ldots\}$ is finite (up to logical equivalence) so $B = \bigwedge_i B_i$ is in L_k. But clearly $p' \models B$ and not $q' \models B$ for each $q' \in \Omega$. Hence $p \models \langle e \rangle B$ but not $q \models \langle e \rangle B$.

(ii) $p \not\uparrow$ but $q \uparrow$. Then $p \models [e]\, True$ but not $q \models [e]\, True$.

(iii) $p \not\uparrow$, $q \not\uparrow$ and $q \stackrel{e}{\Longrightarrow} q'$ but, for all p', $p \stackrel{e}{\Longrightarrow} p'$ implies $p' \not\lesssim_k q'$. Just as in case (i), we find a formula $B \in L_k$ such that $p' \models B$ for each p' such that $p \stackrel{e}{\Longrightarrow} p'$, but not $q' \models B$. (The only difference is that B is now a finite disjunction, not conjunction.) It then follows that $p \models [e]\, B$ but not $q \models [e]\, B$. ⊠

It is worth remarking that the forward implication of the theorem holds without assuming either of the conditions.

It is natural to try to find a more expressive L for which the theorem would hold without conditions. From the proof it is clear that infinite conjunctions and disjunctions would suffice, but we would like a finitary language. It is not clear, however, whether we could then have $\models$ an effective relation.

On the other hand, a simple modification of the preorder $\underset{\sim}{\sqsubset}$ does yield the characterisation theorem without conditions. It is due to Hennessy and Plotkin [HP] who employed the idea for other purposes.

Let $F \subseteq E$ be finite. Let $\underset{\sim}{\sqsubset}^F$ be the preorder defined as before but restricting e to belong to F; also, let $L^F \subseteq L$ be those formulae containing $\langle e \rangle$ only when $e \in F$, and let $AFF^F(p) = AFF(p) \cap L^F$. Then our theorem immediately yields

$$p \underset{\sim}{\sqsubset}^F q \quad \text{iff} \quad AFF^F(p) \subseteq AFF^F(q)$$

Thus, if we choose to define the preorder $\underset{\sim}{\sqsubset}^O$ by

$$p \underset{\sim}{\sqsubset}^O q \quad \text{iff, for all finite } F \subseteq E,\ p \underset{\sim}{\sqsubset}^F q$$

we obtain easily the following

<u>Corollary</u> $\quad p \underset{\sim}{\sqsubset}^O q \quad \text{iff} \quad AFF(p) \subseteq AFF(q)$.

The simplicity of L, with the results in [HP] and this corollary, is some evidence that $\underset{\sim}{\sqsubset}^O$ is more natural - at least, more mathematically tractable - than $\underset{\sim}{\sqsubset}$.

6. <u>Affirmation is effective</u>

We now construct an interpretation of $\overset{e}{\Longrightarrow}$ and $\uparrow$ under which affirmation $\models$ is an effective relation. We start with a finer structure over P and E than that of Section 2. Assume a set $\{\xrightarrow{e} \mid e \in E\}$ of image-finite relations, and a single image-finite relation $\xrightarrow{o}$, over P.

We further assume that time is discrete, and that $p \xrightarrow{e} p'$ represents an instantaneous e-action of p , an <u>acceptance</u> of experiment e , which is only possible if e is attempted, while $p \xrightarrow{o} p'$ represents <u>silent</u> instantaneous action, not accepting an experiment, but possible under any attempted experiment. ($\xrightarrow{o}$ corresponds to $\xrightarrow{\tau}$ in [Mil].)

The <u>derivation tree</u> of an agent p has arcs labelled in $E \cup \{o\}$, e.g.

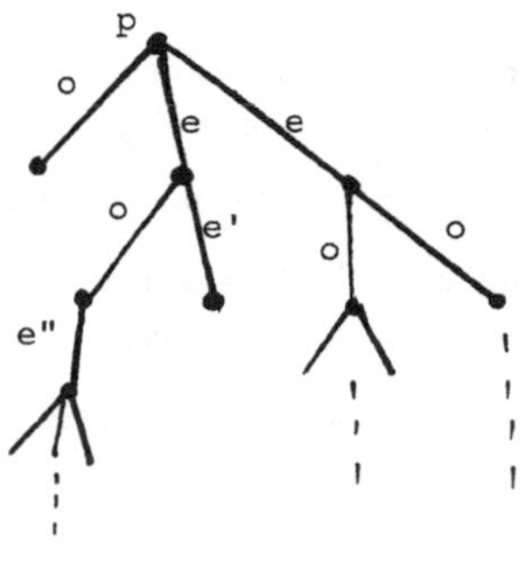

where each node represents an agent, and each arc a possible action. The tree may be infinite or infinitely branching, but its restriction to any finite subset of $E \cup \{o\}$ is finitely branching; in particular, its restriction to $\{e,o\}$ for any $e \in E$ is finitely branching, and includes all possible courses of action under an attempted e-experiment.

We now define over P the binary relations

$$\Longrightarrow \;=\; \xrightarrow{o}{}^{*} \text{, the transitive reflexive closure of } \xrightarrow{o} \text{,}$$

$$\overset{e}{\Longrightarrow} \;=\; \Longrightarrow \xrightarrow{e} \text{, acceptance of } e \text{ possibly preceded by silent action,}$$

and the predicate

$$\uparrow \;=\; \xrightarrow{o}{}^{\omega} \text{, infinite silent action.}$$

Thus we have an interpretation of the structure of Section 2.

(Note: we could have chosen $\overset{e}{\Longrightarrow} = \Longrightarrow \xrightarrow{e} \Longrightarrow$, allowing silent action to follow the acceptance of e; this would be more consistent with the treatment in [Mil] and [HP]. The same results are then obtainable, but the present choice allows a slightly simpler treatment.)

Under a trial, by which we mean an attempt of some duration $n \leq \omega$ to observe some e, an agent p proceeds by an indeterminate selection of either a silent or an e-action, at each instant $i < n$, until $\xrightarrow{e}$ is selected or no selection is possible or instant n is reached. More precisely, an n-outcome of p under e may be any of the following:

(i) Accept with result q if $p \xrightarrow{o}{}^{i} p' \xrightarrow{e} q$, $i < n$.

(ii) Reject if $p \xrightarrow{o}{}^{i} p'$, $i < n$, and p' has no silent action and no e-action.

(iii) Unsettled if $p \xrightarrow{o}{}^{n} p'$ (i.e. $p\uparrow$, if $n = \omega$)

We wish to know that, by varying the ambient ("weather") conditions, an experimenter can cause any possible outcome of a trial. To this end we postulate a set W of instantaneous ambient conditions, exactly one of which holds at each instant; we also assume that, for each e and each p possessing at least one silent or e-action, there exists a surjective function from W to the set of such actions of p, determining one such action under an attempt to observe e.

(Note: the assumptions imply that the image of each p under $\xrightarrow{o}$ or $\xrightarrow{e}$ is not only finite but bounded by the size of W. It is possible to relax the assumptions a little by allowing the size of W to be finite but variable with time.)

It follows easily that each $c \in W^n$ determines an n-outcome of p under e, and that every such outcome is so determined. Hence for given e

(a) $p \overset{e}{\Longrightarrow} q$ iff, for some $n < \omega$ and some $c \in W^n$, c determines the outcome accept with result q.

Moreover, using König's Lemma,

(b) $p\hat{\uparrow}$ iff , for some $n < \omega$, no $c \in W^n$ determines the outcome unsettled.

The experimental set-up is as follows. Given p as a black box, the experimenter may choose any e , and any $c \in W^*$, which determines both the duration and the outcome of a trial. An outcome accept with result q (the e-button goes down) means that he now has available a black box q on which to conduct further trials. The green light serves to distinguish the outcomes reject and unsettled. Black boxes are replicable, so he may conduct many trials on the same p with different e and c .

Theorem For any p , AFF(p) may be effectively enumerated by experiment.

Proof. It is enough to show by induction on k that if $A \in L_k$ then $p \models A$, if true, may be ascertained by finite experiment.

This is trivial for $k = 0$. At $k + 1$, it is enough to consider A of form $\langle e \rangle B$ or $\neg \langle e \rangle B$.

(i) Assume $p \models \langle e \rangle B$, $B \in L_k$. Then, for some q , $p \overset{e}{\Longrightarrow} q$ and $q \models B$. Hence, by (a) and by enumerating W^* , we may enumerate by experiment all outcomes of p under e (a possibly infinite set) of the form accept with result q. By interleaving this enumeration with attempts to ascertain $q \models B$ for each such q , at least one such attempt will succeed, ascertaining $p \models \langle e \rangle B$.

(ii) Assume $p \models \neg \langle e \rangle B$, $B \in L_k$. Then $p \not\models \langle e \rangle B$, so $p\hat{\uparrow}$ and, for all q, $p \overset{e}{\Longrightarrow} q$ implies $q \models \neg B$. From (b), by enumerating W^* we may both ascertain $p\hat{\uparrow}$ and collect all black boxes q (necessarily a finite set) for which $p \overset{e}{\Longrightarrow} q$. Then, by induction, $q \models \neg B$ may be ascertained for all such q by finite experiment; thus $p \models \neg \langle e \rangle B$ is ascertained.

⊠

7. Alternative interpretations

Our experimental set-up was rather simple. We assumed that a single trial consisted of a persistent attempt to observe some e . There are several variations which do not disturb our results; for example, an iterative trial in which some sequence of experiments is attempted, with acceptance as soon as one is accepted, can be considered as a sequence of our more primitive trials. Further, a disjunctive trial in which some finite subset of E is simultaneously attempted, with acceptance when one is accepted, yields no difficulty; effectiveness then depends on our remark that the derivation tree restricted to any finite subset of $E \cup \{o\}$ is finitely branching, allowing König's lemma to be applied. (Choosing the empty subset amounts to allowing only silent action by the agent.) By such refinements we come closer to considering the experimenter himself as an agent, communicating with his object agent by experiment.

Interesting problems arise when we reconsider the divergence predicate $\Uparrow$. A first refinement is to replace it by a family $\{\Uparrow e \mid e \in E\}$ of predicates, where $p\Uparrow e$ means "p may diverge under an e-trial of infinite duration". What we have done amounts to assuming that $\Uparrow e$ is independent of e. Now the first part of this paper is entirely unaffected by the refinement; we have set out Sections 3-5 so that everything makes sense when $\Uparrow, \not\Uparrow$ are textually replaced by $\Uparrow e$, $\not\Uparrow e$, so affirmation still characterizes behaviour. In contrast, the <u>effectiveness</u> of affirmation is strongly dependent on the <u>interpretation</u> of $\Uparrow e$. A possible interpretation, in which $\Uparrow e$ does depend upon e, is to assume a <u>finite delay</u> or <u>weak fairness</u> property. In terms of the action relations of Section 6, this may be expressed as follows:

$p\Uparrow e$ iff there is an infinite derivation $p \xrightarrow{o} p_1 \xrightarrow{o} p_2 \xrightarrow{o} \ldots$
such that infinitely often p_i has no e-action.

In other words, if in an infinite trial an agent can almost always accept e, then it <u>will</u> do so.

Immediately we lose effectiveness, since the essential property (b) in Section 6 no longer holds. The term <u>weak fairness</u> is due to D. Park; his <u>strong fairness</u> property, gained by exchanging "infinitely often" and "almost always" above, also leads to loss of effectiveness. We have not treated effectiveness fully formally, but these observations accord with the result of A. Chandra [Cha] that, in the presence of the finite delay property, the input/output relation computed by a nondeterministic machine is in general highly non-computable, and can be any set in Σ_1^1. It is very likely that a more detailed study of our affirmation relation will produce a similar result, even when E is finite.

References

(LNCSn stands for Vol n, Lecture Notes in Computer Science, Springer-Verlag)

[Cha] Chandra, A. (1978) "Computable Nondeterministic Functions", Proc. 19th Ann. Symposium on Foundations of Computer Science, IEEE, pp.127-131.

[HM] Hennessy, M. and R. Milner (1980) "On observing nondeterminism and concurrency", Proc. 7th ICALP, Amsterdam, LNCS 85, pp. 299-309

[HP] Hennessy, M. and G. Plotkin (1980) "A term model for CCS", Proc. 9th MFCS, Poland, LNCS 88, pp. 261-274

[Mil] Milner, R. (1980) "A Calculus of Communicating Systems", LNCS92.

[Pra] Pratt, V. (1976) "Semantical Considerations on Floyd-Hoare Logic", Proc. 17th Ann. Symposium on Foundations of Computer Science, IEEE, pp.109-121.

TREES AS DATA AND FILE STRUCTURES

J. Nievergelt, Informatik, ETH, CH-8092 Zurich

ABSTRACT

Trees have been important data structures since the mid-fifties when the first list processing applications and languages were developed. When Knuth systematized the accumulated knowledge about data structures in his 1968 book on Fundamental Algorithms, he devoted half the space to tree structures. During the seventies, data structures based on trees were extended to files on secondary storage and to multidimensional problems such as multi-key access. Trees also became the dominant data structure for many algorithms in the field of concrete complexity, because they are the only structures known that guarantee an O(log n) worst case bound on sequential and random access, insertion and deletion on linearly ordered sets. Recent events indicate, however, that in the coming decade the predominance of trees as all-round data structures may be challenged by address computation techniques. Various refinements and generalizations of the old programming trick called hashing have caused an unexpected extension of the domain of applicability of address computation techniques to dynamic files and multi-key access.
This paper surveys the history of data structures used for tables, files or record management systems: structures designed for efficient retrieval, insertion and deletion of single records characterized by key values chosen from linearly ordered sets. By describing a few milestones and highlights, I attempt to identify trends and speculate on future developments.

CONTENTS

1 The emergence of trees as all-round data structures

The very first data structures used on computers during the late forties were based on sequential access (e.g. input and output from paper tape) and address computation: programs to access arrays are described in the venerable report "Planning and coding of problems for an electronic computing instrument" written by Goldstine and von Neumann in 1947 [GN 47]. Arrays and sequential files were firmly entrenched as the dominant data structures when the first high-level programming language was developed in 1954: FORTRAN supported these two structures, and no others, with such statements as DIMENSION and READ TAPE, WRITE TAPE, REWIND and BACKSPACE.

It is easy to understand why the software pioneers of the first decade did not look beyond the two types of data structures array and sequential file. Central memories (electrostatic storage tubes, drums, and later core) had such a limited storage capacity that any structure that requires space for pointers would have been dismissed as a luxury. And the only secondary storage media available, tapes and card decks, can only be accessed efficiently in a sequential manner.

In the mid-fifties certain applications in artificial intelligence or heuristic programming were studied that require dynamic data structures - structures whose extent and shape changes at run-time, not just the values stored within given memory locations. Newell, Shaw and Simon pioneered the development of list processing techniques and designed the first list processing language, IPL-II (e.g.[NS 56]).

The simplest lists are linear: stacks, queues, circular lists, with one-way or two-way links. Access and manipulation routines are short, but access times are long - O(n) -, so that linear lists are only practical for small data collections. Arbitrarily linked lists are often used to reflect explicitly any relationship that may exist between two objects of the real world being modeled by the data - such as in a semantic net. They are suitable for static structures, but too complex to allow efficient insertions and deletions.
In between these extremes there are two categories of list structures that strike an elegant compromise between generality and efficiency: trees and dags (directed acyclic graphs). Both of these data structures play an important role in the first high-level list processing language, LISP (McCarthy, [MC 60]), based on recursion.
Trees (in this article they are always rooted and ordered) directly represent the structure of successive invocations of recursive definitions or functions. Dags permit a compact representations of trees that have many isomorphic subtrees, something that may easily occur with recursive programming. Many techniques developed for trees could be extended to dags - but the study of data structures essentially bypassed dags, presumably because the phenomenon of many isomorphic subtrees is uncommon in the main application of tree structures: to represent linearly ordered sets. Thus, trees have

emerged as THE class of list structures which are most widely used, and are understood best from a theoretical point of view. Knuth deserves much of the credit for this state of affairs. He collected and classified the accumulated knowledge on data structures in his books on Fundamental Algorithms [Kn 68] and Sorting and Searching [Kn 73]. Of all the data structures considered, trees claim the lion's share of the space.

The main use of trees in programming is as a structure for organizing what is variously called a (dynamic) table, file, or record management system: A collection of records on which the operations: retrieval (by name), insertion and deletion must performed with equal efficiency. The above three are the most basic operations. Sometimes a fourth operation is important: given a record, find the next one in a predetermined linear order defined on the space from which the key values are drawn. Other operations, such as retrieval by rank or percentile, merge or split, are less frequently required; when they are, they can often be realized efficiently in a data structure that is efficient for all four operations mentioned earlier: Find, Insert, Delete, Next (but not necessarily in one that supports only the three basic file operations).

Almost all data structures known lead to operation times of the following three orders: O(1), O(log n), O(n). Structures with Find in O(1) tend to have Next in O(n), and vice-versa. Trees are the only structures that guarantee random access as well as sequential access (in order) to a dynamic file all in time O(log n) (see, for example [Ni 74]). While they are not optimal in any one respect, they perform acceptably with respect to all of them, and thus have emerged as the all-round compromise data structure. Before this claim could be substantiated, however, a lot of interesting theoretical and practical work had to be done. The most important aspect of this work, a guaranteed O(log n) worst case bound on each of the operations Find, Next, Insert and Delete, will be reviewed in the next section. This achievement, first accomplished by Adelson-Velskii and Landis [AL 62], was instrumental in motivating researchers to keep extending the domain of application of tree structures. The elegant tree transformations developed to keep dynamic trees balanced gave researchers the confidence that, with sufficient cleverness, they could always invent special classes of trees and matching algorithms to solve any reasonable problem. In view of the difficulties encountered later (section 3) this expectation is no longer shared by experts. But while it lasted, it lead to many new developments during the seventies. Let us mention the following three: the use of trees as file structures for secondary storage, for concurrent access by several transactions to a common data base, and for access to records by means of several keys.

During the seventies, data structures based on trees were extended to files on secondary storage. At first, binary search trees with one record per node were adapted to a paging environment, where a page capable of holding many records is transferred as a unit between central memory and disk. The problem then becomes that of paginating the tree efficiently, so that no path from the root of the tree to a leaf crosses many page boundaries. Such "dichromatic trees" with branches of two kinds (those that can be considered internal to a page and thus cause a brief delay, others crossing page boundaries and causing long delays) have recently been studied again [GS 78]. The most influential approach, however, the B-trees of Bayer and McCreight [BM 72], abandoned early the tradition of storing one record per node, appropriate to central memory, and introduced multiway trees, with an arbitrary number of records per node.

Whereas data structures for central memory are usually accessed by only one process at a time, files on secondary storage must often be made accessible to many processes concurrently. The application may require this, as for example in a distributed information or reservation system; or the underlying operating system may give the writer of a file system no control over the sequencing of page accesses, thus interleaving different file operations. Thus concurrent access to tree-structured data became an important research topic (see, for example [BS 77], or [KL 80]). The resulting protocols turn out to be complex and costly beyond expectation – a first indication that a point of diminishing returns may have been reached in the ever expanding domain of application of trees as data structures.

As a consequence of the shifting mode of computer operation in many applications from batch processing (which leads to sequential processing of data) to transaction processing (which requires random access), multi-key access to files has become more important. Many kinds of multidimensional trees have been designed to cope with this problem – a comprehensive bibliography [EL 80] lists hundreds of publications on this topic. Here again the solutions found so far to the problem of providing multi-key access to dynamic files are not as elegant or efficient as the corresponding approaches to the one-dimensional case.

Before these shortcomings became apparent, however, trees were used successfully as the most popular data structure for designing algorithms with a good worst case behavior. The major theoretical blemish (also of practical importance), had been that a balanced tree will gradually become skewed under random insertions and deletions, and may even degenerate to a linear list under biases as they tend to occur in practice (such as inserting presorted sequences). This shortcoming was eliminated by various classes of balanced trees. Trees are still the only data structures known that guarantee an O(log n) worst case bound on random and sequential access, insertion and deletion of a single element chosen from a linearly ordered set. The next section describes some highlights of this development.

2 Balancing: the quest for an O(log n) worst case bound

Many useful data processing operations on trees, such as search, insertion, deletion, merging and splitting can be performed along a single path from the root to a leaf. This discovery of the sixties must be counted as a major achievement of the budding field of data structures. Since trees usually have a height proportional to the logarithm of the number of nodes, it implies that the most basic data processing operations can all be performed "in logarithmic time". Of course, a couple of assumptions must be made: the storage medium must permit random access, and the tree must not be degenerate. In the sixties the first assumption had become true thanks to central memories large enough to encourage list processing, and to disks as on-line mass storage that relegated tapes to archival storage. The second assumption used to be insured by periodically dumping a file that had become too "skewed" because of holes left by deletions or overflow chains caused by clustered insertions, and rebuilding it in a balanced shape.

Adelson-Velskii and Landis [AL 62] made a fundamental practical as well as theoretical contribution by defining a class of trees later called AVL trees or height-balanced trees. A tree is height-balanced iff, for every node, the heights of its two subtrees differ by at most 1. Although an AVL tree may look sparse and skewed, its height is clearly logarithmic in the number of nodes; in fact, it is at most 44% larger than that of a complete binary tree with the same number of nodes. The novelty of AVL trees came from insertion and deletion algorithms that include rebalancing, that is: if a tree violates the AVL condition after insertion or deletion, a sequence of local transformations (rotations or double rotations) along the root-to-leaf path restores the AVL condition.

AVL trees demonstrated an instance of the following schema for designing a class C of balanced trees:

1) define a balance condition such that every tree in C has height in O(log n)
2) invent rebalancing algorithms that transform in logarithmic time any tree in C disturbed by a single insertion or deletion back into C.

Are there other classes of balanced trees? [NR 73] introduced weight-balanced trees, where the balance condition is on the number of nodes rather than on the height, and the same local transformations (rotations and double rotations) can be used to design a logarithmic rebalancing algorithm. Differences are:

- the user can set a parameter to control the maximal degree of skewedness he wishes to allow,
- insertion or deletion AND rebalancing can normally be done in a single pass from root to leaf.

Hopcroft introduced 2-3 trees [AHU 74], where the balance condition is on the fan-out: each node has either 2 or 3 sons, and all leaves are at the same depth. The B-trees of Bayer and McCreight [BM 72] are a generalization of 2-3 trees suitable for secondary storage: the number of sons of each node (except the root which needs more freedom) is constrained to the range m to 2m-1. The parameter m is chosen so that a full node fills a page or disk block; a worst case disk space utilization of 50% (about 2/3 on average) is quite acceptable. B-trees became widely used in practice, and many commercial file systems are based on them, under a variety of names such as VSAM. Many modifications of the original B-tree scheme are surveyed in [Co 79]. Many other classes of balanced trees have been invented, but it is unlikely that any new class can be found that performs significantly better. We have simply learned that balanced trees have a common range of performance that is quite independent of the particular variety. Inventing new balanced trees has reached the point of diminishing returns. It is more useful to understand the existing classes well, in particular w.r.t. their average performance, and to study properties common to all such classes, or at least to a large number of them.

As an example of the former, [BM 78] proves that, in weigth-balanced trees, the average number of rebalancing operations per insertion or deletion is bounded by a constant. In most classes of balanced trees it may happen that an update (usually the deletion of a leaf) requires each node along the search path to be rebalanced. Theoretically, this is absorbed by the O(log n) bound; in practice, it is a nuisance. Fortunately, simulation shows that rebalancing occurs rarely, typically less than one transformation per update. The reason is that a rebalancing operation improves the shape of the entire tree, so that subsequent insertions or deletions are less likely to cause further rebalancing. [BM 78] gives a theoretical basis to an empirical observation. Recently [Ol 81] a new class of balanced trees has been discovered where the number of transformations needed to rebalance a tree is bounded by a constant in the worst case (three rotations suffice in all cases).

As an example of properties common to several classes of trees, Guibas and Sedgewick [GS 78] have introduced a "dichromatic framework" for balanced trees that allows them to treat AVL-trees as well as B-trees under a common scheme. It reduces all trees to binary, but distinguishes edges internal to a macro-node from edges between them. This generalization of known classes of balanced trees allows them to design update-and-rebalancing algorithms that work in a single top down pass.

The multiplicity of balancing techniques, each one tailored to a specific class of trees, has also yielded to "general dynamization techniques": approaches that allow the transformation of static data structures to dynamic ones. The general idea is to replace one large static structure by a sequence of structures of the same kind, of increasing size. Operations on the entire structure are then decomposed into operations on the individual structures. The first use of this technique appears to be [Wi 78]. It has since been studied by several authors, e.g. [Be 79a], [SB 79], [LM 80], [Me 80].

In summary, the fact that efficient balancing techniques exist for many classes of trees was perhaps the single most important reason for the popularity of data structures based on comparative search trees. As the domain of application of these data structures was gradually extended to files on secondary storage that require fast random access, support concurrent operation and multi-key access, the acrobatics required for balancing a dynamic file under a growing number of constraints became more difficult, less elegant and less efficient. Where and how the inherent limitations of tree-structured lists began to show up is the subject of the next section.

3 Serious limitations: mass storage, concurrency, multidimensional data

Comparative search trees for dynamic sets derive their strength from the ease of modification of list structures. Lists consist of pointer chains. These are traversed quickly when they reside in central memory - with a delay of the order of a microsecond per pointer. But they may lead to unacceptable delays when they cross page boundaries and cause a disk access that consumes tens of milliseconds - 10**4 times longer. Computers are being used more and more interactively - in such applications as CAD, information or reservation systems, and programming. Imagine a text processing system where the spelling of every word is checked against a dictionary as soon as it has been typed. The dictionary resides on disk, the typist does not want to be slowed down by spelling checks. One important design goal for an interactive system is to respond instantaneously to trivial user requests, such as "show me the next item of information". "Instantaneously" means, by human standards, 100 milliseconds. This leaves time for at most one disk access. If a data collection is large, and organized as a tree implemented by pointers, random access to one item is likely to require 3 or more disk accesses. This is because a page or disk block might contain anywhere from a dozen to perhaps a hundred separating key values and pointers. With a fan-out of 100, it takes a tree of depth 3 to address a million key values. Most likely an additional disk access is involved in going from the key value found to the corresponding record. As the performance of any file structure is measured by the number of disk accesses, lists are subject to definite performance limitations.

Another limitation of trees as file structures shows up starkly when concurrent access to data is allowed. Every node in a tree is the sole entry point to the entire subtree rooted at that node, and thus is a potential bottleneck. Ideally, two processes that access data located in different units of storage, such as pages, should be able to proceed unhampered by each other. In trees this is not the case, as a split or merge caused by one process may propagate to other parts of the tree. Several papers ([BM 77], [KL 80], [El 80]) provide evidence that concurrent access to tree structures requires elaborate protocols to insure integrity of the data structure. The best attempt appears to be update-and-rebalancing algorithms that work on a single pass from root to leaf [NR 73], [GS 78]. This is possible when only successful insertions and deletions are considered, but is problematic when an insertion of an already present element is attempted, or a deletion of an absent element. But in any case, concurrency of one process that changes a list structure with others that traverse that structure requires a potentially elaborate and costly locking activity.

Multi-key access is a third area in which comparative search trees have not yielded structures as elegant as in the single key case. Many variations have been tried, all based on the idea that each node is a separator with respect to one of the key-fields (dimensions). Multidimensional binary search trees, also called k-d trees, are one specific example [Be 79]. The main shortcoming of multidimensional trees is that they have no elegant rebalancing algorithms, at least not in the worst case. Instead of O(log n) we need an O(n**(1-1/k)) bound for insertion and deletion, where k is the number of dimensions or key fields. The general dynamization techniques mentioned in section 2 may help in solving the rebalancing problem for multidimensional trees, but practical experience is still lacking.

The inherent difficulties mentioned above suggest that in the coming decade the dominance of trees (of the comparative search type) as all-round data structures may be challenged. As the next two sections attempt to show, data structures based on address computation are not subject to the particular limitations discussed here. Their traditional difficulty has been the implementation of dynamic data structures, and we will see how this is being overcome.

4 Radix trees: organize the embedding space rather than the set of given objects

Much more attention has been given to comparative search than to digital search. Perhaps this imbalance has its roots in the historical fact that memory has been the bottleneck in data processing applications much more often than CPU time. Comparative search techniques economize memory, whereas digital search uses it lavishly in order to achieve a fast look-up.

Let S be a (typically infinite or astronomically large) set of possible key values, F $\subset$ S a much smaller set (but often still large compared to central memory capacity) that we must store. Comparative search techniques require space for just the elements in F, whereas the natural implementations of digital search lead to data structures that reserve space for each element in S. Since S is too large, some compromise results.

Radix trees implemented as tries [Fr 60] are such a compromise. S = A* is assumed to be the set of strings over a finite alphabet A. The branches emanating from each node of a radix tree T are identified with the letters of A. Thus each string in S is represented uniquely as a path from the root to some node of T, or equivalently, S is in 1:1 correspondence with the set of nodes of T. The file content F can be represented by a single presence bit per node. Many data compression schemes are applicable to radix trees, such as the following. Obviously, only a top part of T that contains all of F must be explicitly represented. Or the letters of the alphabet A may be encoded as strings over {0,1}; in the resulting binary radix tree, long paths with only one-way branching can be compressed by using the "Patricia trick" (see, e.g. [Kn 73]).

Radix trees and their compression schemes can be refined and analyzed further if the relationship between the size N of the key space S and the number n of keys to be stored is known. In several applications, such as sparse Gaussian elimination or LR parsing, large but sparse static tables must be stored with N = O(n**2). Tarjan and Yao [TY 79] present a method for storing static tables based on radix trees and table compression by displacement techniques. Their data structure achieves a worst case access time of O(log N/ log n), which is O(1) for N = a polynomial in n, with only O(n) words of storage. Their model allows a word to store log N bits, that is one key value. This is the assumption that is commonly made for comparative search techniques. Balanced trees also use storage O(n), but achieve only an O(log n) access time bound. Thus we have an example where radix trees are provably superior to comparative search trees, at least in a theoretical setting. The following quotation from [TY 79] fits the theme of this paper. "The algorithms we have discussed make use of array storage; it seems that they cannot be implemented using only list structures as storage. Thus they indicate a difference in power between random access machines and pointer machines."

The difference between comparative search and radix search is just an instance of a much more general principle of algorithm design, which says that most algorithmic problems can be approached in two entirely different ways: either by organizing the set of objects to be processed, or by organizing the space in which these objects are embedded.

I think computer science has given too little attention to algorithms based on organizing the embedding space of the objects to be processed. The historical reason for this is clear, as stated at the beginning of this section: embedding spaces tend to be large, and representing them explicitly bears the danger of excessive memory requirements. But available memory sizes are steadily growing, and, even more important, we may learn new ways of compressing large embedding spaces, of mapping them homomorphically into a much smaller memory. In some cases, the IDEA of organizing the embedding space is all the algorithm designer needs - it is just a paradigm in the programmer's mind, not a data structure in the computer's memory. Let's look at an example.

I remember the impact that Warnock's algorithm [Wa 69] for hidden-surface elimination had on my thinking. The problem is to draw a picture, as an observer would see it, of a pile of 3-dimensional objects, given by coordinates of points and equations of surfaces. Earlier algorithms examined an object at a time, asking whether it was visible or obscured by other objects in front of it. The complexity of this examination is easily imagined. Warnock turned the problem around. He started by considering the screen, the space in which the picture to be drawn is embedded. He then looked at the pile of objects, asking whether, by chance, it might be very simple to draw a picture of this pile. Two reasons why it might be simple are: the pile, as seen through the window that corresponds to the screen, is empty; or, there is a surface which is clearly in front of all other objects throughout this window. The situation is usually not simple, so we divide the screen into four quadrants and recursively apply the algorithm to draw the picture for each quadrant. In the real world where coordinates are real numbers this procedure might never end, but a graphics screen is a finite embedding space with, say, 1024*1024 pixels, so the recursion will stop at the latest at depth 10, when the limit of resolution has been reached. A striking idea that makes a conceptually hard problem easy to understand.

Let us present a second example; here the organization of the embedding space leads to a tree explicitly stored in memory. Consider a set of intervals I1, I2, . . , where each interval I is a pair of numbers ((l, r), $0 < l < r < 1$. Given a query point $0 < P < 1$, we would like to list rapidly all intervals I that cover P, that is, those with $l <= P <= r$. A direct comparative search approach to this problem fails because useful left-to-right orders on a set of intervals are partial, not total: it is not always meaningful, for two arbitrary distinct intervals, to say that one is to the left of the other. McCreight [Mc 80] solved the problem by organizing an embedding space for these intervals.

Represent an interval (l,r) by a point in the left-upper half of the unit square, and a query point P on the diagonal (see figure 1). The intervals that cover P lie in the shaded quadrant with lower right corner P. How can we pre-process the set of intervals so that the answer to arbitrary queries can be listed fast?

We organize all the possible queries, that is, all the points of the range [0 . . 1], as a binary tree: root = 1/2, its sons 1/4 and 3/4, etc. In practice the number of points in the space is finite, and the potential depth of this tree is manageable. Figure 2 shows this tree: its nodes are squares that indicate all the intervals stored at this node.

These intervals are stored as lists ordered in two ways: by l-value and by r-value. A query P is answered as the example in figure 3 shows: all intervals stored at the root are listed by l-value until the first one is encountered that does not cover P; then all intervals stored at the node 1/4 are listed by r-value until the first that does not cover P; and so on down the tree. The work required is O(s + log r), where s is the size of the answer, i.e. the number of

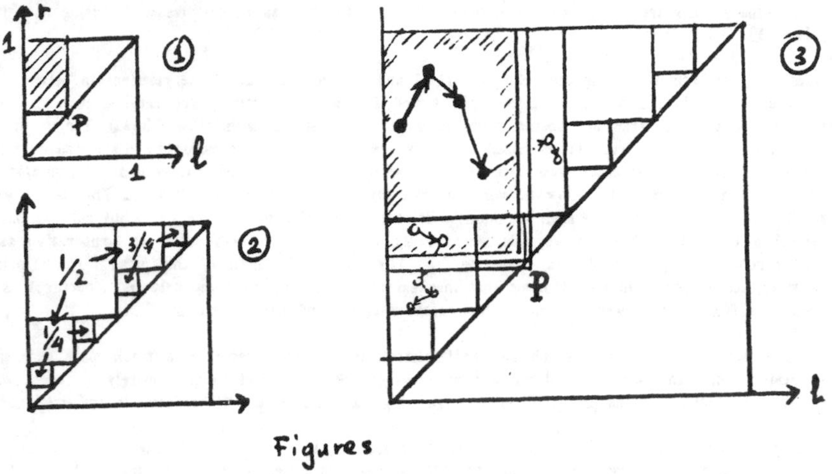

Figures

intervals that cover P, and r is the resolution of our space, i.e. the number of discrete points. The skeleton of this algorithm is given below. Let T(l,r) be the (sub-)tree that corresponds to the interval (l,r); thus T(0,1) is the entire tree; L(T,P) is read as "list all intervals stored in T that cover query P":

```
L(T,P):
  if T is empty then quit
                   else distinguish the following 3 cases:
  P = (l+r)/2 :        list intervals stored at root of T, quit

                     ( lists intervals stored at root of T whose
  P < (l+r)/2 :     <  l-value is < P;
                     ( L(left subtree of T, P);

                     ( list intervals stored at root of T whose
  P > (l+r)/2 :     <  r-value is > P;
                     ( L(right subtree of T, P);
```

In this example, the approach of organizing the embedding space led to a radix search tree, which imposed a left-to-right order on the objects. No such order is apparent when the problem is formulated as a comparative search problem.

5 The resurgence of address computation

Lists and address computation are fundamentally different techniques for implementing data structures: in the former, structural relationships are encoded by means of pointers, in the latter by formulas in the program code. Trees are usually implemented as lists. The inclusion of a section on address computation techniques (hashing, scatter storage) in a survey paper on trees therefore requires justification.

There is no a priori reason why trees should be implemented as lists. A static, regularly shaped tree is more efficiently encoded by address computation than by pointers, just as a rectangular array is. The overhead caused by pointers is justified only when we need to represent irregularly shaped or dynamic trees. In the past only a few examples were known where address computation provides an efficient representation of sets that are either irregularly shaped but static, or dynamic but regularly shaped. Perfect hash functions, tailor-made to map a set whose elements are known a priori into a small address space, illustrate the first case ([Sp 77], [Ci 80]). The second case is illustrated by the data structure called "heap", used in heapsort or as an implementation of a priority queue. A heap is an almost complete binary tree (all leaves are at the same depth level or at adjacent levels) whose structure is given by the following formula: the node with index i has sons with indices 2*i and 2*i+1.

Until recently heaps were an isolated example where address computation had been used to represent a set with an order constraint imposed on the elements – "light" elements are near the root, "heavy" ones at the bottom. Traditional hash tables are used to represent unstructured sets of objects, whose elements are accessed by name only, not by means of relationships such as predecessor or successor (see, however, [AK 75] for an attempt to impose some order on hash tables).

The failure of address computation to preserve a natural order that usually exists on the space of key values was closely tied to its inability to handle dynamic sets efficiently. I will mention a practical and a theoretical example to illustrate the difficulty of combining dynamic sets and address computation. First, the practical one. Anyone who has studied the problem of deletion in a hash table with open addressing will agree that the textbook solutions are ad hoc fixes. An element cannot simply be deleted since this might break collision sequences. But rearranging collision sequences so as to fill in the holes created by deletion is very time consuming. And the alternative of leaving deleted elements in place (marking them "deleted"), so as preserve the continuity of collision sequences, steadily lengthens the latter, so that Finds become slow. In short, deletion in conventional hash tables has no elegant solution.

Second, an example amenable to theoretical analysis: the problem of embedding extendible arrays in memory [Ro 78]. Even such a limited notion of extendibility as appending rows at the bottom or columns at the far right of a 2-dimensional array leads provably to low storage utilization.

In summary, two decades of intensive study of data structures led to a great variety of lists, particularly trees, but only a few structures based on address computation: arrays, fixed-size hash tables, and heaps – all of which exhibit a limited dynamic behavior.

Recent research by Fagin et al [FNPS 79], Larson [La 78,80], and Litwin [Li 78,80] has shown that data structures based on address computation can be efficient for storing dynamic sets. In order to understand how these recent approaches succeeded in making address computation competitive in a field where balanced trees had been used almost exclusively for a decade, it is instructive to realize that tree-structured lists and hash tables are not as unrelated as one might think: a sequence of small changes that produce several well-known and useful data structures will transform a tree that uses comparative search into a hash table. Let us explain how.

A comparative search tree is usually pictured as having the key values along with their associated data stored inside the nodes. When these trees represent files stored on disk, however, where following a pointer from node to node may cause a 50 msec disk access delay, it is inefficient to waste valuable space in the top nodes for data that does not help the search. Hence one stores only key values, and puts the entire record (key and associated data) into nodes at the bottom of the tree. The key values in the top nodes now serve only as separators – they need not identify any records. Hence they don't need to be key values of records to be stored, but can be chosen to be convenient – for example by being short, or equally spaced. A particularly convenient set of separators is the hierarchy of "radix key values" – in the binary system the set 1/2, 1/4, 3/4, 1/8, 3/8, etc. Our comparative search tree has now become a radix tree – a tree with an a priori known regular structure. Such a tree need not be represented by pointers – it can be embedded in memory in a fixed way, whereupon the address of each node can be obtained by evaluating a polynomial. Thus we have arrived at an address computation scheme similar to arrays.

A practical problem had to be solved before address computation became usable for general purpose data structures: how to compress a sparsely and perhaps unevenly filled array or radix tree so that, first, memory utilization is reasonable, and second, address computation remains simple despite a dynamically changing compression scheme.

Extendible hashing [FNPS 79] solves this problem by interposing a large directory address space between the key space and the physical address space. The directory address space is an infinite radix tree; as much of its top as is needed at any time is the current directory. The directory may become large, but as it is stored as an array, the address of the page needed to retrieve a given key value is easily computed. Extendible hashing "almost guarantees" that two disk accesses suffice to access any record: first, to the correct page of the directory, second, to the correct data bucket. This is a major advantage over trees, where 3 or 4 disk accesses are required to retrieve a record in a large file.

Is it possible to design a scheme that will "almost guarantee" a single disk access to retrieve any record by its key value? The answer appears to be YES. Litwin's linear hashing ([Li 80], see also [La 80]) uses no directory. Given a key x, we compute the address of a bucket wherein the record identified by x is found with high probability. In return for avoiding a directory, linear hashing has to reintroduce overflow chains - but thanks to a dynamically varying address formula that adapts to the content of the file, these chains are much shorter than they would be for fixed size hash files.

Address computation has given us an effective solution to a problem that tree structured files have only approached: how to manage large dynamic sets on secondary storage with 1 to 2 disk accesses on the average for each find, insert, or delete In addition, address computation techniques may solve the other two thorny problems of section 3: concurrency and multikey access. The following two examples support this statement.

Because address computation structures lack pointer chains along which structural changes may propagate, concurrent access protocols are simpler than those of trees. [HN 81] describes such a protocol for extendible hash files.

Extendible "hashing" with an identity hash function, but with the mechanism for gracefully maintaining a dynamic partition, generalizes fairly directly to multikey access. The grid file [NSH 81] allows efficient answers to multidimensional range queries (including partially specified queries) on dynamic sets stored on disk.

In addition to promising solutions to some of the fundamental problems such as concurrency and multikey access, address computation techniques have been the object of significant theoretical developments, such as the notion of a universal class of hash functions [CW 77]: roughly, a class that contains a good hash function for any set of key values. And of interesting experiments, such as the search for perfect hash functions mentioned earlier.

A comparison of the recent literature on data structures based on trees with those based on address computation reveals a striking fact: whereas trees are being embellished with frills on top of the basic concepts developed a decade ago, address computation has achieved some conceptual breakthroughs. I urge practical designers of data structures to study these, and to experiment with the modern descendants of one of the programmers' favorite tricks - the old hashing technique.

ACKNOWLEDGMENT

I am grateful to H. Hinterberger and H. Olivie' for helpful comments on an earlier draft of this paper, and to A. Muller for helping me with his formatter.

REFERENCES

[AHU 74]
Aho, A. V., Hopcroft, J. E. and Ullman, J. D.
The design and analysis of computer algorithms,
Addison-Wesley, 1974.

[AL 62]
Adelson-Velskii, G. M. and Landis, Ye. M.
An algorithm for the organization of information (in Russian),
Dokl. Akad. Nauk SSSR, Vol 146, 263-266, 1962.

[AK 75]
Amble, O. and Knuth, D. E.
Ordered hash tables, Computer J., Vol 18, 135-142, 1975.

[Be 79a]
Bentley, J. L.
Decomposable searching problems,
Inform. Proc. Letters, Vol 8, No 5, 244-251, 1979.

[Be 79b]
Bentley, J. L.
Multidimensional binary search trees in database applications
IEEE Trans. Software Engr., Vol 5, No 4, 333-340, July 1979.

[BM 72]
Bayer, R. and McCreight, E. M.
Organization and maintenance of large ordered indexes,
Acta Informatica, Vol 1, 173–189, 1972.

[BS 77]
Bayer, R. and Schkolnick, M.
Concurrency of operations on B-trees,
Acta Informatica, Vol 9, 1–21, 1977.

[BM 78]
Blum, N. and Mehlhorn, K.
On the average number of rebalancing operations in weight-balanced trees, Theor. Comp. Sci, 1978.

[CW 77]
Carter, J. L. and Wegman, M.
Universal classes of hash functions,
Report RC 6687, IBM Yorktown Heights, 1977.

[Ci 80]
Cichelli, R. J.
Minimal perfect hash functions made simple,
Comm. ACM, Vol 23, No 1, 17–19, Jan 1980.

[Co 79]
Comer, D.
The ubiquitous B-tree,
ACM Computing Surveys, Vol 11, No 2, 121–138, June 1979.

[EL 80]
Edelsbrunner, H. and van Leeuwen, J.
Multidimensional algorithms and data structures (Bibliography)
Bulletin of the EATCS, 1980.

[FNPS 79]
Fagin, R., Nievergelt, J., Pippenger, N., and Strong, H. R.,
Extendible hashing – a fast access method for dynamic files,
ACM Trans. Database Systems, Vol 4, No 3, 315–344, Sep 1979.

[Fr 60]
Fredkin, E.
Trie memory, Comm. ACM, Vol 3, 490–500, 1960.

[GN 45]
Goldstine, H. H. and von Neumann, J.
Planning and coding of problems for an electronic computing instrument, Part II, Vol 1, 1947;
reprinted in A. H. Taub (ed.), "John von Neumann – Collected Works", Vol 5, Pergamon Press, 1963.

[GS 78]
Guibas, L. and Sedgewick, R.
A dichromatic framework for balanced trees, Proc.
19-th Annual Symp. Found. of Computer Sci., 8–21, IEEE, 1978.

[HN 81]
Hinterberger, H. and Nievergelt, J.,
Concurrent access control in extendible hash files,
(submitted)

[Kn 68], [Kn 73]
Knuth, D. E.
The art of computer programming, Addison-Wesley,
Vol 1, Fundamental Algorithms, 1968,
Vol 3, Sorting and Searching, 1973.

[KL 80]
Kung, H. T. and Lehman, P. L.
Concurrent manipulation of binary search trees,
ACM Trans. Database Sys, Vol 5, No 3, 354-382, Sep 1980.

[La 78]
Larson, P.
Dynamic hashing, BIT, Vol 18, 184-201, 1978.

[La 80]
Larson, P.
Linear hashing with partial expansions,
Proc. 6-th Conf. on Very Large Databases, Montreal, Oct 1980.

[LM 80]
van Leeuwen, J. and Maurer, H. A.
Dynamic systems of static data structures,
Univ. Graz, Institut Informationsver. Report 42, Jan 1980.

[Li 78]
Litwin, W.
Virtual hashing: a dynamically changing hashing,
Proc. 4-th Conf. Very Large Databases, Berlin, 1978, 517-523.

[Li 80]
Litwin, W.
Linear hashing: a new tool for file and table addressing,
Proc. 6-th Conf. on Very Large Databases, Montreal, Oct 1980.

[Mc 80]
McCreight, E. M.
Efficient algorithms for enumerating intersecting intervals
and rectangles, XEROX PARC Report CSL-80-9, 1980.

[Me 78]
Mehlhorn, K.
Arbitrary weight changes in dynamic trees,
RAIRO, Th CS

[Me 80]
Mehlhorn, K.
Lower bounds on the efficiency of static to dynamic transforms
of data structures,
Univ. Saarland Report 1980.

[NR 73]
Nievergelt, J. and Reingold, E. M.
Binary search trees of bounded balance,
SIAM J. Computing, Vol 2, No 1, 33-43, Mar 1973.

[Ni 74]
Nievergelt, J.
Binary search trees and file organization,
ACM Computing Surveys, Vol 6, No 3, 195-207, Sep 1974.

[NSH 81]
Nievergelt, J., Sevcik, K. and Hinterberger, H.
The grid file: a dynamic multikey access structure,
(in preparation)

[NS 56]
Newell, A. and Simon, H. A.
The logic theory machine - A complex information processing system,
IRE Trans. Information Theory, Vol IT-2, No 3, 61-79, Sep 1956.

[Ol 81]
Olivie', H. J.
Half-balanced binary search trees,
Report 81-01, IHAM, B-2000 Antwerp, 1981.

[Ro 78]
Rosenberg, A. L.
Storage mappings for extendible arrays,
in R. T. Yeh (ed.), Current Trends in Programming Methodology,
Vol IV: Data Structuring, Ch 10, Prentice-Hall 1978.

[SB 79]
Saxe, J. B. and Bentley, J. L.
Transforming static data structures to dynamic structures,
20-th IEEE Symp. Foundations of Computer Scsi., 148-168, 1979.

[Sp 77]
Sprugnoli, R.
Perfect hashing functions: a single probe retrieving method
for static sets, Comm. ACM, Vol 20, No 11, 841-850, Nov 1977.

[TY 79]
Tarjan, R. E. and Yao, A. C.-C.
Storing a sparse table,
Comm. ACM, Vol 22, No 11, 606-611, Nov 1979.

[Wa 69]
Warnock, J. E.
A hidden-surface algorithm for computer generated half-tone
pictures, Univ. Utah, Dept. Computer Sci. TR 4-15, 1969.

[Wi 78]
Willard, D. E.
Balanced forests of h-d trees as a dynamic data structure,
Harvard Univ., Aiken Computer Lab Report TR-23, 1978.

INFINITARY RELATIONS

Maurice NIVAT
Laboratoire d'Informatique Théorique et Programmation
I.N.R.I.A.
Domaine de Voluceau
Rocquencourt
78150 LE CHESNAY
FRANCE

I - INTRODUCTION

If one considers the set of finite behaviours of a process p as a subset of the free monoïd generated by the finite alphabet of actions A, let us denote it $HR^{\star}(p)$, one is lead to extend it to infinity to include infinite behaviours which are infinite words on the alphabet A. The set $HR^{\omega}(p)$ of such infinite behaviours is, in the normal case of a process p which has the finite non determinism property, linked to $HR^{\star}(p)$ by the formula :

$$HR^{\omega}(p) = \{u \in A^{\omega} \mid \forall n \in \mathbb{N} \quad u[n] \in HR^{\star}(p)\}$$

In other words an infinite sequence of actions $u \in A^{\omega}$ is an infinite behaviour of p if and only if the initial segment of length n of u (denoted u[n]) is a finite behaviour of p for all n.

And we can reformulate this by writing simply :

$$HR^{\omega}(p) = \mathrm{Adh}\ (HR^{\star}(p))$$

by using the notion of adherence as it is defined in [3]. The adherence of a finitary language $L \subseteq A^{\star}$ is the set of cluster points of L in the natural metric topology on $A^{\infty} = A^{\star} \cup A^{\omega}$.

Now if a finite number k of processes $p_1,\dots,p_k$ behave simultaneously respecting some synchronisation condition S, a finite behaviour of the system $(\vec{p}, S)$ thus formed is a k-uple of words $<f_1,\dots,f_k>$ where for all $i \in [k]$, $f_i \in HR^{\star}(p_i)$.

In [7] it is suggested that a general form of a synchronisation condition is the pair $<p_S, \vec{\varphi}>$ of a synchronizing process p_S and a multimorphism $\vec{\varphi} = <\varphi_1,\dots,\varphi_k>$ where for all $i \in [k]$ φ_i maps the alphabet A of p_S into $A_i \cup \{\varepsilon\}$ where A_i is the alphabet of actions of p_i.

Then $HR^{\star}(\vec{p}, S)$ is exactly the set of all k-uples $<f_1,\dots,f_k>$ which lie in $HR^{\star}(p_1) \times \dots \times HR^{\star}(p_k)$ and satisfy the synchronisation condition :

$$\exists\, g \in HR^{\star}(p_S) \ :\ \forall\, i \in [k] \quad \varphi_i(g) = f_i$$

We write :

$$HR^{*}(\vec{p},S) = (HR^{*}(p_1) \times \ldots \times HR^{*}(p_k)) \cap \vec{\varphi}\,(HR^{*}(p_S))$$

A question is immediately raised which is to define the infinite behaviours of the system $(\vec{p}, S)$. Given a finitary relation $R \subset A_1^{*} \times \ldots \times A_k^{*}$ can we define such a thing as the adherence of R and write $HR^{\omega}(\vec{p},S) = Adh(HR^{*}(\vec{p},S))$?

It is immediate to see that this question cannot be answered as easily in the case of relations as in the case of languages though it is of the utmost importance for the study of systems of synchronized processes. The present paper is devoted to the mathematical problem of extending to infinity finitary relations : it should be read as a sequel of [6] and a companion paper to [7] . It is an essential part of a theory of infinite computations and synchronization which the author tries to build in close collaboration with André Arnold [1].

II - LEFT FACTORS AND ENTENDABILITY

For words in A^∞ the notion of left factors is well-known. We define it in the following way :

- if $f = f(1) \dots f(n)$ is a word of length n in A^* and $p \in \mathbb{N}$
  $f[o] = \varepsilon$ where ε is the empty word
  $f[p] = f(1) \dots f(p)$ for $1 \le p \le n$
  $f[p] = f[n] = f$ for $n \le p$

- if $u = u(1)\ u(2) \dots u(n) \dots$ is an infinite word in A^ω and $p \in \mathbb{N}$
  $u[o] = \varepsilon$
  $u[p] = u(1) \dots u(p)$ for $p \ge 1$

Then for all $\alpha \in A^\infty$ we define

$$FG(\alpha) = \{\alpha[p] \mid p \in \mathbb{N}\} \text{ which is a subset of } A^*$$

The product on A^∞ is defined by :

- if $f \in A^n$ and $g \in A^p$, fg is the word of length n+p given by :
  $(fg)(\ell) = f(\ell)$ for $\ell \le n$
  $(fg)(\ell) = g(\ell-n)$ for $n+1 \le \ell \le n+p$

- if $f \in A^n$ and $u \in A^\omega$, fu is the infinite word given by
  $(fu)(\ell) = f(\ell)$ for $\ell \le n$
  $(fu)(\ell) = u(\ell-n)$ for $\ell \ge n+1$

- if $u \in A^\omega$ and $\alpha \in A^\infty$: $u\,\alpha = u$

The relation of extendability is then defined by :

$$\alpha \le \beta \iff \exists\, \gamma : \alpha\,\gamma = \beta$$

We say that α is extendable into β iff $\alpha \le \beta$.

Clearly $\alpha \le \beta \iff \alpha \in FG(\beta)$ or $\alpha = \beta \in A^\omega$.

Some equalities and equivalences hold

$$FG(FG(\alpha)) = FG(\alpha) \text{ (where } FG(FG(\alpha)) = \cup \{FG(\beta) \mid \beta \in FG(\alpha)\}$$
$$\alpha \le \beta \iff FG(\alpha) \subseteq FG(\beta)$$
$$\alpha = \beta \iff FG(\alpha) = FG(\beta)$$
$$\alpha \in A^* \iff \text{card } (FG(\alpha)) < \infty$$

Define an FG-set as any set $L \subseteq A^\infty$ which is totally ordered by the relation $\le$. In other words :

$$L \text{ is an FG-set} \iff \forall\, \alpha,\beta \in L \text{ either } \alpha \le \beta \text{ or } \beta \le \alpha.$$

Property 1 : For every FG-set $L \subseteq A^\infty$, $L \neq \emptyset$, there exists a unique word $\alpha \in A^\infty$ such that :

$$L \subseteq FG(\alpha) \text{ and for all } \beta \quad L \subseteq FG(\beta) \Rightarrow \alpha \le \beta$$

This word α is denoted Sup (L) and characterized by :

$$FG(Sup(L)) = FG(L) = \cup \{FG(\alpha) \mid \alpha \in L\}.$$

We assume the above property to be well-known and introduce a few more useful notations :

$$\alpha \in FG(\beta) \Rightarrow \alpha \in A^* \text{ and } \exists\,!\ \gamma : \ \alpha\gamma = \beta$$

This unique element γ is denoted $(\beta{:}\alpha)$.

If $L \subseteq A^\infty$ is an infinitary language and $f \in A^*$ a finite word :

$$(L{:}f) = \{(\beta{:}f) \mid \beta \in L \text{ and } f \le \beta\}$$

Clearly we have :

$$f \in L \iff \varepsilon \in (L{:}f)$$

L is an FG-set implies $(L{:}f)$ is an FG-set and if $(L{:}f) \neq \emptyset$

$$f\ Sup\ (L{:}f) = Sup(L)$$

This last assertion comes from the fact that :

$$FG(L{:}f) = (FG(L){:}f)$$

In this paragraph we extend the well-known notions just recalled to multiwords and relations.

If $A_1,\ldots,A_k$ are alphabets, it is convenient to denote $\mathcal{A} = A_1^\infty \times \ldots \times A_k^\infty$. Elements in $\mathcal{A}$ are called multiwords (or k-words if we wish to specify the arity) : they are k-uples of the form :

$$\vec{\alpha} = \langle\alpha_1, \ldots, \alpha_k\rangle \quad \text{where } \forall\, i \in [k] \;\; \alpha_i \in A_i^\infty$$

We write also $\alpha_i = \pi_i(\vec{\alpha})$.

The multiword $\vec{\alpha}$ is finite iff $\pi_i(\vec{\alpha}) \in A_i^*$ for all $i \in [k]$ and if $\vec{\alpha}$ is finite we define its maximal length

$$|\vec{\alpha}| = \max\, \{\pi_i(\vec{\alpha}) \mid i \in [k]\}$$

The multiword $\vec{\alpha}$ is infinite iff $\pi_i(\vec{\alpha}) \in A_i^\omega$ for at least one i.
Its maximal length $|\vec{\alpha}|$ is then set to be infinite $|\vec{\alpha}| = \infty$.
The minimal length of a multiword is :

$$||\vec{\alpha}|| = \min\, \{\pi_i(\vec{\alpha}) \mid i \in [k]\}$$

This minimal length is finite unless $\pi_i(\vec{\alpha}) \in A_i^\omega$ for all $i \in [k]$
If this is the case $\vec{\alpha}$ is said to be totally infinite.

We denote :

$$\mathcal{A}^{fin} = \{\vec{\alpha} \in \mathcal{A} \mid |\vec{\alpha}| < \infty\}$$
$$\mathcal{A}^{inf} = \{\vec{\alpha} \in \mathcal{A} \mid |\vec{\alpha}| = \infty\}$$
$$\mathcal{A}^{tinf} = \{\vec{\alpha} \in \mathcal{A} \mid ||\vec{\alpha}|| = \infty\}$$

One multiplies multiwords component wise :

$$\vec{\alpha}\,\vec{\beta} = \langle\alpha_1\,\beta_1, \ldots, \alpha_k\,\beta_k\rangle$$

and denote $\vec{\varepsilon}$ the multiword each component of which is ε so that :

$$\vec{\varepsilon}\,\vec{\alpha} = \vec{\alpha}\,\vec{\varepsilon} = \vec{\alpha} \quad \text{for all } \vec{\alpha}.$$

We first define the relation of entendability :

$$\vec{\alpha} \le \vec{\beta} \iff \exists\, \vec{\gamma} \quad \vec{\alpha}\,\vec{\gamma} = \vec{\beta}$$
$$\iff \forall\, i \in [k] \quad \alpha_i \le \beta_i .$$

We denote for all $\vec{\alpha} \in \mathcal{R}$:

$$PP(\vec{\alpha}) = \{\vec{f} \in \mathcal{R}^{fin} \mid \vec{f} \leq \vec{\alpha}\}$$

We clearly have :

$$PP(\vec{\alpha}) = FG(\alpha_1) \times \ldots \times FG(\alpha_k)$$

and if $\vec{f} \in PP(\vec{\alpha})$ there exists a unique $\vec{\beta} \in \mathcal{R}$ denoted $(\vec{\alpha}:\vec{f})$ such that $\vec{\alpha}\,\vec{\beta} = \vec{\alpha}(\vec{\alpha}:\vec{f}) = \vec{\alpha}$.

Obviously $\pi_i(\vec{\alpha}:\vec{f}) = (\alpha_i:f_i)$ for all $i \in [k]$.

If we now define a PP-set as an relation $R \subseteq \mathcal{R}$ which is totally ordered by $\leq$ we can state a property which is very similar to property 1.

Property 2 : For every non empty PP-relation $R \subseteq \mathcal{R}$ there exists a unique multiword $\vec{\alpha} \in \mathcal{R}$ such that :

$$R \subseteq PP(\vec{\alpha}) \text{ and for all } \vec{\beta} \in \mathcal{R}\ R \subseteq PP(\vec{\beta}) \Rightarrow \vec{\alpha} \leq \vec{\beta}$$

This $\vec{\alpha}$ is denoted Sup(R) and is characterized by :

$$PP(Sup(R)) = PP(R) = \cup\, \{PP(\vec{\alpha}) \mid \vec{\alpha} \in R\}.$$

Proof : If R is a PP-set then for all $i \in [k]$
$\pi_i(R) = \{\pi_i(\vec{\alpha}) \mid \vec{\alpha} \in R\}$ is an FG-language and $R \neq \emptyset \Leftrightarrow \pi_i(R) \neq \emptyset$ for all i.
$Sup(R) = \langle Sup(\pi_1(R)), \ldots, Sup(\pi_k(R))\rangle$ has all the desired properties. □

Many rather obvious identities and equivalences can be stated :

$$PP(PP(\vec{\alpha})) = PP(\vec{\alpha})$$

$$\vec{\alpha} \leq \vec{\beta} \Leftrightarrow PP(\vec{\alpha}) \subseteq PP(\vec{\beta})$$

$$\vec{\alpha} = \vec{\beta} \Leftrightarrow PP(\vec{\alpha}) = PP(\vec{\beta})$$

$\vec{\alpha} = Sup\ (PP(\vec{\alpha}))$ ($PP(\vec{\alpha})$ is clearly a non empty PP-set)

$\vec{f}\ Sup\ (R:\vec{f}) = Sup\ (R)$ if R is a PP-set and $R:\vec{f} \neq \emptyset$

A major difficulty when dealing with multiwords and relations is that we have to distinguish between the two relations "$\vec{\alpha}$ is extendable into $\vec{\beta}$" and the relation we now define "$\vec{\alpha}$ is a left factor of $\vec{\beta}$". For $\vec{\alpha} \in \mathcal{R}$ define

$$\vec{\alpha}[p] = \langle\alpha_1[p], \ldots, \alpha_k[p]\rangle$$

$$FG(\vec{\alpha}) = \{\vec{\alpha}[p] \mid p \in \mathbb{N}\}$$

It is no longer true that :

$$\vec{\alpha} \le \vec{\beta} \iff \vec{\alpha} \in FG(\vec{\beta}) \text{ or } \vec{\alpha} = \vec{\beta} \in \mathscr{A}^{inf}$$

(one can provide immediate exemples).

We have the following inclusions identities and implications :

$$FG(FG(\vec{\alpha})) = FG(\vec{\alpha}) \tag{1}$$

Proof : For all $\vec{\alpha} \in \mathscr{A}$, n, p $\in$ IN

$$(\vec{\alpha}[n])\,[p] = \vec{\alpha}[\min(n,p)] \qquad \square$$

$$FG(\vec{\alpha}) = FG(\vec{\beta}) \iff \vec{\alpha} = \vec{\beta} \tag{2}$$

Proof : For all $i \in [k]$ we have :

$$\pi_i(FG(\vec{\alpha})) = FG(\pi_i(\vec{\alpha}))$$

for if $f_i \in FG(\alpha_i)$ and $n = |f_i|$ then $\pi_i(\vec{\alpha}[n]) = \alpha_i[n] = f$ which proves one inclusion. The reverse inclusion is obvious.

Then $FG(\vec{\alpha}) = FG(\vec{\beta})$ implies for all $i \in [k]$

$$\pi_i(FG(\vec{\alpha})) = \pi_i(FG(\vec{\beta}))$$

which implies $FG(\pi_i(\vec{\alpha})) = FG(\pi_i(\vec{\beta}))$.

Thus for all $i \in [k]$ $\pi_i(\vec{\alpha}) = \pi_i(\vec{\beta})$ and $\vec{\alpha} = \vec{\beta}$ $\square$

$$FG(\vec{\alpha}) \subseteq FG(\vec{\beta}) \Rightarrow \vec{\alpha} \le \vec{\beta} \tag{3}$$

The proof is the same as above.

The interesting fact is that the reverse is not true and we can only state if we define for all R, R' $\subseteq \mathscr{A}$

$$R \le R' \iff \forall \vec{\alpha} \in R \ \exists \vec{\beta} \in R' \quad \vec{\alpha} \le \beta$$

$$\vec{\alpha} \le \vec{\beta} \iff FG(\vec{\alpha}) \le FG(\vec{\beta}) \tag{4}$$

Proof : $\alpha_i \le \beta_i \Rightarrow \alpha_i[n] \le \beta_i[n]$ for all n.

Whence $\vec{\alpha} \le \vec{\beta}$ > $\vec{\alpha}[n] \le \vec{\beta}[n]$ for all n.

Conversely $FG(\vec{\alpha}) \le FG(\vec{\beta})$ implies.

$\pi_i(FG(\vec{\alpha})) \le \pi_i(FG(\vec{\beta}))$ for all i since $f_i \in \pi_i(FG(\vec{\alpha})) \Rightarrow f_i = \pi_i(\vec{f}\,)$ for some $\vec{f} \in FG(\vec{\alpha})$.

There exists then $\vec{g} \in FG(\vec{\beta})$ such that $\vec{f} \leq \vec{g}$ and clearly $f_i = \pi_i(\vec{f}) \leq \pi_i(\vec{g}) \in \pi_i(FG(\vec{\beta}))$.
For all i we then have $FG(\pi_i(\vec{\alpha})) \leq FG(\pi_i(\vec{\beta}))$ which obviously implies $FG(\pi_i(\vec{\alpha})) \subseteq FG(\pi_i(\vec{\beta}))$ since $f \leq g \in FG(\pi_i(\vec{\beta})) \Rightarrow f \in FG(g) \subseteq FG(\pi_i(\vec{\beta}))$.
Eventually for all i $\pi_i(\vec{\alpha}) \leq \pi_i(\vec{\beta})$ and $\vec{\alpha} \leq \vec{\beta}$. □

For all $\vec{\alpha} \in \mathscr{X}$ $FG(\vec{\alpha})$ is a non empty PP set and we have $\vec{\alpha} = \text{Sup}\ (FG(\vec{\alpha}))$. (5)

An immediate consequence is that $PP(\vec{\alpha}) = PP(FG(\vec{\alpha})) = FG(PP(\vec{\alpha}))$.

Proof. Clearly for all n, p

$$\vec{\alpha}[\min(n,p)] \leq \vec{\alpha}[\max(n,p)]$$

Suppose $\vec{\beta} = \text{Sup}(FG(\vec{\alpha}))$. Then for all i :

$$\pi_i(\vec{\beta}) = \text{Sup}\ (\pi_i(FG(\vec{\alpha}))) = \text{Sup}\ (FG(\pi_i(\vec{\alpha}))) = \pi_i(\vec{\alpha}). \quad \square$$

III - CONVERGING SEQUENCES, ADHERENCES AND CENTERS

In A^∞ there exists a very natural metric topology defined by the distance $d : A^\infty \times A^\infty \to \mathbb{R}_+$ given by :

$$d(\alpha,\beta) = 2^{-\min\{n|\alpha[n] \neq \beta[n]\}} \text{ if } \alpha \neq \beta$$

$$d(\alpha,\beta) = 0 \text{ if } \alpha = \beta$$

A^∞ equiped with the metric d and the topology induced by d is a complete metric space which is also compact.

It may be useful to recall a few properties of this d-topology. The sequence α_n, $n \in \mathbb{N}_+$ converges iff it is d-Cauchy, i.e. satisfies :

$$\forall\, n \in \mathbb{N}\ \exists\, N \in \mathbb{N} \quad p,q \geq N \Rightarrow \alpha_p[n] = \alpha_q[n]$$

Its limit $\alpha_o = \lim \alpha$ is then the unique word in A^∞ such that :

$$\forall\, n \in \mathbb{N}\ \exists\, N \in \mathbb{N} \quad p \geq N \Rightarrow \alpha_p[n] = \alpha_o[n]$$

And we can see that $\lim \alpha$ is a finite word $\alpha_o \in A^*$ only if the sequence α_n is stationary :

$$\forall\, n \in \mathbb{N}\ \exists\, N \in \mathbb{N} \quad p \geq N \Rightarrow \alpha_p = \alpha_o$$

A property which will play a role in the sequel is that :

<u>Property 3</u> : From every sequence α_n, $n \in \mathbb{N}_+$ of words in A^∞ one can extract a converging subsequence.

<u>Proof</u> : The statement means that one can find an infinite subset N' of $\mathbb{N}$ such that α_n, $n \in N'$ converges, i.e. satisfies the d-Cauchy condition :

$$\forall\, n\ \exists\, N \quad p,q \geq N \text{ and } p,q \in N' \Rightarrow \alpha_p[n] = \alpha_q[n]$$

Let us remark immediately that the convergence of α_n, $n \in \mathbb{N}_+$ implies the convergence of α_n, $n \in N'$ for all infinite subset N' of $\mathbb{N}_+$.

Two cases arise :

- The sequence of lengths $|\alpha_n|$, $n \in \mathbb{N}_+$ is ultimately bounded i.e. satisfies :

$$\exists N \in \mathbb{N}_+, M \in \mathbb{N} \ \forall n \geq N \quad |\alpha|_n \leq M$$

Then for all $n \geq N$, α_n belongs to the finite set :

$$A^{\leq M} = \{f \in A^* \mid |f| \leq M\}$$

And certainly there exists $f_o \in A \leq M$ such that :

$$N' = \{n \geq N \mid \alpha_n = f_o\} \text{ is infinite.}$$

Then α_n, $n \in N'$ converges towards f_o.

- The sequence $|\alpha_n|$ is not ultimately bounded i.e.

$$\forall N \in \mathbb{N}_+, M \in \mathbb{N} \quad \exists n \geq N \quad |\alpha_n| > M$$

Define for all $m \in \mathbb{N}_+$:

$$E_m = \{f \in A^m \mid \text{card } \{n \mid f \leq \alpha_n\} = \infty\}$$

This set is for all m finite and non empty : finite for $E_m \subseteq A^m$ which is finite and non empty for there are infinitely many n's for which $|\alpha_n| \geq m$ and thus has a left factor in A^m. Whence certainly at least one $f \in A^m$ is a left factor of α_n for infinitely many n's.

Obviously if $f \in E_{m+1}$, $f[m] \in E_m$.

By Koenig's lemma one can find a sequence f_m, $m \in \mathbb{N}_+$ such that $f_m \in E_m$ and $f_m = f_{m+1}[m]$ for all m.

This sequence obviously converges towards $u \in A^\omega$ such that $u[m] = f_m$ for all m.

Now we can build an infinite strictly increasing sequence of integers $n_1, n_2, \ldots, n_m, \ldots$ such that for all m $f_m \leq \alpha_{n_m}$ and $n_m < n_{m+1}$. The sequence α_{n_m} converges towards u. □

It will be quite convenient to use increasing sequences of words. In fact any increasing sequence α_n, $n \in \mathbb{N}_+$ such that :

$$\forall n \quad \alpha_n \leq \alpha_{n+1} \text{ is converging to Sup } \{\alpha_n\}.$$

Closed subsets : an infinitary language $L \subseteq A^\infty$ is closed iff it contains the limits of sequences α_n, $n \in \mathbb{N}_+$ where $\alpha_n \in L$. From the definition of $\alpha_o = \lim \alpha_n$ it is clear that $\alpha_o \in A^* \Longleftrightarrow \alpha_o \in L^{fin} = L \cap A^*$. And $\alpha_o \in A^\omega \Longleftrightarrow \forall n \exists p\ \alpha_o[n] = \alpha_q[n]$ for all $q > p$. It follows that all the limits of sequences of words in L are finite or belong to $Adh(L) = \{u \in A^\omega \mid FG(u) \subseteq FG(L)\}$. And $u \in Adh(L) \Rightarrow \forall n \exists \alpha_n \in L$ $u[n] = \alpha_n[n] \Rightarrow u = \lim \alpha_n$.

We can state the :

Property 4 : The infinitary language $L \subseteq A^\infty$ is closed iff :

$$Adh(L) \subseteq L$$

The topological closure of L i.e. the smallest closed language containing L is $\bar{L} = L \cup Adh(L)$.

The set Adh(L) is called the adherence of L in [3]. Many properties of this adherence have been established in the same papers : a major tool to study the adherence is the center of a language defined as the set of left factors of the adherence :

$$\underline{L} = FG(Adh(L))$$

The center is characterized by :

$$\underline{L} = FG(L^{inf}) \cup \{f \in FG(L^{fin}) \mid (L:f) \text{ is infinite}\}$$

In this writing $L^{inf} = L \cap A^\omega$, $L^{fin} = L \cap A^*$.

And we have :

$$Adh(L) = Adh(\underline{L})$$

$$Adh(L_1) = Adh(L_2) \Longleftrightarrow \underline{L_1} = \underline{L_2}$$

IV - ADHERENCES AND CENTERS OF RELATIONS

The cartesian product $\mathcal{A} = A_1^\infty \times \ldots \times A_k^\infty$ can be equiped with the distance d : $\mathcal{A}\times\mathcal{A} \to \mathbb{R}_+$ given by :

$$d(\vec{\alpha},\vec{\beta}) = 2^{-\min\{n \mid \vec{\alpha}[n] \neq \vec{\beta}[n]\}} \text{ if } \vec{\alpha} \neq \vec{\beta}$$

$$= 0 \text{ if } \vec{\alpha} = \vec{\beta}$$

Clearly :

$$d(\vec{\alpha},\vec{\beta}) = \max\{d(\alpha_i,\beta_i) \mid i \in [k]\}$$

<u>Proof</u> : If $\vec{\alpha} \neq \vec{\beta}$ and n is the smallest integer such that $\vec{\alpha}[n] \neq \vec{\beta}[n]$ we have :

$$\vec{\alpha}[n-1] = \vec{\beta}[n-1] \text{ and for at least one i}$$

$$\pi_i\ (\vec{\alpha}[n]) \neq \pi_i\ (\vec{\beta}[n]).$$

Thus for all i $d(\alpha_i,\beta_i) \leq 2^{-n}$ and for a least one i $d(\alpha_i,\beta_i) = 2^{-n}$. □

The topology induced by d on $\mathcal{A}$ is the product topology.

Thus the sequence $\vec{\alpha}_n$, $n \in \mathbb{N}$ of multiwords converges iff and only if it is d-Cauchy i.e. :

$$\forall n\ \exists N \quad p,q > N \Rightarrow \vec{\alpha}_p[n] = \vec{\alpha}_q[n]$$

and this happens iff for all i :

$$\pi_i(\vec{\alpha}_n),\ n \in \mathbb{N} \text{ is d-Cauchy}$$

If $\vec{\alpha}_n$, $n \in \mathbb{N}$ is d-Cauchy it converges towards $\vec{\alpha}_o$, which is unique, such that :

$$\forall n\ \exists N \quad p > N \Rightarrow \vec{\alpha}_o[n] = \vec{\alpha}_p[n]$$

and if $\vec{\alpha}_o = \lim \vec{\alpha}_n$ we have :

$$\pi_i\ (\lim \vec{\alpha}_n) = \lim\ (\pi_i(\vec{\alpha}_n)) \text{ for all i.}$$

More precisely $\vec{\alpha}_n$ converges iff $\pi_i(\vec{\alpha}_n)$ converges for all i and $\lim\ (\vec{\alpha}_n) = \langle \lim \pi_1(\vec{\alpha}_n), \ldots, \lim \pi_k(\vec{\alpha}_n)\rangle$.

We can entend property 3 and state.

Property 5 : From any sequence $\vec{\alpha}_n$, $n \in \mathbb{N}$ of multiwords. We can extract a converging subsequence.

Proof : We extract the converging subsequence :

$$\pi_1 (\vec{\alpha}_n),\ n \in N_1 \text{ from } \pi_1 (\vec{\alpha}_n)$$

Then we can extract $\pi_2 (\vec{\alpha}_n)$, $n \in N_2$ from the sequence $\pi_2 (\vec{\alpha}_n)$, $n \in N_1$: this means that we can find an infinite subset N_2 of N_1 such that :

$$\pi_2 (\vec{\alpha}_n),\ n \in N_2 \text{ converges.}$$

By induction we thus build $N_1 \supseteq N_2 \supseteq \dots \supseteq N_k$ such that $\pi_i(\vec{\alpha}_n)$, $n \in N_i$ converges for all i.

By a remark made above $N_k \subseteq N_i$ and N_k infinite implies that $\pi_i(\vec{\alpha}_n)$, $n \in N_k$ converges.

Thus $\vec{\alpha}_n$, $n \in N_k$ converges. □

Closed relations

Let $R \subseteq \mathcal{A}$ be an infinitary relation.

It is closed iff it contains all the limits of converging sequences of multiwords in R.

As in the case of languages $\vec{\alpha}_n$, $n \in \mathbb{N}$ converges towards $\vec{\alpha}_o \in \mathcal{A}^{fin}$ iff $\vec{\alpha}_n = \vec{\alpha}_o$ for all sufficiently large n's and this implies $\vec{\alpha}_o \in R^{fin} = R \cap \mathcal{A}^{fin}$.

And the set of infinite limits :

$$\{\vec{\alpha}_o \in \mathcal{A}^{inf} \mid \vec{\alpha}_o = \lim \vec{\alpha}_n,\ \alpha_n \in R\}$$

is equal to :

$$Adh(R) = \{\vec{\alpha} \in \mathcal{A}^{inf} \mid FG(\vec{\alpha}) \subseteq FG(R)\}$$

We call Adh(R) the adherence of R.

Property 6 : The infinitary relation R is closed (in the d-topology) iff $Adh(R) \subseteq R$. The topological closure of R is $\bar{R} = R \cup Adh(R)$.

As in the case of languages a major tool to study adherences and closed relations is the notion of center.

The center of a relation $R \subseteq \mathcal{A}$ is $\underline{R} = FG(Adh(R))$. We first characterize the center of a finitary relation.

<u>Property 7</u> : If $R \subseteq \mathcal{A}^{fin}$ is finitary

$$\underline{R} = \{\vec{f} \in \mathcal{A}^{fin} \mid \text{card}\ \{\vec{g} \in R \mid \vec{f} \in FG(\vec{g})\} = \infty\}$$

Proof : We denote $FG^{-1}(\vec{f})$ the set of all $\vec{g}$ such that

$$\vec{f} \in FG(\vec{g}).$$

Suppose first $\vec{f} \in FG(\vec{\alpha})$ for some $\vec{\alpha} \in Adh(R)$. For all n $\vec{\alpha}[n] \in FG(R)$ i.e. for all n there exists $\vec{g}_n \in R$ such that $\vec{\alpha}[n] = \vec{g}_n[n]$.

The fact that $\vec{\alpha} \in \mathcal{A}^{inf}$ implies that :

$|\vec{\alpha}[n]| = n$ whence $|\vec{g}_n| \geq n$ and the set of $\vec{g}_n$'s is infinite.

If $|\vec{f}| = p$, $\vec{f} = \vec{\alpha}[p] = (\vec{\alpha}[n])\ [p]$ for all $n \geq p$.

Thus $\vec{f} = (\vec{g}_n[n])\ [p] = \vec{g}_n[p]$ for all $n \geq p$. And $FG^{-1}(\vec{f}) \cap R$ is infinite.

Conversely suppose $FG^{-1}(\vec{f}) \cap R$ is infinite : this implies $FG^{-1}(\vec{f}) \cap FG(R)$ is infinite since $R \subseteq \mathcal{A}^{fin} \Rightarrow R \subseteq FG(R)$.

Consider then the following sequence of sets indexed by $m \geq n = |\vec{f}|$:

$$E_m = \{\vec{g} \in FG(R) \mid \vec{f} \in FG(\vec{g}) \text{ and } |\vec{g}| = m\}$$

This set is for all m finite (obviously) and non empty for $FG^{-1}(\vec{f}) \cap FG(R)$ implies the existence of $\vec{g}$ in this set with arbitrarily large $|\vec{g}|$. And if $\vec{f} = \vec{g}[n]$, $|\vec{g}| \geq m$ then $\vec{f} = (\vec{g}[m])\ [n]$ whence $\vec{g}[m] \in E_m$ since $|\vec{g}[m]| = m$.

Now if $\vec{g} \in E_{m+1}$ clearly $\vec{g}[m] \in E_m$.

We can find, by Koenig's lemma, $\vec{g}_m \in E_m$ such that for all m, $\vec{g}_m = \vec{g}_{m+1}[m]$.

The increasing sequence $\vec{g}_m$ converges towards some $\vec{\alpha} \in Adh(R)$ since $\vec{\alpha}[m] = \vec{g}_m \in FG(R)$ for all m.

And we have $\vec{f} = \vec{\alpha}[n] \in FG(Adh(R))$. □

One could have been tempted to write :

$$\underline{R} = \{\vec{f} \in FG(R) \mid (R:\vec{f}) \text{ is infinite}\}.$$

This is false as proved by the exemple :

$$R = \{<a^n, b^{2n}> \mid n \in \mathbb{N}\}$$

$$FG(R) = \{<a^n, b^p> \mid n \leq p \leq 2n\}$$

Clearly $R : <a^n, b^p>$ is infinite for all n, $n \leq p \leq 2n$ for :

$$<a^n, b^p> <a^q, b^{2n+2q-p}> = <a^{n+q}, b^{2n+2q}> \in R$$

The adherence of R is $<a^\omega, b^\omega>$ and the center is

$$\underline{R} = <a^n, b^n>.$$

One can check that $FG^{-1}(<a^n, b^p>) \cap R$ is infinite iff $n = p$: indeed $<a^n, b^n> \in FG(<a^m, b^{2m}>)$ for all $m \geq n$ and $<a^n, b^p>$, $n < p \leq 2\,n$ is a left factor of $<a^n, b^m>$, $p \leq m \leq 2n$. □

Now the center of an infinitary relation is :

$$\underline{R} = FG(R^{inf}) \cup \underline{R^{fin}} \text{ where :}$$

$$R^{inf} = R \cap \mathcal{R}^{inf} \text{ and } R^{fin} = R \cap \mathcal{A}^{fin}.$$

The properties of the center we stated for languages are preserved :

$$Adh(R) = Adh(\underline{R})$$

$$Adh(R_1) = Adh\ (R_2) \iff \underline{R_1} = \underline{R_2}.$$

We also have for every finitary relation R

$$Adh(R) \neq \emptyset \iff \underline{R} \neq \emptyset \iff R \text{ is infinite.}$$

Other obvious properties are :

$$Adh(R) = Adh(FG(R))$$

$$\underline{R} = \underline{FG(R)} = FG(\underline{R})$$

And the following exemple proves that Adh (PP(R)) is usually strictly greater than Adh(R) :

$$R = \{<a^n, b^n> \mid n \in \mathbb{N}\}$$

$$\mathrm{Adh}(R) = \{<a^{\omega}, b^{\omega}>\}$$

$$\mathrm{PP}(R) = a^* \times b^*$$

$$\mathrm{Adh}(\mathrm{PP}(R)) = a^* \times \{b^{\omega}\} \cup \{a^{\omega}\} \times b^* \cup \{<a^{\omega}, b^{\omega}>\}$$

<u>Property 8</u> : $\mathrm{Adh}(\mathrm{PP}(R)) = \widehat{\mathrm{PP}}(\mathrm{Adh}(R)) \cap \mathcal{R}^{inf}$.

Where $\widehat{\mathrm{PP}}\ (\hat{\beta}) = \{\hat{\alpha} \mid \hat{\alpha} \leq \hat{\beta}\}$

Proof : One inclusion is obvious.

To prove the reverse inclusion we remark that :

$\vec{f} \in \widehat{\mathrm{PP}}(R)$ and $|\vec{f}| = n$ imply that there exists

$\vec{g} \in \mathrm{FG}(R)$ such that $|\vec{g}| = n$ and $\vec{f} \leq \vec{g}$.

Indeed : $\vec{f} \in \mathrm{PP}(R) \Rightarrow \exists\ \vec{\alpha} \in R : \vec{f} \leq \vec{\alpha}$.

But then : $\vec{f}[n] = \vec{f} \leq \vec{\alpha}[n] \in \mathrm{FG}(R)$.

Suppose : $\vec{\alpha} \in \mathrm{Adh}(\mathrm{PP}(R))$.

Consider then the sequence of sets :

$$E_n = \{\vec{g} \in \mathrm{FG}(R) \mid \vec{\alpha}[n] \leq \vec{g} \text{ and } |\vec{g}| = n\}$$

These sets being finite and non empty and satisfying :

$$\forall n \quad \vec{g} \in E_{n+1} \Rightarrow \vec{g}[n] \in E_n$$

we can find, by Koenig's lemma, an infinite sequence g_n with for all n $g_n \in E_n$ and $g_n = g_{n+1}[n]$. This sequence converges to β such that :

$$g_n = \beta[n] \text{ for all } n$$

This limit β belongs to Adh(R) and from $\vec{\alpha}[n] \leq \vec{g}_n \leq \vec{\beta}$ for all n we deduce :

$$\vec{\alpha} = \lim \vec{\alpha}[n] \leq \beta \text{ whence } \vec{\alpha} \in \widehat{\mathrm{PP}}(\mathrm{Adh}(R)). \quad \square$$

We can make precise a remark made above.

<u>Property 9</u> : <u>PP(R)</u> = FG(Adh(PP(R))) is precisely the set of all $\vec{f}$ such that $R:\vec{f}$ is infinite.

<u>Proof</u> : One inclusion is obvious. The second one is one more straight foward application of Koenig's lemma. □

V - OPERATIONS ON RELATIONS AND CLOSEDNESS

V.1. Union and product

In [ ] we have studied the topological closure of the union and product of two languages and the star of a language.

There are many more operations on relations and in this paragraph, we shall establish a number of formulae to compute the topological closure of the result of these operations in terms of the topological closure of the relations on which they are performed.

The simplest cas is the union which gives rise to :

Property 10 : $\overline{R_1 \cup R_2} = \overline{R_1} \cup \overline{R_2}$.

Proof : The union of two closed relations is closed.
Thus $R_1 \subseteq \overline{R_1}$ and $R_2 \subseteq \overline{R_2}$ imply :

$$R_1 \cup R_2 \subseteq \overline{R_1} \cup \overline{R_2} \quad \text{and} \quad \overline{R_1 \cup R_2} \subseteq \overline{R_1} \cup \overline{R_2}$$

Conversely $R_1 \subset R_1 \cup R_2$ and $R_2 \subseteq R_1 \cup R_2$ imply :

$$\overline{R_1} \subseteq \overline{R_1 \cup R_2} \quad \text{and} \quad \overline{R_2} \subseteq \overline{R_1 \cup R_2}$$

for the closure is obviously increasing $R \subseteq R' \Rightarrow \bar{R} \subseteq \bar{R}'$ □

As concerns the product componentwise i.e.

$$R_1 R_2 = \{\vec{\alpha} \vec{\beta} \mid \vec{\alpha} \in R_1 \quad \vec{\beta} \in R_2\}$$

we have the same formula as in the case of languages.

Property 11 : $\overline{R_1 R_2} = \overline{R_1}\ \overline{R_2}$

Proof : In the case of languages $L_1, L_2 \subseteq A^*$, we had proved :

$$Adh(L_1 L_2) = Adh(L_1) \cup L_1^{fin} Adh(L_2)$$

whence :

$$\overline{L_1 L_2} = L_1 L_2 \cup \mathrm{Adh}(L_1 L_2)$$

$$= (L_1 \cup \mathrm{Adh}(L_1))(L_2 \cup \mathrm{Adh}(L_2)) = \overline{L_1}\,\overline{L_2}$$

since :

$$(L_1 \cup \mathrm{Adh}(L_1))(L_2 \cup \mathrm{Adh}(L_2)) = L_1 L_2 \cup \mathrm{Adh}\, L_1 \cup L_1^{fin}\, \mathrm{Adh}(L_2)$$

for :

$$L_1^{inf}\, \mathrm{Adh}(L_2) = L_1^{inf}\, L_2 = L_1^{inf} \subseteq L_1 L_2$$

We cannot really use the same method for relations as for languages since the formula $FG(L_1\ L_2) = FG(L_1) \cup L_1^{fin}\ FG(L_2)$ is not valid for relations.

We give here a proof which uses as few theorems from topology as possible. It relies on the :

Lemma 1 : If α_n, $n \in \mathbb{N}_+$ and β_n, $n \in \mathbb{N}_+$ are two sequences of words in A^∞ such that the sequence :

$$\gamma_n = \alpha_n\, \beta_n \text{ converges towards } \gamma_o.$$

There exists an infinite subsets N' of $\mathbb{N}$ such that α_n, $n \in N'$ and β_n, $n \in N'$ are both convergent and their limits α_o, β_o satisfy $\alpha_o\, \beta_o = \gamma_o$.

Proof : Suppose $\gamma_o = \lim \gamma_n$ is finite $\gamma_o \in A^*$.
We know that γ_n is stationary i.e.

$$\exists N \quad n \geq N \Rightarrow \gamma_n = \gamma_o$$

For all $n \geq N$ we can thus write :

$$\langle \alpha_n, \beta_n \rangle \in \{\langle f_1, f_2 \rangle \mid f_1\, f_2 = \gamma_o\}$$

Since the set of binary factorisations of γ_o is finite, there exist $\langle f_1, f_2 \rangle$ such that $f_1\, f_2 = \gamma_o$ and $\langle \alpha_n, \beta_n \rangle = \langle f_1, f_2 \rangle$ for infinitely many n's say all $n \in N'$ where N' is an infinite subset of $\mathbb{N}_+$. Clearly then α_n, $n \in N'$ converges towards f_1 and β_n, $n \in N'$ converges towards f_2.

Suppose now γ_o is infinite : two case arise :

1) $|\alpha_n|$ is ultimately bounded i.e. $|\alpha_n| \leq M$ for all sufficiently large $n \in \mathbb{N}$. We have $\forall p\ \exists n_p\ \ n \geq n_p \Rightarrow \gamma_n[p] = \gamma_o[p]$ and since $\gamma_n = \alpha_n\ \beta_n (\alpha_n\ \beta_n)[p] = \gamma_o[p]$. But $|\alpha_n| \leq M$ whence for all $p \geq M$ and $n \geq n_p$:

$$\gamma_n[p] = \alpha_n \cdot \beta'_n \quad \text{with} \quad \beta'_n \leq \beta_n$$

For infinitely many n's $\alpha_n = f \in A \leq M$, write :

$$\alpha_n = f \text{ and } \gamma_o[p] = f\ \beta_n[p-|f|]$$

Clearly if N' is this infinite set of n's

α_n, $n \in N'$ converges towards f and

β_n, $n \in N'$ converges towards $\gamma_o : f$

2) The sequence $|\alpha_n|$ is not ultimately bounded. We can extract α_n, $n \in N'$ which converges towards $u \in A^\omega$.

From $\alpha_n \leq \gamma_n$ for all n we can conclude $\lim \alpha_n \leq \lim \gamma_n$ whence $u \leq \gamma_o$ which implies $u = \gamma_o$. We then entract from β_n, $n \in N'$ any converging sequence β_n, $n \in N''$ and clearly both α_n, $n \in N''$ and β_n, $n \in N''$ converge with $\lim \alpha_n \lim \beta_n = u \lim \beta_n = u = \gamma_o$. □

By a simple induction we extend this lemma to sequences of multiwords and this proves the inclusion :

$$\overline{L_1\ L_2} \subseteq \overline{L_1}\ \overline{L_2}$$

The reverse inclusion comes from the simple fact :

<u>Lemma 2</u> : If α_n, $n \in \mathbb{N}_+$ and β_n, $n \in \mathbb{N}_+$ converge towards α_o and β_o then $\alpha_n\ \beta_n$, $n \in \mathbb{N}$ converges towards $\alpha_o\ \beta_o$. The same holds for sequences of multiwords.

The proof is straight forward. □

We can now give a formula to compute $Adh(R_1\ R_2)$:

$$Adh(R_1\ R_2) = Adh(R_1)\ R_2 \cup R_1\ Adh(R_2) \cup Adh(R_1)\ Adh(R_2)$$

This comes from :

$$\begin{aligned} Adh(R_1\ R_2) &= \overline{R_1\ R_2}^{\,inf} \\ &= ((R_1 \cup Adh(R_1))\ (R_2 \cup Adh(R_2)))^{inf} \end{aligned}$$

5.2. Projections and slices of relations

Let $\vec{\alpha} \in \mathcal{A} = A_1^\infty \times \ldots \times A_k^\infty$ and I be a subset of $[k] = \{1,2,\ldots,k\}$ which can always be written $I = \{i_1,\ldots,i_\ell\}$ with $i_1 < i_2 < \ldots < i_\ell$.

Then $\pi_I(\vec{\alpha})$ is the multiword in $A_{i_1}^\infty \times \ldots \times A_{i_\ell}^\infty$ given by :

$$\pi_I(\vec{\alpha}) = <\pi_{i_1}(\vec{\alpha}),\ldots,\pi_{i_\ell}(\vec{\alpha})>$$

π_I is a projection of $\mathcal{A}$ onto $\mathcal{A}_I = A_{i_1}^\infty \times \ldots \times A_{i_j}^\infty$. We shall need an operation of recomposition.

If I,J are disjoint subsets of [k], we denote I + J their union and if $\vec{\alpha} \in \mathcal{A}_I$, $\vec{\beta} \in \mathcal{A}_J$ we denote $\vec{\alpha} \times \vec{\beta}$ the multiword in $\mathcal{A}_{I+J}$ given by :

$$\begin{aligned} \pi_i(\vec{\alpha}\times\vec{\beta}) &= \pi_i(\vec{\alpha}) \text{ if } i \in I \\ &= \pi_i(\vec{\beta}) \text{ if } i \in J \end{aligned}$$

These operations are entended to relations :

$$\pi_I(R) = \{\pi_i(\vec{\alpha}) \mid \vec{\alpha} \in R\}$$

$$R_1 \times R_2 = \{\vec{\alpha} \times \vec{\beta} \mid \vec{\alpha} \in R_1 , \vec{\beta} \in R_2\} \text{ if}$$

$R_1 \subseteq \mathcal{A}_I$, $R_2 \subseteq \mathcal{A}_J$ where I and J are disjoint subsets of [k].

We now define slices of a relation :

Let $\vec{\alpha}$ be an element of $\mathcal{A}_I$ for some non empty subset I of [k]. We denote $R(I,\vec{\alpha})$, and call slice of R along $\vec{\alpha}$, the relation in $\mathcal{A}_{[k]\setminus I}$ given by :

$$\vec{\beta} \in R(I,\vec{\alpha}) \iff \vec{\beta} \times \vec{\alpha} = \vec{\alpha} \times \vec{\beta} \in R.$$

We have some obvious identities.

For all non empty subset I of [k]

$$R = \cup \{R(I,\vec{\alpha}) \times \{\vec{\alpha}\} \mid \vec{\alpha} \in \mathcal{A}_I\}$$

and since obviously :

$$R(I,\vec{\alpha}) \neq \emptyset \iff \vec{\alpha} \in \pi_I(R)$$

we can rewrite this identity :

$$R = \cup \{R(I,\vec{\alpha}) \times \{\vec{\alpha}\} \mid \vec{\alpha} \in \pi_I(R)\}.$$

A relation is said to be decomposable iff there exist a partition of [k] in $[k] = I_1 + \ldots + I_\ell$, where $I_1, \ldots, I_\ell$ are non empty pairwise disjoint subsets such that :

$$R = \pi_{I_1}(R) \times \pi_{I_2}(R) \times \ldots \times \pi_{I_\ell}(R)$$

A cartesian relation is a relation R which is equal to the product of its components $R = \pi_1(R) \times \ldots \times \pi_k(R)$.

We establish a few easy properties which will be useful later on :

Property 12 : $\overline{\pi_I(R)} = \pi_I(\bar{R})$.

Proof : Suppose $\vec{\alpha}_n$, $n \in \mathbb{N}_+$, $\alpha_n \in R$ converges towards $\vec{\alpha}_o$ i.e. $\forall p\ \exists N \quad n \geq N \Rightarrow \alpha_o[p] = \alpha_n[p]$.

Then $\pi_I(\vec{\alpha}_n)$ converges towards $\pi_I(\vec{\alpha}_o)$ since $\pi_I(\vec{\alpha}[p]) = (\pi_I(\vec{\alpha}))\,[p]$ for all p.

Conversely if $\vec{\beta}_n$, $n \in \mathbb{N}_+$, $\vec{\beta}_n \in \pi_I(R)$ converges towards $\vec{\beta}_o$ we consider for all n some $\vec{\alpha}_n \in R$ such that $\vec{\beta}_n = \pi_I(\vec{\alpha}_n)$.

From the sequence of multiwords $\vec{\alpha}_n$, $n \in \mathbb{N}_+$ we can entract a converging subsequence $\vec{\alpha}_n$, $n \in N'$ for some infinite subset N' of $\mathbb{N}_+$.

Suppose $\vec{\alpha}_n$, $n \in N'$ converges towards $\vec{\alpha}_o$: it is easy to see that $\pi_I(\vec{\alpha}_o) = \vec{\beta}_o$ since $\pi_I(\vec{\alpha}_o) = \lim \pi_I(\vec{\alpha}_n) = \lim \vec{\beta}_n$. □

Property 13 : $\overline{R_1 \times R_2} = \overline{R_1} \times \overline{R_2}$

Proof : It is immediate if one remarks that for all $\vec{\alpha} \in \mathcal{R}_{I+J}$ and $p \in \mathbb{N}$:

$$\vec{\alpha}[p] = (\pi_I(\vec{\alpha}))\,[p] \times (\pi_J(\vec{\alpha}))\,[p]\ . \ \square$$

In order to characterize the closure of slices we have to introduce the notion of a limit of a sequence of sets or relations.

If R_n, $n \in \mathbb{N}_+$ is a sequence of relations, its limit is defined by :

$$\lim R_n = \{\lim \vec{\beta}_n \mid \vec{\beta}_n \text{ converges and } \forall n \;\; \vec{\beta}_n \in R_n\}$$

Then we can state :

Property 14 : $\bar{R}(I,\vec{\alpha}_o) = \lim R(I,\vec{\alpha}_n)$ for all sequence $\vec{\alpha}_n$ converging towards $\vec{\alpha}_o$ and satisfying $\vec{\alpha}_n \in \pi_I(R)$ for all n.

Proof : Let $\vec{\beta}_o \in \bar{R}(I,\alpha_o)$ i.e. $\vec{\alpha}_o \times \vec{\beta}_o$ is the limit of some sequence $\vec{\gamma}_n$, $\vec{\gamma}_n \in R$.

We can write $\vec{\gamma}_n = \pi_I(\vec{\gamma}_n) \times \pi_J(\vec{\gamma}_n)$ if $I+J = [k]$. Clearly $\pi_J(\vec{\gamma}_n)$ converges towards $\vec{\beta}_o$ and for all n :

$$\pi_J(\vec{\gamma}_n) \in R(I, \pi_I(\vec{\gamma}_n)).$$

The reverse inclusion is also immediate :

Suppose $\vec{\beta}_n \in R(I,\vec{\alpha}_n)$ for all n and $\vec{\alpha}_n \to \vec{\alpha}_o$. Suppose $\vec{\beta}_n$ converges towards $\vec{\beta}_o$. Then $\vec{\alpha}_n \times \vec{\beta}_n$ converges towards $\vec{\alpha}_o \times \vec{\beta}_o \in \bar{R}$ and thus $\vec{\beta}_o \in \bar{R}(I,\vec{\alpha}_o)$. □

An immediate corollary is :

Corollary 1 : If $\alpha_o \in \mathcal{A}^{fin}$ $\bar{R}(I,\vec{\alpha}_o) = \overline{R(I,\vec{\alpha}_o)}$

Proof : $\vec{\alpha}_n \to \vec{\alpha}_o$ implies that $\vec{\alpha}_n$ is stationary whence $R(I,\vec{\alpha}_n)$ is stationary, equal to $R(I,\vec{\alpha}_o)$ for all sufficiently large n's. □

5.3. Infinite stars of relations

The star of the relation $R \subseteq \mathcal{A}$ is defined by

$$R^* = \cup \{R^n \mid n \in \mathbb{N}\}$$

where $R^o = \{\vec{\varepsilon}\}$ by definition and for all n :

$$R^{n+1} = R\, R^n$$

R^ω is the set of all infinite products of the form

$$\vec{\alpha} = \vec{\alpha}_1\, \vec{\alpha}_2 \ldots \vec{\alpha}_n \ldots \text{ where } \vec{\alpha}_i \in R\backslash\{\vec{\varepsilon}\}$$

The infinite product $\vec{\alpha}$ is defined without any difficulty as the limit of the increasing sequence $\vec{\beta}_n = \vec{\alpha}_1,\ldots, \vec{\alpha}_n$.

An obvious remark is that $R^{\omega} \subseteq \mathcal{A}^{inf}$ since for all n there exists $i \in [k]$ $\pi_i(\vec{\alpha}_n) \neq \varepsilon$ implies that there exists $i \in [k]$ such that $\pi_i(\vec{\alpha}_n) \neq \varepsilon$ for an infinite number of n's. Whence $\pi_i(\vec{\alpha}) \in A_i^{\omega}$. The infinite star R^{∞} is defined as $R^{\infty} = R^{\star} \cup R^{\omega}$.

<u>Theorem 1</u> : $\overline{R^{\infty}} = \overline{R}^{\infty}$.

<u>Proof</u> : We first establish the inclusion $\overline{R}^{\infty} \subseteq \overline{R^{\infty}}$. By induction, using property 10, we get for all $n \in \mathbb{N}$:

$$\overline{R^n} = \overline{R}^n$$

and this suffices to prove $\overline{R}^{\star} \subseteq \overline{R^{\infty}}$.

We now prove $\overline{R}^{\omega} \subseteq \overline{R^{\infty}}$.

For this we consider $\vec{\alpha} = \vec{\alpha}_1 \vec{\alpha}_2 \dots \vec{\alpha}_n \dots$ with $\vec{\alpha}_n \in \overline{R}\backslash\{\vec{\varepsilon}\}$. We can write for all n $\vec{\alpha}_n = \lim \vec{\beta}_p^{(n)}$ where $\vec{\beta}_p^{(n)}$ is a converging sequence of elements in R.

We claim that $\vec{\alpha} = \lim \vec{\gamma}_n$ if we define $\vec{\gamma}_n = \vec{\beta}_n^{(1)} \dots \beta_n^{(n)}$. Let us consider one component $\pi_i(\vec{\alpha})$: two cases may happen :

1) $\pi_i(\vec{\alpha}_1), \dots, \pi_i(\vec{\alpha}_m)$ are finite and $\pi_i(\vec{\alpha}_{m+1})$ is infinite. Then $\pi_i(\vec{\alpha}) = \pi_i(\vec{\alpha}_1) \dots \pi_i(\vec{\alpha}_m)\ \pi_i(\vec{\alpha}_{m+1})$.

   The sequences $\pi_i(\vec{\beta}_p^{(j)})$ are stationary fo all $j = 1,\dots,m$ and there exists $N \in \mathbb{N}_+$ such that $n \geq N$ implies :

   $$\pi_i(\vec{\beta}_n^{(j)}) = \pi_i(\vec{\alpha}_j) \text{ for all } j = 1,\dots,m$$

   We also write that $\pi_i(\vec{\beta}_p^{(m+1)}) \to \pi_i(\vec{\alpha}_{m+1})$:

   $$\forall p \ \exists N_p \quad n \geq N_p \Rightarrow \pi_i(\vec{\beta}_n^{(m+1)})[p] = \pi_i(\vec{\alpha}_{m+1})[p]$$

   If $q = \Sigma\ \{|\pi_i(\vec{\alpha}_j)| \mid j = 1,\dots,m\}$ then :

   $$\forall p,\ n \geq \max(N, N_{p+q}) \Rightarrow \pi_i(\vec{\gamma}_n)[p+q] = \pi_i(\vec{\alpha})[p+q]$$

   This clearly implies $\pi_i(\vec{\gamma}_n) \to \pi_i(\vec{\alpha})$.

2) Suppose $\pi_i(\vec{\alpha}_m)$ is finite for all $m \in \mathbb{N}_+$. All the sequences $\pi_i(\vec{\beta}_n^{(m)})$ are stationary.

   For all p there exists m_p such that :

   $$\pi_i(\vec{\alpha})[p] \leq \pi_i(\vec{\alpha}_1) \dots \pi_i(\vec{\alpha}_{m_p})$$

But then there exists N_p such that $n \geq N_p$ implies :

$$\pi_i(\vec{\beta}_n^{(j)}) = \pi_i(\vec{\alpha}_j) \text{ for all } j = 1,\ldots,m_p$$

Whence for all $n \geq N_p$, $\pi_i(\vec{\gamma}_n)[p] = \pi_i(\vec{\alpha})[p]$. □

In order to establish the reverse inclusion $\overline{R^\infty} \subseteq \bar{R}^\infty$. We shall make an induction on k. The result is known for k = 1 that is in the case of languages.

Property 15 : For all $L \subset A^\infty$ $\overline{L^\infty} = \bar{L}^\infty$.

Proof :

$$(L^\infty)^{fin} = (L^{fin})^*$$
$$(L^\infty)^{inf} = (L^{fin})^* L^{inf} \cup (L^{fin})^\omega$$
$$(\bar{L})^{fin} = L^{fin}$$
$$(\bar{L})^{inf} = \mathrm{Adh}(L)$$

From these identities we compute both sides of the equation :

$$(\overline{L^\infty})^{fin} = (L^\infty)^{fin} = (L^{fin})^* \text{ which is equal to :}$$
$$(\bar{L}^\infty)^{fin} = (\bar{L}^{fin})^* = (L^{fin})^*$$
$$(\overline{L^\infty})^{inf} = \mathrm{Adh}(L^\infty) = \mathrm{Adh}(FG(L^\infty))$$

But $FG(L^\infty) = FG(L^*) = (L^{fin})^* FG(L)$ and we can use formulae established in [ ].

$$\mathrm{Adh}(FG(L^\infty)) = (L^{fin})^* \mathrm{Adh}(L) \cup (L^{fin})^\omega$$

And this is equal to :

$$(\bar{L}^\infty)^{inf} = (\bar{L}^{fin})^* \bar{L}^{inf} \cup (\bar{L}^{fin})^\omega$$
$$= (L^{fin})^* \mathrm{Adh}(L) \cup (L^{fin})^\omega \quad □$$

We assume now that $\overline{R^\infty} \subseteq \bar{R}^\infty$ has been proved for all $k' < k$ and consider $\vec{\alpha} = \lim \vec{\alpha}_n$ in $\overline{R^\infty}$.

A very simple case is when $\vec{\alpha}$ has a finite component :

$$\pi_i(\vec{\alpha}) = f \in A_i^*$$

The sequence $\pi_i(\vec{\alpha}_n)$ is stationary and there exists thus $N \in \mathbb{N}$ such that $n \geq \mathbb{N} \Rightarrow \pi_i(\vec{\alpha}_n) = f$. Equivalently $n \geq \mathbb{N} \Rightarrow \vec{\alpha}_n \in R^\infty(i,f) \times \{f\}$. This implies $\vec{\alpha} \in \overline{R^\infty(i,f) \times \{f\}} = \overline{R^\infty(i,f)} \times \{f\}$

We can write $R^{\infty}(i,f)$ as the finite union of all the products of the form $R(i,\varepsilon)^{\infty}\ R(i,f_1)\ R(i,\varepsilon)^{\infty} \dots R(i,f_\ell)\ R(i,\varepsilon)^{\infty}$ where $f_1,\dots,f_\ell \in A_i^+$ are such that $f = f_1,\dots,f_\ell$.

The closure of $R^{\infty}(i,f)$ is the union of the closures of these products that is the union of $\overline{R(i,\varepsilon)^{\infty}}\ \overline{R(i,f_1)} \dots \overline{R(i,f_\ell)}\ \overline{R(i,\varepsilon)^{\infty}}$

By induction we know that :

$$\overline{R(i,\varepsilon)^{\infty}} \subseteq \overline{R(i,\varepsilon)}^{\infty} = \bar{R}(i,\varepsilon)^{\infty}$$

whence, since $\overline{R(i,f_j)} = \bar{R}(i,f_j)$.

$\overline{R(i,\varepsilon)^{\infty}}\ \overline{R(i,f_1)} \dots \overline{R(i,f_\ell)} - \overline{R(i,\varepsilon)^{\infty}}$ is contained in $\bar{R}(i,\varepsilon)^{\infty}\ \bar{R}(i,f_1)\dots \bar{R}(i,f_\ell)\ \bar{R}(i,\varepsilon)^{\infty}$. The last product is obviously contained in $\bar{R}^{\infty}(i,f)$.

Consider now the case $\pi_i(\vec{\alpha})$ is infinite for all $i \in [k]$ where $\vec{\alpha} = \lim \vec{\alpha}_n$.

We can find $\vec{\beta}_o^{(1)}$ and $\vec{\gamma}_o^{(1)}$ such that :

$$\vec{\alpha}[1] \leq \vec{\beta}_o^{(1)},\ \vec{\beta}_o^{(1)} \in \bar{R}^{\infty}\ \bar{R},\ \vec{\gamma}_o^{(1)} \in \overline{R^{\infty}}$$

and :

$$\vec{\alpha} = \vec{\beta}_o^{(1)}\ \vec{\gamma}_o^{(1)}.$$

From $\vec{\alpha}[1] = \vec{\alpha}_n[1]$ for all sufficiently large n's we deduce that there exists a factorization of $\vec{\alpha}_n$ in $\vec{\alpha}_n = \vec{\beta}_n\ \vec{\gamma}_n$, $\vec{\beta}_n \in R^{\ell_n}$, $\vec{\gamma}_n \in R^{\infty}$ such that $\vec{\alpha}_n[1] = \vec{\beta}_n[1]$.

In fact we suppose that ℓ_n is minimum, i.e. we consider $\vec{\alpha}_n = \vec{\beta}_n\ \vec{\gamma}_n$ where $\vec{\beta}_n \in R^{\ell_n}$, $\vec{\gamma}_n \in R^{\infty}$, $\vec{\alpha}_n[1] = \vec{\beta}_n[1]$ and for all factorizations of $\vec{\beta}_n$ in $\vec{\beta}'_n$, $\vec{\beta}''_n$, $\vec{\beta}'_n \in R^{\ell_n - 1}$, $\vec{\beta}''_n \in R$, $\vec{\alpha}_n[1] \neq \vec{\beta}'_n[1]$.

Since α_n converges we can extract convergent subsequences, β_n, $n \in N'$, γ_n, $n \in N'$ which converge towards $\vec{\beta}_o^{(1)}$ and $\vec{\gamma}_o^{(1)}$ such that $\vec{\alpha} = \vec{\beta}_o^{(1)}\ \vec{\gamma}_o^{(1)}$.

Two things may happen ℓ_n, $n \in N'$ is ultimately bounded by M.

Then β_n, $n \in N'$ converges towards $\beta_o^{(1)} \in \overline{R^{\leq M}}$ which is obviously equal to $\bar{R}^{\leq M}$.

We can write $\beta_o^{(1)} \subseteq \bar{R}^{\infty}\ \bar{R}$ since $\vec{\varepsilon} \in \bar{R}^{\infty}$ and certainly since $\beta_o^{(1)}[1] = \vec{\alpha}[1]$, $\beta_o^{(1)} \subset \bar{R}^{\ell}$ for some $\ell \geq 1$. ℓ_n is not ultimately bounded.

For all β_n consider $\vec{\beta}_n = \vec{\beta}'_n\ \vec{\beta}''_n$ with $\vec{\beta}'_n \in R^{\ell_n - 1}$ and $\vec{\beta}''_n \in R$.

We certainly have, since $\vec{\alpha}[1] \neq \beta'_n[1]$, $\pi_i(\beta'_n[1]) = \varepsilon$ for some i which implies :

$$\exists i \ \pi_i(\beta'_n[1]) = \varepsilon \text{ for infinitely many n's in N'}.$$

Then $\vec{\beta}'_n$, for these n's in N", belongs to :

$$R^{\ell_n - 1}(i,\varepsilon) \times \varepsilon$$

And we can write the limit of $\vec{\beta}_n = \vec{\beta}'_n \vec{\beta}''_n$ as the product of two converging subsequences of $\vec{\beta}'_n$ and $\vec{\beta}''_n$ i.e. as $\vec{\beta}_o^{(1)'} \vec{\beta}_o^{(1)''}$ where $\vec{\beta}_o^{(1)'} = \lim \vec{\beta}'_n$ belongs to $R^\infty(i,\varepsilon) \times \varepsilon$ which is contained by induction in $\overline{R}^\infty(i,\varepsilon) \times \{\varepsilon\}$ itself contained in $\overline{R}^\infty$. $\vec{\beta}_o^{(1)''} = \lim \vec{\beta}''_n$ belongs to $\bar{R}$ since $\vec{\beta}''_n \in R$. We have proved what was annonced.

Repeating the process we can find a sequence :

$$\vec{\beta}_o^{(1)} \vec{\beta}_o^{(2)} \dots \vec{\beta}_o^{(n)} \dots \text{ such that for all } n \in \mathbb{N}_+$$

$$\vec{\alpha}[n] = (\vec{\beta}_o^{(1)} \dots \vec{\beta}_o^{(n)})[n]$$

The sequence $\vec{\beta}_o^{(1)} \dots \vec{\beta}_o^{(n)}$ obviously converges towards $\vec{\alpha}$ and we have written $\vec{\alpha}$ as an infinite product $\vec{\alpha} = \vec{\beta}_o^{(1)} \dots \vec{\beta}_o^{(n)} \dots$ of elements in $\overline{R}^\infty \bar{R}$.

To get our result it then suffices to prove the :

Lemma 3 : For all R $(R^\infty)^\infty = R^\infty$.

Proof : We know that $(R^\infty)^{fin} = (R^{fin})^*$ and $(R^\infty)^{inf} = (R^{fin})^* R^{inf} \cup (R^{fin})^\omega$. And we compute :

$$((R^\infty)^\infty)^{fin} = ((R^\infty)^{fin})^* = ((R^{fin})^*)^* = (R^{fin})^* = (R^\infty)^{fin}$$

$$\begin{aligned} ((R^\infty)^\infty)^{inf} &= ((R^\infty)^{fin})^* (R^\infty)^{inf} \cup ((R^\infty)^{fin})^\omega \\ &= ((R^{fin})^*)^* [(R^{fin})^* R^{inf} \cup (R^{fin})^\omega] \cup ((R^{fin})^*)^\omega \\ &= (R^{fin})^* R^{inf} \cup (R^{fin})^\omega = (R^\infty)^{inf}. \quad \square \end{aligned}$$

Exemple : The following exemple shows that we cannot really simplify our proof. We take :

$$\alpha_n = \langle a,\varepsilon\rangle^n \langle \varepsilon,b\rangle^n, \quad n \in \mathbb{N}_+$$

This sequence obviously converges towards $\langle a^\omega, b^\omega\rangle = \vec{\alpha}$.

But $\vec{\alpha}_n = \vec{\beta}_n \vec{\gamma}_n$ and $\vec{\alpha}[1] = \langle a,b\rangle \leq \vec{\beta}_n[1]$ implies $\vec{\beta}_n = \langle a,\varepsilon\rangle^n \langle \varepsilon,b\rangle \in R^{n+1}$.

Writing $\vec{\beta}_n = \vec{\beta}'_n \vec{\beta}''_n$ with $\vec{\beta}'_n = <a,\varepsilon>^n$, $\vec{\beta}''_n = <\varepsilon,b>$ we get :

$$\vec{\beta}_o^{(1)} = \lim \vec{\beta}_n = \lim \vec{\beta}'_n \lim \vec{\beta}''_n = <a^\omega,\varepsilon> <\varepsilon,b> = <a^\omega,b>$$

The factorization of $\vec{\alpha}$ we can obtain from $\vec{\alpha} = \lim \vec{\alpha}_n$ is :

$$\vec{\alpha} = <a^\omega,b> <a^\omega,b> \ \ldots <a^\omega,b> \ldots$$

From theorem 1 we can derive a formula to compute the adherence of the star a relation which is exactly the same as the formula to compute the adherence of the star a language.

Corollary 2 : $\mathrm{Adh}(R^*) = (R^{fin})^* \, \mathrm{Adh}(R^\infty) \cup (R^{fin})^\omega$

Proof :

$$\mathrm{Adh}(R^*) = (\overline{R^*})^{inf} = (\overline{R^\infty})^{inf} = (\bar{R}^\infty)^{inf}$$

$$= (\bar{R}^{fin})^* \, (\bar{R}^{inf}) \cup (\bar{R}^{fin})^\omega$$

$$= (R^{fin})^* \, \mathrm{Adh}(R) \cup (R^{fin})^\omega \qquad \square$$

5.4. Composition of relations

Let $I+J_1$ and $I+J_2$ be subsets of $[k]$ such that $I+J_1 \cap I+J_2 = \emptyset$.

We can compose over I two relations $R_1 \subseteq \mathcal{R}_{I+J_1}$ and $R_2 \subseteq \mathcal{R}_{I+J_2}$ to get :

$$R_1 \circ_I R_2 = \{\vec{\alpha} \times \vec{\gamma} \mid \vec{\alpha} \in \mathcal{R}_{J_1} \ \vec{\gamma} \in \mathcal{R}_{J_2} \ \exists \vec{\beta} \in \mathcal{R}_I : \vec{\alpha} \times \vec{\beta} \in R_1 \text{ and } \vec{\beta} \times \vec{\gamma} \in R_2\}$$

Example : The following example shows that $\overline{R_1 \circ_I R_2} = \bar{R}_1 \circ \bar{R}_2$ is not usually true.

Take $J_1 = \{1\}$, $I = \{2\}$, $J_2 = \{3\}$:

$$R_1 = <a^n, (b^{2n})^p \mid n, p \in \mathbb{N}_+>$$

$$R_2 = <b(b^{2n})^p, c^n \mid n, p \in \mathbb{N}_+>$$

The composition over I.

$R_1 \circ_2 R_2$ is clearly empty since $\pi_2(R_1) \cap \pi_2(R_2) = \emptyset$.

But : $\bar{R}_1 = R_1 \cup a^+ \times \{b^\omega\} \cup \langle a^\omega, b^\omega\rangle$

$\bar{R}_2 = R_1 \cup \{b^\omega\} \times c^+ \cup \langle b^\omega, c^\omega\rangle$

And $\bar{R}_1 \circ R_2$ contains $a^+ \times c^+ \cup \langle a^\omega, c^\omega\rangle$. □

<u>Theorem 2</u> : If $FG(R_1) \subseteq R_1$ and $FG(R_2) \subseteq R_2$ then $\overline{R_1 \circ_I R_2} = \bar{R}_1 \circ_I \bar{R}_2$.

<u>Proof</u> : The inclusion $\overline{R_1 \circ_I R_2} \subseteq \bar{R}_1 \circ \bar{R}_2$ ... is always true.

Every $\vec{\delta} = \vec{\alpha} \times \vec{\gamma} \in \overline{R_1 \circ_I R_2}$ is the limit of a séquence $\vec{\delta}_n = \vec{\alpha}_n \times \vec{\gamma}_n$ where $\vec{\alpha}_n \times \vec{\gamma}_n \in R_1 \circ_I R_2$. Thus for all n there exists $\vec{\beta}_n \in \mathcal{A}_I$ such that :

$$\vec{\alpha}_n \times \vec{\beta}_n \in R_1 \quad \text{and} \quad \vec{\beta}_n \times \vec{\gamma}_n \in R_2$$

We can extract from $\vec{\beta}_n$ a converging subsequence, $\vec{\beta}_n$, $n \in N'$ with $\vec{\beta}$ as a limit.

Clearly the sequences $\vec{\alpha}_n \times \vec{\beta}_n$, $n \in N'$ and $\vec{\beta}_n \times \vec{\gamma}_n$, $n \in N'$ converge towards respectively $\vec{\alpha} \times \vec{\beta}$ and $\vec{\beta} \times \vec{\gamma}$. Since $\vec{\alpha} \times \vec{\beta} \in \bar{R}_1$ and $\vec{\beta} \times \vec{\gamma} \in \bar{R}_2$, we have : $\vec{\delta} = \vec{\alpha} \times \vec{\gamma} \in \bar{R}_1 \circ_I \bar{R}_2$.

Conversely assume $FG(R_i) \subseteq R_i$ for $i = 1,2$. Consider $\vec{\delta} = \vec{\alpha} \times \vec{\gamma} \in \bar{R}_1 \circ_I \bar{R}_2$.

These exists $\vec{\beta}$ such that $\vec{\alpha} \times \vec{\beta} \in \bar{R}_1$ and $\vec{\beta} \times \vec{\gamma} \in \bar{R}_2$. We can write :

$$\vec{\alpha} \times \vec{\beta} = \lim (\vec{\alpha}_n \times \vec{\beta}_n) \text{ where } \forall n \ \ \vec{\alpha}_n \times \vec{\beta}_n \in R_1$$

$$\vec{\beta} \times \vec{\gamma} = \lim (\vec{\beta}'_n \times \vec{\gamma}_n) \text{ where } \forall n \ \ \vec{\beta}'_n \times \vec{\gamma}_n \in R_2$$

Then for all $p \in \mathbb{N}$ there exists N such that :

$$n \geq N \Rightarrow \vec{\alpha}_n[p] = \vec{\alpha}[p],\ \vec{\gamma}_n[p] = \vec{\gamma}[p]$$

$$\text{and} \qquad \vec{\beta}_n[p] = \vec{\beta}[p] = \vec{\beta}'_n[p].$$

Whence $\vec{\alpha}[p] \times \vec{\beta}[p] = \vec{\alpha}_n[p] \times \vec{\beta}_n[p] = (\vec{\alpha}_n \times \vec{\beta}_n)[p]$ is in $FG(R_1) \subseteq R_1$ and similarly $\vec{\beta}[p] \times \vec{\gamma}[p]$ is in R_2.

The sequence $\vec{\alpha}[p] \times \vec{\gamma}[p]$, $p \in \mathbb{N}$ obviously converges towards $\vec{\delta} = \vec{\alpha} \times \vec{\gamma}$ and is composed of elements of $R_1 \circ_I R_2$, where $\vec{\delta} \in \overline{R_1 \circ_I R_2}$. □

<u>Corollary</u> : For all relations R_1, R_2

$$\mathrm{Adh}(R_1) \circ \mathrm{Adh}(R_2) \cup \mathrm{Adh}(R_1) \circ FG(R_2) \cup FG(R_1) \circ \mathrm{Adh}(R_2) = \mathrm{Adh}(FG(R_1) \circ FG(R_2)).$$

Proof :

$$\begin{aligned}\mathrm{Adh}(FG(R_1) \circ FG(R_2)) &= (\overline{FG(R_1) \circ FG(R_2)})^{\inf} \\ &= (\overline{FG(R_1)} \circ \overline{FG(R_2)})^{\inf} \\ &= ((\bar{R}_1 \cup FG(R_1)) \circ (\bar{R}_2 \cup FG(R_2)))^{\inf} \\ &= (\bar{R}_1)^{\inf} \circ (\bar{R}_2 \cup FG(R_2)) \cup (\bar{R}_1 \cup FG(R_1))(\bar{R}_2)^{\inf} \quad \square\end{aligned}$$

CONCLUSION

We have obtained a number of results concerning the topological closure of infinitary relations : in practice, at least for modeling the synchronization of concurrent processes, we shall use mainly infinitary rational relations. A forthcomming paper of the same author is devoted to theim definition and properties. The author has had very helpful discussions with A. Arnold, L. Boasson, F. Boussinot, G. Roncairol and G. Ruggin.

BIBLIOGRAPHY

[1] A. ARNOLD and M. NIVAT
Metric interpretations of infinite trees and semantics of non deterministic recursive programs. Theor. Comp. Sci., Vol. 11 (1980), 181-205.

[2] J. BEAUQUIER and M. NIVAT
Application of formal language theory to problems of security and synchronization , in Formal Language Theory (R. Book, éd.) Academic Press, New York, 1980.

[3] L. BOASSON and M. NIVAT
Adherences of languages, Jour. Comp. Syst. Sci., Vol. 20 (1980), 285-309.

[4] S. EILENBERG
Automata, Languages and Machines, Vol. A, Academic Press, New York, 1974.

[5] M. NIVAT
Systèmes de transition permanents et équitables, Research Report n° 2577, Laboratoire Central de Recherches Thomson-CSF, Orsay, 1980.

[6] M. NIVAT
Infinitary languages (to appear).

[7] M. NIVAT
Synchronization et multimorphismes (to appear).

FULL APPROXIMABILITY OF A CLASS OF PROBLEMS OVER POWER SETS

G. AUSIELLO, A. MARCHETTI SPACCAMELA, M. PROTASI

1. INTRODUCTION

The aim of this paper is to discuss methods for the full approximation of combinatorial problems and to study the full approximability of a class of NP-complete optimization problems defined over a lattice. Most combinatorial optimization problems can be naturally defined as optimization problems over lattices according to the ground algebraic structure of the set of feasible solutions. For example the problem of graph colouring can be viewed as an optimization problem over a partition lattice, the problem of minimum spanning tree is an optimization problem over a matroid, ecc.. In [AMP] a large class of optimization problems was formalized as the class of max-subset problems over power sets and some basic properties of these problems were studied. Despite its simple characterization the class of max-subset problems is indeed sufficiently general to include problems with very different properties with respect to approximability. In fact this class includes many problems which are known to be non fully approximable and, at the same time, practically all known examples of fully approximable NP-complete problems.

The existence of good approximations to the solution of hard optimization problems has been studied by several authors [S,GJ,JK,L, etc.]. What is more interesting for the development of our work is that 1) the techniques used in proving the full approximability of a problem are essentially based on variations of dynamic programming, 2) generally single problems (and not classes of problems) have been shown to be fully approximable.

In particular as regards 2) many difficulties arise when trying to find general natural conditions for the approximability and despite of the interest for this type of results few steps have been made in thid direction ([PM],[KS]).

In order to establish a connection between good approximability of hard problems and the intrinsic combinatorial properties which characterize such problems it is useful to restrict ourselves to con-

---

G. Ausiello - Istituto di Automatica, Università di Roma, Roma.
A. Marchetti Spaccamela - IASI - CNR, Roma.
M. Protasi - Istituto Matematico, Università dell'Aquila, L'Aquila.

sidering max-subset problems and the properties of the set of their feasible solutions.

In the whole we can say that three possible research areas are worth-while of beeing pursued: 1) to find new simple methods of full approximation, 2) to give general conditions for the full approximability of a class of problems, 3) to introduce new approximate algorithms of lower complexity for problems which are already known to be fully approximable.

In this paper we will be concerned with points 1) and 2). In fact, in par. 3 we will consider a new method for showing the full approximability. Its computational complexity will be studied and its advantages with respect to the classical schemes will be also shown. Instead in par. 4 we will give a sufficient condition for the full approximability of a subclass of max-subset problems which is based on the structural properties of the set of feasible solutions and which is verified by the most important problems which are known to be fully approximable.

2. A FULLY POLYNOMIAL APPROXIMATION SCHEME

Given an NP-complete optimization problem *A* with measure m the following definitions capture the concept of good approximability.

DEFINITION 2.1. A is an *ε-approximate algorithm* for *A* if, given any instance $x \in A$, we have

$$\left| \frac{m^*(x) - m(A(x))}{m^*(x)} \right| \leq \varepsilon$$

where $m^*(x)$ is the measure of the optimal solution of the instance x.

DEFINITION 2.2. A problem *A* is said to be

a) *polynomially approximable* if given any $\varepsilon > 0$ there exists an ε-approximate algorithm for *A* which runs in polynomial time;

b) *fully polynomially approximable* if *A* is polynomially approximable and there exists a polynomial q such that, given any ε, the running time of the ε-approximate algorithm is bounded by $q(|x|, 1/\varepsilon)$.

DEFINITION 2.3. A constructive method that, for any given ε, provides the corresponding polynomial ε-approximate algorithm A_ε is said to be a *polynomial approximation scheme* (PAS). Besides if, for every ε, the running time of A_ε is bounded by $q(|x|, 1/\varepsilon)$ for some polynomial q we say that the scheme is a *fully polynomial approximation scheme*.

As we said in the introduction the main aim of this paper is to characterize optimization problems belonging to the class of max-subset problems which are fully approximable and hence we will only consider fully polynomial approximation schemes for this class of problems.

DEFINITION 2.4. A NP *max-subset problem* A over an alphabet Σ is a quadruple

$$A = \langle \text{INPUT}, F, \pi, m \rangle \text{ where}$$

INPUT : is a polynomially decidable subset of Σ^* (set of *instances*)

F : INPUT $\rightarrow$ P (Σ^*) is a polynomially computable mapping that to every input x associates a finite set of (encodings of) *objects*

π : is a polynomially decidable *property* of subsets of F(x)

m : F (INPUT) $\rightarrow$ N (where F (INPUT) = $\bigcup_{x \in \text{INPUT}} P(F(x))$ is the *measure* that, given x, associates a non negative integer to every subset of F(x).

DEFINITION 2.5. Given an instance x of a NP max-subset problem A,

1) the *search space* of x is the lattice L_x of the powerset P(F(x)) under inclusion

2) the set of *feasible solutions* of x is the subsemilattice SOL(x), which is formed by the elements of L_x which satisfy π

3) the *optimal solutions* of x are the elements of SOL(x) for which m is maximal.

The definitions can be extended to minimization problems by inverting the lattice ordering.

Examples: graph problems = Max-clique, Min node cover, node deletion, arc deletion, max-subgraph

: set problems = Max-set packing, min-set covering, min hitting set

: mathematical programming = Max-knapsack, max subset sum

: problems of scheduling theory

: problems on matroids and independent systems

(For the definitions of the above problems see [G,J]).

As a detailed example let us consider the 0-1 knapsack problem

$$\max \textstyle\sum_i c_i x_i \text{ subject to } \sum_i a_i x_i \le b \qquad x_i \in \{0,1\}$$

In this case we have:

INPUT = (2n+1)-tuples of positive integers $\langle c_1,\ldots,c_n;\ a_1,\ldots,a_n;\ b\rangle$

$F(\langle c_1,\ldots c_n;\ a_1,\ldots,a_n;\ b\rangle) = \{c_1,\ldots,c_n\}$

$\pi(\{c_{j_1},\ldots,c_{j_K}\}) \Leftrightarrow a_{j_1} + \ldots + a_{j_K} \leq b$

$m(\{c_{j_1},\ldots,c_{j_K}\}) = \sum_{i=1}^{K} c_{j_i}$

The fundamental technique for constructing fully polynomial approximation schemes are all based on the classic dynamic programming scheme. This scheme, in the case of max subset problems, can be so summarized

```
L:= Ø;
for all items i in F(x) do
        for all sets S_j in L do
            if S_j ∪ {i} satisfies π
            then
              begin insert S_j ∪ {i} in L;
                    eliminate dominated elements
              end
        end
end.
```

take the best solution in L.

It is easy to see that the number of steps of the algorithm is proportional to the number of items in F(x) times the length of the list L.

Clearly variations of this scheme are obtained by considering different conditions of dominance between elements.

In the case of knapsack we can define the following dominance rule:

Given two sets S' and S" in L we say that S' is *dominated* by S"

if $\sum_{i\in S'} c_i \leq \sum_{i\in S''} c_i$ and

$$\sum_{i\in S'} a_i \geq \sum_{i\in S''} a_i.$$

Clearly the elimination of S' does not introduce any error.

Therefore we can obtain the following exact algorithm for the knapsack problem:

Algorithm A_1

```
L:= Ø;
for i = 1 to n do
    for all sets S_j in L do
        if  Σ_{j∈S_j} a_j + a_i ≤ b
          then
            begin L:= L ∪ (S_j ∪ {i})
                  eliminate all S' ∈ L
                  such that ∃S" ∈ L
                  Σ_{j∈S'} c_j ≤ Σ_{j∈S"} c_j
                  and
                  Σ_{j∈S'} a_j ≥ Σ_{j∈S"} a_j
            end
    end
end
take the best solution in L.
```

To evaluate the complexity of the above algorithm it is sufficient to note that, at each step, the number of solutions contained in the list L is less than

$\min(b, \sum_{j=1}^{n} a_j , \sum_{j=1}^{n} c_j)$. So with a suitable implementation of the elimination step it is not hard to see that the complexity of algorithm A_1 is $O(n \cdot \min(b, \sum_{j=1}^{n} a_j , \sum_{j=1}^{n} c_j))$ (i.e. exponential in the size of the input).

It is also possible to obtain the elements of the optimal solution without increasing the overall complexity of the algorithm (see [L]).

In order to achieve a fully polynomial approximation scheme the first technique which was used for finding an approximate solution to the knapsack problem was based on scaling all coefficients a_i by a factor $k = \varepsilon \cdot c_{MAX}/n$.

This technique is summarized by the following algorithm

Algorithm A_2

```
    for j = 1 to n do
        c'_j = c_j / k
    end;
```

Apply algorithm A_1 taking as input

$(c'_1 \ldots, c'_n; \; a_1 \ldots, a_n; \; b)$

take the best solution and multiply its value for k.

If $m(A_2(x))$ is the value of the approximate solution we have that

$$m^*(x) - m(A_2(x)) \leq n \cdot k$$

On the other side we can assume (w.l.o.g.) that

$$m^*(x) \geq c_{MAX}.$$

It follows that

$$\frac{m^*(x) - m(A_2(x))}{m^*(x)} \leq \frac{n \cdot k}{c_{MAX}} = \varepsilon$$

As regards the running time we have that the complexity of the algorithm is $O(n \cdot (\sum_{j=1}^{n} c'_j))$. Due to the scaling we have that

$$\sum_{j=1}^{n} c'_j \leq \frac{n \cdot c_{MAX}}{k} = \frac{n^2}{\varepsilon}$$

So the overall complexity is $O(\frac{n^3}{\varepsilon})$.

Algorithm A_2 can be improved in several different ways obtained by Ibarra and Kim [IK] and Lawler [L], lowering the complexity to $O(n \lg \frac{1}{\varepsilon} + \frac{1}{\varepsilon^4})$.

3. DIFFERENT FULLY POLYNOMIAL APPROXIMATIONS SCHEMES

The fully polynomial approximation scheme described in Par. 2, although very useful for many problems, suffers some drawbacks.

In fact in order to find the fully polynomial approximation scheme we need to know good bounds to m^* and this is a severe limitation to the generality of the method as it can be easily seen if we simply switch from max knapsack to min knapsack problems.

Another limitation of this scheme is that it cannot be applied for solving other NP-complete optimization problems such as the product knapsack problem, which instead can be shown to be fully approximable by other methods.

Due to these facts the search for general full approximation schemes has been pursued with the aim of finding results which,despite a slight loss in efficiency could be applied to a broader class of problems and that could provide some insight in the properties of fully approximable problems and in their characterization.

The first attempt to provide such a general scheme was the condensation algorithm due to Moran [M]. With respect to the dynamic programming scheme (A_1) the elimination step is performed by eliminating more partial solutions and therefore introducing an error.

More precisely we say that S" dominates S'

$$\text{if } (1-\delta) \sum_{i \in S'} c_i \leq \sum_{i \in S''} c_i \text{ and}$$

$$\sum_{i \in S'} a_i \geq \sum_{i \in S''} a_i$$

where $\delta = \min\{\varepsilon^2, \frac{1}{n^2}\}$, the condensing parameter, is the relative error introduced in the dominance test. As there is a propagation of the error then the total relative error is at most $n\delta \leq \varepsilon$. Moreover the running time, as analyzed by Moran, is $O(\max\{|x^4|, |x^2|/\varepsilon^2\})$ when applied to variants of max subset-sum and max subset-product problems.

A different approach which leads to a more efficient algorithm is based on the technique of *variable partitioning* (as opposed to the constant partitioning technique introduced by Sahni [S]). This method is based on the partitioning of the range of the measure into intervals of exponentially increasing size and on an elimination rule which preserves only one solution for every interval.

To allow a better understanding of the advantages of this approach the method and the results will be given for the 0/1 knapsack and the 0/1 product knapsack. It can be immediately extended to other fully approximable problems.

More in detail the method is as follows.

Let R be the range of the possible values of the measure. In a general NP-complete max-subset problem, and therefore in our cases R is smaller than $2^{p(|x|)}$ for some polynomial p and as we will see the whole development of the algorithm allows us to refer only to this general bound without requiring any more precise estimate of a bound for m^*. The range R is then partitioned into K intervals $[0,m_1)$, $[m_1,m_2),\ldots[m_{K-1},m_K)$ where $m_i = (1+\varepsilon/n)^i$. Let us denote T_i the i-th interval.

The elimination rule for the 0/1 knapsack is the following:
Given two sets S' and S", S' is *dominated* by S"

$$\text{if } \sum_{i \in S'} c_i \in T_i, \quad \sum_{i \in S''} c_i \in T_j, \quad j \geq i \text{ and}$$

$$\sum_{i \in S'} a_i \geq \sum_{i \in S''} a_i$$

Clearly changing the sums in products we have the elimination rule for the 0/1 product knapsack.

In every interval there will be at most one feasible solution and hence, at each iteration, we will have, at most K elements in the list.

THEOREM 2.1. The variable partitioning method provides a fully polynomial apparoximation scheme for the 0/1 knapsack and the 0/1 product knapsack.

PROOF. The error that may result by using this algorithm may be bounded as follows. At stage i at most the error $\Delta_i = m_i - m_{i-1}$ may arise; in the worst case this error may happen at every stage. Since there are n stages and since $\Delta_i < \Delta_{i+1}$ we have that $|m^*(x)-m(A_\varepsilon(x))| \leq n\Delta_{i_{MAX}}$ where i_{MAX} is such that $m_{i_{MAX}-1} \leq m^*(x) < m_{i_{MAX}}$. From the above inequalities we deduce that the overall error is

$$\left|\frac{m^*(x)-m(A_\varepsilon(x))}{m^*(x)}\right| \leq \frac{n[(1+\frac{\varepsilon}{n})^{i_{MAX}} - (1+\frac{\varepsilon}{n})^{i_{MAX}-1}]}{(1+\frac{\varepsilon}{n})^{i_{MAX}-1}} = \varepsilon$$

As for as the complexity is concerned, the number of steps of the given algorithm is as usual a function of n and the length of the list L. In this case the number of solutions which may be preserved in L is equal to the number of intervals K which should satisfy the following inequalities

$$(1 + \frac{\varepsilon}{n})^K \leq 2^{p(|x|)}$$

$$K \log(1 + \frac{\varepsilon}{n}) \leq p(|x|)$$

$$K \leq \frac{p(|x|)}{\log(1 + \frac{\varepsilon}{n})}$$

Hence with a suitable implementation the complexity of the method is

$$O(n \cdot \frac{p(|x|)}{\log(1 + \frac{\varepsilon}{n})})$$

Therefore in the case of knapsack we have that the range R is bounded by $n \cdot c_{MAX}$ and therefore in this case we have a complexity

$$O(n \cdot \frac{\log n + \log c_{MAX}}{\log(1+\varepsilon/n)})$$

while in the case of product knapsack we obtain

$$O(n^2 \cdot \frac{\log c_{MAX}}{\log(1 + \varepsilon/n)})$$

QED

The complexity of the method could be improved in two directions: a) from a general point of view using together the variable partitioning with Sahni's fixed partitioning b) for a single problem, exploiting some particular features. For istance some ideas by Ibarra, Kim and Lawler for the knapsack could also be applied in our case.

However we will not describe these results further because they are obvious extensions and because in this paper we are interested in the general characteristics of the scheme and in defining conditions which guarantee its applicability.

4. A SUFFICIENT CONDITION FOR THE FULL APPROXIMABILITY OF MAX SUBSET PROBLEMS

The results shown in the preceding paragraph suggest to introduce an abstract characterization of the condition of dominance that allows the elimination of feasible solutions and to establish on this basis a condition of full approximability for max-subset problems.

For this purpose we have to require that the satisfaction of the property π by a feasible solution of a max subset problem is "measured" by a function f (which generalizes the concept of occupancy as it appears in knapsack problems).

DEFINITION 3.1. A max subset problem A is said to be *regular* if there exists a polynomially computable set function f with integer value such that the following conditions hold:

1. for every $S \in P(F(x))$ $f(S) \leq 0$ iff $\pi(S)$
2. $\forall S$ $f(\phi) \leq f(S)$, $m(\phi) \leq m(S)$
3. $\forall S_1, S_2$ and any disjoint S_3

$$f(S_1) \leq f(S_2) \rightarrow f(S_1 \cup S_3) \leq f(S_2 \cup S_3)$$

$$m(S_1) \leq m(S_2) \rightarrow m(S_1 \cup S_3) \leq m(S_2 \cup S_3)$$

4. $\forall S_1, S_2$ and disjoint S_3

$$m(S_1) \geq m(S_2) \rightarrow \frac{m(S_1)}{m(S_2)} \geq \frac{m(S_1 \cup S_3)}{m(S_2 \cup S_3)}$$

In [AMP] two weaker properties of max-subset problems were introduced, namely the *hereditarity* of the property

$$\forall S_1, S_2 \quad S_1 \subseteq S_2 \rightarrow (\pi(S_2) \rightarrow \pi(S_1))$$

and *monotonicity* of the measure

$$\forall S_1, S_2 \quad S_1 \subseteq S_2 \rightarrow m(S_1) \leq m(S_2)$$

PROPOSITION 3.1. *A regular* max-subset problem has the properties of hereditarity and monotonicity.

PROOF. Let $S_1 = \phi$.

Since property 2 of definition 3.1 guarantees $f(\emptyset) \leq f(S_2)$ and $m(\emptyset) \leq m(S_2)$ for every S_2, then, given any S_2 and S_3, property 3 of definition 3.1 implies

$$m(S_3) \leq m(S_2 \cup S_3) \qquad \text{(monotonicity)}$$

and $\quad f(S_3) \leq f(S_2 \cup S_3)$

that is $\pi(S_2 \cup S_3) \rightarrow \pi(S_3)$ (hereditarity)

QED

On the other side the property of regularity is indeed strictly stronger than hereditarity and monotonicity.

PROPOSITION 3.2. Max-clique is hereditary and monotone but is not regular.

PROOF. The fact that max-clique is hereditary and monotone is trivial. On the other side whatever f we choose there will be instances of the problem and sets S_1, S_2, S_3 of nodes such that S_1 and S_2 are nodes of complete subgraphs, $S_1 \cup S_3$ corresponds to a complete subgraph, $S_2 \cup S_3$ corresponds to a non complete subgraph. Then we would have $f(S_1) = a$, $f(S_2) = b$ for some negative a and b. W.l.o.g. let $b \leq a$. At the same time $f(S_1 \cup S_3) = c \leq 0$ while $f(S_2 \cup S_3) > 0$. So we would have $f(S_2) < f(S_1)$ but $f(S_2 \cup S_3) > f(S_1 \cup S_3)$.

QED

Examples of problems that satisfy the definition of regularity are knapsack, product knapsack, some scheduling problems. The fact that all these problems are also known to be fully approximable is not surprising because we can prove the following theorem:

THEOREM 3.3. A regular NP-complete max-subset problem is fully approximable.

PROOF. Let us consider the following algorithm based on Moran's approach

```
δ = min(1/n , ε);
L:= ∅;
for all items i in F(x) do
        for all sets S_j in L do
            if S_j ∪ {i} satisfies π
              then
                  begin insert S_j ∪ {i} in L;
                   eliminate all elements S' ∈ L
                   for which there exists S" ∈ L
                   such that
                     (1 - δ/n)m(S') ≤ m(S") and
                      f(S') ≥ f(S")
              end
        end
end
take the best solution on L.
```

As regards the analysis of the error we observe that, at each step, the error introduced by eliminating S' and keeping S" is at most $\frac{\delta}{n}$. By property 3 of definition 3.1 we have that for each subset T (disjoint with S' and S") if $S' \cup T$ is feasible then also $S'' \cup T$ is feasible. By property 4 we have that if $m(S'') \leq m(S')$ then $\frac{m(S'' \cup T)}{m(S' \cup T)} \geq \frac{m(S'')}{m(S')} \geq 1 - \frac{\delta}{n}$; in the other case by property 3 we have that $m(S'' \cup T) \geq m(S' \cup T)$. Hence there is a propagation of the error introduced at each step; since there are n steps the total error is $n \cdot \frac{\delta}{n} \leq \varepsilon$.

The computational complexity of the algorithm is, with a suitable implementation, $0(n\, \ell)$ if ℓ is the maximum number of partial solutions in the list L. As there exists a polynomial $p(|F(x)|)$ such that $m(F(x)) \leq 2^{p(|F(x)|)} = 2^{p'(x)}$ we have that $\ell = 0(\lg_{1+\delta/n} m(F(x))) \leq$ $\leq 0(\frac{n}{\delta}\, p'(n))$. Therefore the complexity of the algorithm is $0(p'(n) \max(h^2, \frac{n}{\varepsilon}))$.

QED

Note that the theorem could also be proved using the method of the variable partitioning.

REFERENCES

[AMP] G.AUSIELLO, A.MARCHETTI SPACCAMELA, M.PROTASI: *Combinatorial problems over power sets*, Calcolo, Vol. 16, n. 4, 1979.

[GJ] M.R.GAREY, D.S.JOHNSON: *Computers and intractability. A guide to the theory of NP-completeness*, Freeman and Company, 1979.

[KS] B.KORTE, R.SCHRADER: *On the existence of fast approximation schemes*, Tech. Rep. 80163-OR, University of Bonn.

[IK] O.H.IBARRA, C.E.KIM: *Fast approximation algorithms for the knapsack and sum subset problems*, J. ACM Vol. 22, n.4, 1975.

[L] E.L.LAWLER: *Fast approximation algorithms for knapsack problems*, Proc. 18th FOCS, Long Beach, 1977.

[M] S.MORAN: *General approximation algorithms for some arithmetical combinatorial problems*, Tech. Rep. 140, Technion, Haifa, 1978.

[PM] A.PAZ, S.MORAN: *NP-optimization problems and their approximations*, Proc. 4th ICALP, Turku, 1977.

[S] S.SAHNI: *Approximate algorithms for the* 0/1 *knapsack problem*, J ACM Vol. 22, n. 1, 1975.

How to compute generators for the intersection of subgroups in free groups

J. Avenhaus, K. Madlener

Kaiserslautern

In this paper we present a breath first tree search to solve combinatorial problems in free groups. Let F be a free group with finitely generated subgroups H and K. It is known that H∩K is again finitely generated (i.e. F has the Howson intersection property). We give a new proof of this fact and present a polynomial time algorithm for computing a set of generators for H∩K. The main algebraic tools are Nielsen type arguments.

The strategy is to search in a tree, thereby cancelling so many subtrees that only a polynomial number of nodes have to be visited. This strategy can be used to give polynomial time decision procedures for some problems, e.g. to decide whether the intersection H∩K is trivial or whether two elements define the same double coset in F. These problems turn out to be polynomial time complete under log-space reducibility.

1. Introduction

Let $S=\{s_1,\dots,s_r\}$ be a finite set and $F=\langle S;\emptyset\rangle$ the free group with basis S. If $\underline{S}=\{s_1,\dots,s_r,s_1^{-1},\dots,s_r^{-1}\}$ is the set of generators and their formal inverses then any element of F can be represented by a finite word over $\underline{S}$. We denote by $\underline{S}^*$ the set of all finite words over $\underline{S}$, by e the empty word, by $|w|$ the length of a word w and by $\equiv$ the identity on $\underline{S}^*$. Then e represents the unit element of F, s_i and s_i^{-1} represent inverse group elements and the multiplication in F corresponds to concatenation in $\underline{S}^*$. For $x,y\in \underline{S}^*$ we write x=y in F, if x and y represent the same element in F.

A subgroup H of F will be given by a set of generators: If $U\subseteq\underline{S}^*$ we denote by $\langle U\rangle$ the subgroup of F generated by U and by $\underline{U}^*$ the set of words $w\equiv u_1\dots u_n$, $u_i\in\underline{U}:=\{u^{\varepsilon}\mid u\in U, \varepsilon=\pm 1\}$, $n\in\mathbb{N}$. For any word $x\in\underline{S}^*$ we have $x\in\langle U\rangle$ iff $x=w$ in F for some $w\in\underline{U}^*$. For $w\in\underline{U}^*$ as above, let $n=|w|_U$ be the

U-length of w. w is called freely reduced in $\underline{U}$, if $u_i \neq u_{i+1}^{-1}$ for all i. For each $x \in \underline{S}^*$ there is a unique freely reduced word $\rho(x)$ with $x = \rho(x)$ in F. It is known that x=y in F iff $\rho(x) \equiv \rho(y)$ [MKS], ρ is linear time computable and the presentation of group elements of free groups by freely reduced words is unique.

If H=<U> and $x \in H$ then x=w in F for some $w \in \underline{U}^*$. This representation for x may not be unique even if one restricts to freely reduced words in $\underline{U}^*$. We say that U generates H freely (or U is independent), if for every $x \in H$ there is a unique $w \in \underline{U}^*$ freely reduced in $\underline{U}$ with x=w in F. An important theorem for free groups, the Nielsen-Schreier Theorem, states that subgroups of free groups are again free groups, so a set of free generators for a subgroup H always exists.

In [AM1] a polynomial time algorithm was presented for computing from $U=\{u_1,\ldots,u_l\}$ a set $V=\{v_1,\ldots,v_m\}$, $m \leq l$, of free generators for <U> based on Nielsen's proof of the subgroup theorem. The set V is a Nielsen-reduced set of generators. Nielsen-reduced sets of generators for a subgroup H have important properties and were used in [AM1, AM2] to give polynomial time algorithms for some decision problems in free groups, such as the generalized word problem (i.e. to decide whether $x \in$<U>), the equality problem and the isomorphism problem for finitely generated subgroups.

In this paper we consider the problem of computing a set of generators for the intersection G=<U> ∩ <V> of two finitely generated subgroups. It is known that this intersection is finitely generated and an explicit bound for the number of generators is known [H,I,LS]. However, no explicit algorithm for computing a set of generators from U and V is given there. We present a polynomial time algorithm for computing a set of generators for G based on a tree search in which nodes are labelled by elements of $\underline{U}^*$ up to certain length. Conditions for the length of the generators and a search strategy are developed from properties of Nielsen reduced sets.

The strategy is also used to decide whether the intersection <U> ∩ <V> is trivial and whether $x \in$ <U>y<V> (i.e. x and y define the same double coset). We develop polynomial time algorithms for these problems. Using a reduction from [AM3] it is shown that the problems are polynomial time complete under log-space reducibility.

For the complexity statements of the algorithms we will use as computing model the usual deterministic Turing machine, which has a finite-state control, a two-way read only input tape, a one way write-only output tape and some two way read-write work tapes. The inputs of the algorithms will be of the type $U=\{u_1,\dots,u_l\}$, $V=\{v_1,\dots,v_m\}$ with $u_i \in \underline{S}^*$ and $|u_i| \leq n$. We measure the time requirements as functions of l,m,n. For $L_1 \subseteq \Sigma^*$, $L_2 \subseteq \Pi^*$, $L_1 \leq L_2$, (L_1 is log-space reducible to L_2) means that there is a log-space computable function $\varphi: \Sigma^* \to \Pi^*$ such that $y \in L_1 \Leftrightarrow \varphi(y) \in L_2$ for all $y \in \Sigma^*$.

We are interested in free groups with a finite but aribitrary large number of generators s_i. In order to state the completness results we have to encode the problems in a finite alphabet Σ. To do so we choose Σ to contain $s, \bar{s}, o, 1$ and represent the s_i and $\bar{s}_i$ by $s\mathfrak{z}(i)$ and $\bar{s}\mathfrak{z}(i)$, respectively, where $\mathfrak{z}(i)$ is i in binary. We will usually give a polynomial time algorithm for the unencoded problem and it will always be easy to see that the algorithm works also in polynomial time for the encoded problem.

2. Nielsen reduction

A lot of subgroup problems in free groups can be solved using properties of Nielsen reduced sets [LS]. Nielsen proved the subgroup theorem (cf.[LS]) by showing that each finite set U can be transformed by a finite sequence of operations into a Nielsen reduced set which freely generates the same subgroup $\langle U \rangle$.

To introduce this concept we define the sets of great prefixes, small prefixes and halves of $U \subseteq \underline{S}^*$ by

$$\mathrm{GPREF}(U) = \{x \mid \exists y : xy \in \underline{U} \text{ and } |x|>|y|\}$$
$$\mathrm{SPREF}(U) = \{x \mid \exists y : xy \in \underline{U} \text{ and } |x|\leq|y|\}$$
$$\mathrm{HALF}(U) = \{x \mid \exists y : xy \in \underline{U} \text{ and } |x|=|y|\}$$

A word x is called an isolated prefix of $\underline{U}$ if there is exactly one y such that $xy \in \underline{U}$ [MKS]. So an isolated prefix of $\underline{U}$ determines an $u \in \underline{U}$ uniquely.

2.1 Definition. A set $U \subseteq \underline{S}^*$ of freely reduced words is Nielsen reduced (N-reduced), if (N1) and (N2) hold:

(N1) If $x \in \mathrm{GPREF}(U)$, then x is an isolated prefix of $\underline{U}$.

(N2) If $x\bar{y}^{1} \in \underline{U}$, $|x|=|y|$, then either x or y is an isolated prefix of $\underline{U}$.

It is easy to see that any N-reduced set is independent and further it has the following important property [MKS]

(N3) If $z_1,\dots,z_p \in \underline{U}$, $z_i \neq z_{i+1}^{-1}$, then $\varrho(z_1 \dots z_p)$ (the free reduction in $\underline{S}^*$) contains a character from any z_i, i.e. there are $x_1,\dots,x_{p+1}$, $y_1,\dots,y_p$ such that $z_i \equiv x_i y_i x_{i+1}^{-1}$, $y_i \neq e$ $(1 \leq i \leq p)$ and

$\varrho(z_1 \ldots z_p) \equiv x_1 y_1 y_2 \ldots y_p x_{p+1}$. In particular, the initial segment of z_1 which remains uncancelled in the free reduction is either a great prefix or a half of z_1 and it is isolated in both cases.

This property enables one to reconstruct the product $z_1 \ldots z_p$ out of the freely reduced word $\varrho(z_1 \ldots z_p)$: The greatest isolated prefix of $\underline{U}$ which is an initial segment of $\varrho(z_1 \ldots z_p)$ determines z_1 uniquely.

A process that transforms a set U into a N-reduced set V such that $\langle U \rangle = \langle V \rangle$ is called a N-reduction. The following theorem is proved in [AM1].

<u>2.2 Theorem</u> A set $U=\{u_1,\ldots,u_p\}$ with $|u_i| \leq n$ can be Nielsen-reduced on a TM in time $O(p^5 n^2)$.

The idea of the proof is to show that a polynomial number of operations of type 1) Delete u_i from U if $u_i = u_j^{\varepsilon}$ ($\varepsilon = \pm 1$) of $u_i \equiv e$

2) Replace u_j by $(u_j u_i^{\varepsilon})$ $i \neq j$, $\varepsilon = \pm 1$

are enough to transform U into a Nielsen reduced set V.

In order to test whether $x \in \langle U \rangle$, U finite, we first transform U into a N-reduced set V. Property (N3) for V can now be used to decide whether $x \in \langle U \rangle = \langle V \rangle$. More precisely, from [AM1] we have the following theorem.

<u>2.3 Theorem</u> Let $V=\{v_1,\ldots,v_m\} \subseteq \underline{S}^*$ be N-reduced, $|v_i| \leq n$ and $x \in \underline{S}^*$ with $|x| \leq t$. There are functions f_V, g_V computable in time $O(tmn)$ with

a) $x = f_V(x) g_V(x)$ in F, $f_V(x) \in \underline{V}^*$, $|f_V(x)|_V \leq t$ and $|g_V(x)| \leq t$.
b) $g_V(x)$ has no prefix $z \in$ GPREF(V)
c) $g_V(x) \in$ SPREF(V), if there is a $y \in \underline{S}^*$ such that $xy \in \langle V \rangle$ and xy is freely reduced.
d) For any $y \in \underline{S}^*$ with $\langle V \rangle y = \langle V \rangle x$, $|g_V(x)| \leq |y|$, and so $x \in \langle V \rangle$ iff $g_V(x) \equiv e$.

The idea for the proof is to split a maximal factor $f_V(x) \in \underline{V}^*$ from the left of x, leaving $g_V(x)$ with $x = f_V(x) g_V(x)$ in F: If w is freely reduced and $w \equiv xw'$, where $x \in$ GPREF(V) $\cup$ HALF(V) is an isolated prefix of maximal length, then there is a unique $z \in \underline{V}$ with $z \equiv x y^{-1}$. We have $w = z \cdot yw'$ in F and the process can be repeated with input yw' until no such isolated prefix x is found or two factors z, z^{-1} appear. Then $f_V(x)$ and $g_V(x) \equiv yw'$ have been computed. Notice, that yw' is freely reduced, $y \in$ SPREF(V) and w' is a suffix of w.

Property c) will be of great importance because it restricts and gives a test for the set of words which are prefixes of words in $\langle V \rangle$.

3. Computing generators for $H \cap K$.

A group is said to have the Howson property if the intersection of any two finitely generated subgroups is again finitely generated. It is known that free groups have the Howson property: If $H=\langle U\rangle$ and $K=\langle V\rangle$ are subgroups of the free group $F=\langle S;\emptyset\rangle$ with cardinalities for U,V $\|V\|=m$ and $\|U\|=l$, respectively, then $G=H\cap K$ is generated by a set W with cardinality $\|W\| \leq 2\cdot(l-1)\cdot(m-1)+1$. There are different proofs of this fact [H,I,LS] but no explicit algorithm for computing the generators is given there.

We present an algorithm for computing from U and V a set W of generators for G in polynomial time.

In the sequel we will use the following notations

$$U=\{u_1,\dots,u_l\} \qquad H=\langle U\rangle \qquad G=H\cap K$$
$$V=\{v_1,\dots,v_m\} \qquad K=\langle V\rangle$$
$$SP(V,U) = SPREF(V)\cdot SPREF(U)^{-1}$$
$$\alpha(V,U)=|SPREF(V)|\cdot|SPREF(U)|$$

Because of Theorem 2.2 we may assume that U,V are N-reduced. We will be interested in solutions $h\in H$, $k\in K$ for the equation $uhv=k$ in F. The following technical Lemma gives information about the length and the structure of such solutions.

<u>3.1 Lemma.</u> Let U,V be N-reduced, $u,v\in \underline{S}^*$ freely reduced such that $\bar{u}^1$ and v have no prefix in GPREF(U). If there is a $w\in \underline{U}^*$, freely reduced in $\underline{U}$ with $uwv \in K$ and $|w|_U \geq \alpha(V,U)$, then there are w',w" with w=w'w" in F, $|w'|_U < 2\cdot\alpha(V,U)$, $|w''|_U<|w|_U$ and $uw''v \in K$.

<u>Proof</u>: Let $w\equiv z_1\dots z_p$, $z_i\in\underline{U}$, $z_i \neq z_{i+1}^{-1}$, $p\geq\alpha(V,U)$ and $uwv\in K$. Since $\bar{u}^1$ and v have no prefix in GPREF(U) and because of property (N3) of N-reduced sets there is a decomposition

$$u\equiv u_o x_o^{-1},\quad z_i\equiv x_{i-1}y_i x_i^{-1},\quad v\equiv x_p v_o$$

with $\rho(uwv)\equiv u_o y_1\dots y_p v_o$, $\rho(uz_1\dots z_i)\equiv u_o y_1\dots y_i x_i^{-1}$ and $x_i\in SPREF(U)$ for $i=o,1,\dots,p$. We have $u_o y_1\dots y_p v_o\in K$ so $w_i :\equiv g_V(u_o y_1\dots y_i)\in SPREF(V)$ by Theorem 2.3(c).

Now $p\geq\alpha(V,U)= |SPREF(V)|\cdot|SPREF(U)|$, so there must exist $o\leq i<j\leq\alpha(V,U)$ such that $w_i\equiv w_j$ and $x_i\equiv x_j$. By Theorem 2.3(a) $u_o y_1\dots y_t=k_t w_t$ in F with $k_t\equiv f_V(u_o y_1\dots y_t)\in K$ for $t=o,1,\dots,p$. This gives

$$uz_1\dots z_i=u_o y_1\dots y_i \bar{x}_i^1 = k_i w_i \bar{x}_i^1 = k_i w_j \bar{x}_j^1$$
$$=k_i k_j^{-1} k_j w_j x_j^{-1} = k_i k_j^{-1} uz_1\dots z_j \text{ in } F$$

It is now easy to verify that the statements of the lemma hold with $w' :\equiv z_1 \ldots z_j (z_1 \ldots z_i)^{-1}$ and $w'' :\equiv z_1 \ldots z_i z_{j+1} \ldots z_p$.

Lemma 2.1 immediately gives that if there is a solution of the equation $uxv \in K$ there must be one with U-length less than $\alpha(V,U)$. It also gives a new proof for the fact that free groups have the Howson property:

3.2 Corollary. Let U,V be N-reduced.

a) There is a set $W \subseteq \underline{U}^*$ such that $\langle W \rangle = H \cap K$ and $|w|_U < 2\alpha(V,U)$ for all $w \in W$.

b) Free groups have the Howson property.

Proof: Clearly, a) implies b). We prove a): If $x \in H \cap K$ then there is a $w \in \underline{U}^*$ freely reduced in $\underline{U}$ such that $x=w$ in F and $w \in K$.

By Lemma 3.2 with $u \equiv e \equiv v$, w is a product $w=w'w''$ in F with $w', w'' \in \underline{U}^*$ and $w'' \in K$, so $w', w'' \in G = H \cap K$. Furthermore, we have $|w'|_U < 2\alpha(V,U)$ and $|w''|_U < |w|_U$. Repeating this argument we get that w is a product $w = w_1 \ldots w_q$ in F with $w_i \in G$, $w_i \in \underline{U}^*$, and $|w_i|_U < 2\alpha(V,U)$. This proves a).

We want to compute a set W of generators for G. In order to do this we organize the set $M = \{w \in \underline{U}^* \mid w \text{ freely reduced in } \underline{U},\ |w|_U < 2\alpha(V,U)\}$ in a tree as follows. The root of T is labelled with the empty word e (unit element of F) and a node with label w has sons wz, $z \in \underline{U}$, such that $wz \in M$. In order to find a set W of generators for G, one could traverse the whole tree and collect all the $w \in \langle V \rangle$. By Theorem 2.2 the test whether $w \in \langle V \rangle$ can be done in polynomial time. Nevertheless, this algorithm takes exponential time, since there are too many tests to be performed and also the set W may become too large. On the other hand, a polynomial number of w's are enough to generate G. This situation seems to be a typical example where nondeterministic computations are more powerfull than deterministic ones. For instance, the special case to decide whether $G \neq \langle e \rangle$, i.e. $G = H \cap K$ is non trivial, can be solved as follows: $G \neq \langle e \rangle$ iff there is a node with label $w \neq e$ such that $w \in \langle V \rangle$. This can be decided nondeterministically in polynomial time simply by guessing a w in the tree T and verifying $w \in \langle V \rangle$.

Using Theorem 2.3 and some book-keeping we can reduce the deterministic search in T drastically and solve both problems in polynomial time on a deterministic TM.

Let < be an order in $\underline{U}$, e.g. $u_1 < \bar{u}_1^1 < \ldots < u_l < \bar{u}_l^1$. This order is extended to M by

$w_1 < w_2 \iff |w_1|_U < |w_2|_U$ or
$|w_1|_U = |w_2|_U$, w_1 comes before w_2 in the lexicographical order on $\underline{U}^*$ defined by $<$.

We will test elements $w \in M$ for $w \in K$ in order $<$, this corresponds to a breadth first search in the tree T. To avoid the full search in T we need some conditions which garantee that some subtrees must not be visited for finding a set W of generators for G. Such conditions are proved in the following Lemma.

3.3 Lemma. Let w,v be nodes in T and $w' \equiv wy$ a descendant of w.

a) If $g_V(w) \notin SP(V,U)$, then $w' \notin K$

b) If $g_V(w) \equiv g_V(v)$, then $wy \in K \iff vy \in K$.

Proof: a) Assume $w' \equiv wy \in K$. Because of (N3) there is a decomposition $w \equiv w_o\bar{x}^1$, $y \equiv xy_o$, such that $\varrho(w') \equiv \varrho(w_o)\,\varrho(y_o)$ and $x \in SPREF(U)$. Theorem 2.3 c) gives $g_V(\varrho(w_o)) \in SPREF(V) \cdot SPREF(U)^{-1} = SP(V,U)$.

b) Again by Theorem 2.3 a), there exsist $k_1, k_2 \in K$ such that $w = k_1 g_V(w)$ and $v = k_2 g_V(v)$ in F. This together with $g_V(w) \equiv g_V(v)$ implies $wy \in K \iff vy \in K$.

Part a) of this Lemma will be used to avoid unsuccessfull search and part b) will be used to keep W small: Suppose $g_V(w) \equiv g_V(v)$, $w < v$ and $wy \in K$. Then $vy \in K$ also, but $vy = v\bar{w}^1 wy$ in F and so $v\bar{w}^1 \in K$. If we put $v\bar{w}^1$ into W, each descendant of v which is a candidate for beeing in W can be represented as a product of $v\bar{w}^1$ and a descendant of w. So we can leave all descendants of v out of our search.

We will traverse T breadth first and maintain in set W and a list L the following information:

$w \in W \Rightarrow w \in G$

$w \in L \Rightarrow$ there is probably a descendant $wy \in K$ of w which is not known to be a product of elements in W.

More precisely, we start with $W = \emptyset$ and $L = \{e\}$. Handling node w in the tree T consists of

- if $g_V(w) \notin SP(V,U)$, then cancel all descendants of w,
- if $g_V(w) \equiv e$, then put w into W and cancel all descendants of w,
- if $g_V(w) \in SP(V,U)$, $g_V(w) \not\equiv e$ then
  a) if there is a $v \in L$ with $g_V(w) \equiv g_V(v)$, then put $w\bar{v}^1$ into W and cancel all descendants of w.
  b) else put w into L.

Notice that this strategy guarantees the following:

1) $\langle W \rangle \leq G$. It remains to show $G \leq \langle W \rangle$.

2) For $v, w \in L$ we have $g_V(w) \not\equiv g_V(v)$ and $g_V(w), g_V(v) \in SP(V,U)$ so $|L| \leq \alpha(V,U)$.

3) The search in T is restricted to descendants of the $w \in L$.

These considerations show that the tree T must not be constructed explicitly, it is enough to store W and L. The k-th element of L is stored as $(L_k', L_k'') = (w, g_V(w))$. The exact algorithm is given below. We use two marks i,j where i points to the actual list element under consideration and j to the last element of the actual list.

<u>Algorithm for computing generators for the intersection of subgroups</u>

```
    Input:  U,V  N-reduced
    Output: W≤U*  a set of generators for G
    Method:
(1)   L1 ← (e,e); W ← ∅;
(2)   i ← 1; j ← 1;
(3)   while  i≤j  do
(4)    begin  for all z∈U with L'i does not end with z̄⁻¹ do
(5)        begin  x ← L'i z; y ← gV(L"i z);
(6)               if y ∈ SP(V,U) then
(7)               if y ≡ e then W ← W ∪ {x}
(8)                         else if ∃ k≤j : L"k ≡ y
(9)                              then W ← W ∪ {xL'k⁻¹}
(10)                             else j ← j+1; Lj ← (x,y)
(11)       end;
(12)       i ← i+1
(13)   end
```

We prove that the algorithm always stops and gives a set of generators for G as output.

<u>3.4 Theorem</u>. Let $U=\{u_1,\dots,u_l\}$, $V=\{v_1,\dots,v_m\} \leq \underline{S}^*$ be N-reduced, $|u_i|, |v_j| \leq n$ and $H=\langle U\rangle$, $K=\langle V\rangle$. Then $G=H \cap K$ is generated by the set W computed by the above algorithm in time $O(l^3m^2n^5)$.

<u>Proof:</u> As mentioned before $j \leq \alpha(V,U)$ because the list L becomes not longer than $\alpha(V,U)$, so the <u>while</u>-loop is executed at most $\alpha(V,U)$ times and the computation always stops. In line (4) there are at most $2\cdot l$ z's to be considered. Line (5) costs $O(mn^2)$, line (6) costs $O(n\cdot\alpha(V,U))$ and line (8) costs $O(n\cdot\alpha(V,U))$. So one pass through the <u>while</u> loop costs $O(l\cdot(mn^2+\alpha(V,U)n))$. Since $\alpha(V,U) \leq (ln+1)\cdot(mn+1) = O(lmn^2)$, the total cost is $O(l^3m^2n^5)$. W has at most $l\cdot\alpha(V,U) = O(l^2mn^2)$ elements and these have U-length less then $2\alpha(V,U)$.

The only thing which remains to be proved is that $G \leq \langle W\rangle$. Let $x \in \underline{U}^* \cap K$, we prove that $x \in \langle W\rangle$. Suppose this is not the case and let $x \in \underline{U}^* \cap K$ be minimal according to our order $<$ with $x \notin \langle W\rangle$. Then $x \neq e$ and $x \equiv z_1 \ldots z_p$, $z_i \in \underline{U}^*$, $z_i \neq z_{i+1}^{-1}$. Let $x_q \in \underline{U}^*$ be a maximal prefix of x which is in the list and $x \equiv x_q z u$, $z \in \underline{U}$, $u \in \underline{U}^*$. Then $0 \leq |x_q|_U < p$ and $g_V(x_q z) \in SP(V,U)$ but $x_q z$ is not added to the list. This leads to two possible cases:

i) $x_q z \in K$ and so $x_q z \in W$ or

ii) $\exists\, x_i \in \underline{U}^*$ in the list with $x_i < x_q z$ and $x_q z x_i^{-1} \in K$ so that $xzx_i^{-1} \in W$.

In case i) $x = x_q z u \in K$ and $x_q z \in K$, so $u \in K$. But $|u|_U < p$ and the minimality of x leads to $u \in \langle W\rangle$ which means $x \in \langle W\rangle$. In case ii) $x = x_q z x_i^{-1} x_i u \in K$ and $x_q z \bar{x}_i^{1} \in K$, so $x_i u \in K$. But $x_i u < x$, which means $x_i u \in \langle W\rangle$. Together with $x_q z x_i^{-1} \in W$ this again gives $x \in \langle W\rangle$. In both cases we have a contradiction to $x \notin \langle W\rangle$ and so $G = \langle W\rangle$.

One can prove a type of minimality condition for W: If $x \in \underline{U}^* \cap K$ is freely reduced, then x is a product of elements $w \in W$ with $|w|_U \leq |x|_U$. So the generators in W are of minimal U-length.

4. Related Problems

Let us reconsider Lemma 3.1. In section 3 this Lemma was used with $u \equiv v \equiv e$ to find an $h \in H$ such that $h \in K$ also. For aribitrarily given u,v it can be used to find all $h \in H$ such that $uhv \in K$, i.e. to solve the equation $uhv = k$ in F, $h \in H$, $k \in K$. We give polynomial time algorithms for testing the solvability of such equations and for representing all solutions. Notice that $uhv = k$ is solvable iff $Hv \cap u^{-1}K \neq \emptyset$. Thus our problem can be stated as an intersection problem for cosets.

Using the notation $H = \langle U\rangle$ and $K = \langle V\rangle$ we are interested in the following problems

$$\begin{aligned} &\text{ICOS}\ (x,y,U,V) &&\Longleftrightarrow Hx \cap yK \neq \emptyset \\ &\text{IRCOS}(x,y,U,V) &&\Longleftrightarrow Hx \cap Ky \neq \emptyset \\ &\text{INT}\ (U,V) &&\Longleftrightarrow H \cap K \neq \langle e\rangle \ \text{(non trivial intersection)} \end{aligned}$$

The problem INT is already solved (implicitly) in section 3. The problem ICOS is equivalent to the problem whether x and y define the same double cost, for ICOS $(x,y,U,V) \Longleftrightarrow Hx \cap yK \neq \emptyset \Longleftrightarrow HxK = HyK$. IRCOS is a subproblem of ICOS, for IRCOS $(x,y,U,V) \Longleftrightarrow$ ICOS (xy^{-1},e,U,V). So it remains to solve ICOS.

Let x,y,U,V be given and define $u :\equiv g_U(y)^{-1}$ and $v :\equiv g_U(x)$. Then by Theorem 2.3 we have

$$\text{ICOS}\ (x,y,U,V) \Longleftrightarrow \exists w \in \underline{U}^* : y^{-1}wx \in K \Longleftrightarrow \exists w \in \underline{U}^* : uwv \in K.$$

Now Lemma 3.1 can be applied: If there is a solution w, then there is a solution w with $|w|_U \leq \alpha(V,U)$. Here again, one way to decide ICOS

(x,y,U,V) is to test all $w\in \underline{U}^*$ with $|w|_U \leq \alpha(V,U)$ for $uwv\in K$. But this costs exponential time. So we repeat the strategy of section 3 and perform a search in the tree T based on Lemma 3.3.

The algorithm below is very similar to that in section 3. The only differences are as follows : We search just one solution $w\in\underline{U}^*$ for $uwv\in K$. The list elements are $(L_i', L_i'') = (w, g_V(uw))$ and the test "$g_V(w)\in K$?" has to be replaced by "$g_V(uw)v\in K$?". So the correctness of the algorithm can be proved exactly as in section 3.

<u>Algorithm for the problem $y^{-1}hx\in K$</u>

```
Input : x,y∈S* U,V N-reduced
Output: Solution h∈U* with y⁻¹hx∈K or no solution

Method:

u← g_U(y)⁻¹; v←g_U(x);
if uv∈K then h←f_U(y)f_U(x)⁻¹ is solution, stop;
u←g_V(u); v← g_V(v⁻¹)⁻¹;
if u∉ SP(V,U) or v⁻¹ ∉ SP(V,U) then no solution, stop;
i← 1; j← 1; L_1← (e,u);
while i≤j do
begin for all z∈U with L'_i does not end with z⁻¹ do
   begin w← g_V(L''_i z);
      if wv∈K then h← f_U(y)·L'_i z·f_U(x)⁻¹ is a solution, stop;
      if w∈ SP(V,U) and w is not jet second component in the list
      then j← j+1; L_j← (L'_i z,w)
   end;
   i← i+1
end;
stop no solution
```

A similar argument as in the proof of Theorem 3.4 gives that this algorithm always stops and gives the correct answer, i.e. a solution if there is one. We summarize the result in the following theorem.

<u>4.1 Theorem</u>. Let $U=\{u_1,\ldots,u_l\}$, $V=\{v_1,\ldots,v_m\}\subseteq\underline{S}^*$ be N-reduced, $|u_i|,|v_j|\leq n$, $H=\langle U\rangle$, $K=\langle V\rangle$ and $|x|,|y|\leq t$. The problem ICOS(x,y,U,V) is solvable in time $O(tn(l+m)+l^3m^2n^5)$.

<u>Proof</u>: We only prove the time bound. The preconditioning (up to the <u>while</u> loop) takes $O(lnt)+O(mnt)+O(mnt)+O(\alpha(V,U)\cdot n)=O(nlt(l+m)+lmn^2))$. The computations in the <u>while</u> loop are only with words of length $O(n)$. The loop is passed at most $\alpha(V,U)$ times and the cost for one pass is

$O(l(mn^2+mn^2+\alpha(V,U)n))=O(l^2mn^3)$. The whole cost is then the bound given in the statement of the Theorem.

Notice, if U,V are fixed and we want only to decide whether $Hx\cap yK\neq\emptyset$ then we have a linear time algorithm in the length of x and y.

<u>4.2 Corollary.</u> The problems

$ICOS(x,y,U,V) \iff Hx \cap yK \neq \emptyset$
$IRCOS(x,y,U,V) \iff Hx \cap Ky \neq \emptyset$
$INT(U,V) \iff H \cap K \neq \langle e\rangle$

are decidable in polynomial time with the same time bound as in Theorem 4.1. If ICOS, IRCOS or INT are true we can compute a representation for $Hx\cap yK$, $Hx\cap Ky$, $H\cap K$, respectively, in polynomial time.

<u>Proof:</u> Let $w\in Hx\cap yK$ be constructed from the solution h of $\bar{y}^1hx\in K$ given by the algorithm of 4.1. Then $Hx\cap yK= Hw\cap wK = (H\cap wK\bar{w}^1)w$. $H\cap wK\bar{w}^1$ is finitely generated and a set of generators can be computed with the algorithm of section 3.
We have $IRCOS(x,y,U,V) \iff ICOS(x\bar{y}^1,e,U,V)$. Let $w\in Hx\cap Ky$ be constructed from the solution h of $hx\bar{y}^1\in K$. Then $Hx\cap Ky = Hw\cap Kw = (H\cap K)w$ and a representation for this intersection can be computed with the algorithm of section 3.
For INT(U,V) we can use the algorithm of section 3 for computing the generators of the intersection or use the algorithm of 4.1 with $u\equiv v\equiv e$ avoiding the trivial solution e.

Corollary 4.2 deals with the question whether two cosets have trivial intersection. As a matter of completeness we study briefly whether two cosets are equal. Let

$EQCOS(x,y,U,V) \iff Hx = yK$
$EQRCOS(x,y,U,V) \iff Hx = Ky$
$EQCOS^*(U,V) \iff \exists x,y: Hx=yK$
$EQRCOS^*(U,V) \iff \exists x,y: Hx=Ky$

<u>4.3 Lemma</u> The problems EQCOS, EQRCOS, $EQCOS^*$ and $EQRCOS^*$ are decidable in polynomial time.

<u>Proof:</u> It is easy to see that

$Hx=yK \iff \bar{x}^1Hx=K \wedge y\bar{x}^1\in H$
$Hx=Ky \iff H=K \wedge y\bar{x}^1\in H$

In [AM1] it was shown that the generalized word problem ($u\in\langle U\rangle$?) and

the equality problem ($\langle U\rangle=\langle V\rangle$?) are solvable in polynomial time. So EQCOS and EQRCOS are decidable in polynomial time.
We have $EQRCOS^*(U,V) \iff H=K$, so $EQRCOS^*$ is decidable in polynomial time.
We have $EQCOS^*(U,V) \iff \exists x: x^{-1}Hx = K$. This problem was shown to be decidable in polynomial time in [AM2].

5. P-completeness of the problems

In the previous section it was shown that all the decision problems considered are in P, the class of problems solvable in polynomial time on a TM. So an upper bound for the complexities is known. In this section we prove that all the problems are P-complete under log-space reducibility.

To do this we use a construction from [AM3] that allows one to reduce any problem in P to a subgroup problem in a free group:

5.1 Fact Let Z be a polynomial time bounded TM. There is a free group $F=\langle S,\emptyset\rangle$ such that for any input y to Z a letter $s(y)\in S$ and a finite set $U(y)\subseteq \underline{S}^*$ are log-space computable with

a) Z accepts $y \iff s(y) \in \langle U(y)\rangle$

b) Z rejects $y \iff U(y) \cup \{s(y)\}$ is independent.

Now we can prove

5.2 Theorem. The problems INT, ICOS, IRCOS, EQCOS, EQRCOS, $EQCOS^*$ and $EQRCOS^*$ are P-complete under log-space reducibility.

Proof: Since all the problems are in P, it is enough to show that any polynomial time recognizable language L can be reduced to each of our problems.
Let L be such a language and Z a polynomial time TM which recognizes L. By Fact 5.1 we have

$$\begin{aligned}
y\in L &\iff INT(U(y),\{s(y)\})\\
&\iff ICOS(s(y),e,U(y),U(y))\\
&\iff IRCOS(s(y),e,U(y),U(y))\\
&\iff EQCOS(s(y),e,U(y),U(y))\\
&\iff EQRCOS(s(y),e,U(y),U(y))\\
&\iff EQCOS^*(U(y),U(y)\cup\{s(y)\})\\
&\iff EQRCOS^*(U(y),U(y)\cup\{s(y)\})
\end{aligned}$$

So L is log-space reducible to each of our problems and hence the problems are P-complete.

References :

[AM1] Avenhaus, J., Madlener, K.: Polynomial time algorithms for the Nielsen reduction and related problems in free groups. Interner Bericht 3o/8o, FB Informatik, Univ. Kaiserslautern, 198o

[AM2] Avenhaus, J., Madlener, K.: Polynomial algorithms for problems in free groups based on Nielsen type arguments. Interner Bericht 31/8o, FB Informatik, Univ. Kaiserslautern, 198o

[AM3] Avenhaus, J., Madlener, K.: P-complete problems in free groups, 5th GI Conference on Theoretical Informatics, Karlsruhe 1981.

[H] Howson, A.G.: On the intersection of finitely generated free groups, J. London Math. Soc. 29, 428-434 (1954)

[I] Imrich, W.: On finitely generated subgroups of free groups Arch. Math. 28, 21-24 (1977)

[LS] Lyndon, R.C., Schupp, P.E.: Combinatorial group theory, Springer Verlag, Berlin 1977

[MKS] Magnus, W., Karrass, A., Solitar, D.: Combinatorial group theory, New York, 1966

ABSTRACT DATA TYPES AND REWRITING SYSTEMS :

APPLICATION TO THE PROGRAMMING OF ALGEBRAIC ABSTRACT DATA TYPES IN PROLOG

Marc BERGMAN**

Pierre DERANSART*

Abstract : The aim of the paper is to present the operational semantics of Algebraic Abstract Data Types (AAT) in terms of rewriting systems and their programming in PROLOG respectively.

An AAT is considered as an interpretor, the semantical actions of which are rewrite rules.

The power of the methodology makes it possible to construct one and many-sorted types, and may be used as an aid for proofs of properties.

This approach leads to a clean, rapid, and accurate programming close to the abstract specification of the type. Proofs by "constructor induction" or Knuth-Bendix algorithm of equational properties valid in initial or final models may be programmed in PROLOG.

Keywords : Abstract Data Type, rewriting systems, operational semantics, PROLOG, proof of equational properties, Knuth-Bendix algorithm.

* INRIA
BP 105
Domaine de Voluceau - Rocquencourt
78153 Le Chesnay Cédex

** Faculté des Sciences de LUMINY
Case 901
F-13288 Marseille

INTRODUCTION

Considering Algebraic Abstract Data Types (AAT) defined by rewriting systems [Hu-Op 80], we show how to program them in the language PROLOG [Ko-74, Ro 75].

For each type in which we are interested, we suppose the existence of a finite set of oriented axioms and introduce the notion of T-reducibility : i.e. the distinction between the total reduction of a term of type T and the local one, via T, if the term is dominated by an operator (defined in T) the codomain of which is different from T.

Combining T-reducibility, sufficient completeness, and consistency we introduce the notion of canonical types, and show that these types are of general purpose. They authorize the construction of a hierarchical library of independant types in the sense that the definition of a new type has no side-effect on the pre-existing ones.

These types may be programmed in PROLOG in a clear, direct and accurate manner.

Such a definition of types may be used for multiple applications including the symbolic manipulations and the proof properties using induction or not. We show that it is possible to use the Knuth-Bendix algorithm in order to prove equational properties valid in the final model of canonical types.

We start by briefly describing the algebraic formalism of the AAT and giving the sufficient conditions for the "sufficient completeness" and "consistency" in terms of rewrite rules (section I), then the programming of AAT in PROLOG is presented (section II) and finally their use is illustrated (section III) as an aid in writing specification, presenting two proof methods in final algebras.

I - ALGEBRAIC ABSTRACT DATA TYPES AS REWRITING SYSTEMS

The description of types has to be made at two levels : the concrete (or user) one and the abstract one which gives the semantics of the type.

This section is devoted to the latter and intends to achieve the semantical encountered problems : definition of object , genericity, independance. The question of overloading operators (for example, "+" used for reals, vectors, matrices...) is supposed to be resolved trough the translation from the concrete syntax to the abstract one by a "types compiler".

The definition given here for the algebraic abstract data types is similar to the Guttag and Horning one [Gu-Ho 78], i.e. a type is specified by :

- S_T a set of sorts (the names of parameter types and $T \in S_T$).
- Σ_T a set of operators with their functionalities. Σ_T is partitioned as the disjointed union $\Sigma_T = I_T \cup O_T$ where I_T is the set of operators the codomain of which is T and O_T is the set of operators the codomain of which is distinct from T (but they must have at least one argument of type T). By abuse of notation, we shall note Σ_T for the set of operators and the signature.
- R_T a finite set of oriented axioms of the form : $\ell_i = r_i$ where ℓ_i is a term $f(e_1,e_2,\ldots,e_n)$, $f \in \Sigma_T$ and r_i may contain conditional expressions. Both ℓ_i and r_i terms belong to the free Σ-algebra $T_{\Sigma_T \cup O_I \cup V}$ constructed with $\Sigma_T \cup O_I \cup V$ where O_I is the set of operators used for writing the axioms but disjoined from Σ_T, V a denumerable set of typed variables.

The variables appearing in the axioms are supposed to be universally quantified.

We shall note :

$\underset{R_T}{\rightarrow}$ the binary relation defined by R_T ;

$\overset{*}{\underset{R_T}{\rightarrow}}$ its reflexive-transitive closure ;

$\underset{R_T}{=}$ its reflexive, symetric and transitive closure.

This definition implies that an "err" element is used for the error treatment because no preconditions are exhibited here. Nevertheless we should consider more powerful axioms of the form :

"Condition $\Longrightarrow \ell_i = r_i$"

but this case will not be studied here.

Example : Type : A ;

syntax : (B is used for Boolean)

$$\Sigma_A \begin{cases} a \to A \\ b \to A \\ f(A) \to B \\ g(A) \to B \end{cases}$$

semantics :

$$R_A \begin{cases} f(a) = vv \\ f(b) = ff \\ g(x) = vv \wedge f(x) \\ g(b) = ff \end{cases}$$

where - $S_A = \{A,B\}$

- $\Sigma_A = I_A \cup O_A$ with $I_A = \{a,b\}$, $O_A = \{f,g\}$
- R_A is the set of rules given in the semantics
- $O_I = \{\wedge\}$

We shall note $\bar{V}_T$ a denumerable set of variables the type of which is distinct from T.

Let R_T be the sub-set of the Σ-algebra $T_{\Sigma_T \cup O_I \cup \bar{V}_T}$ whose terms are dominated by an operator belonging to Σ_T.

We shall say that R_T is the set of reducible terms in T or T-reducible terms.

In contrast with them, the set of terms which have no operator belonging to Σ_T, namely $T_{O_I \cup \bar{V}_T}$ will be named the set of primitive terms for T or T-primitive. We shall note it P_T ; the types of primitive terms will be named the primitive types (or sorts) for T.

The last part of $T_{\Sigma_T \cup O_I \cup \bar{V}_T}$ contains the terms which have a subterm T-reducible.
In order to avoid some confusion we shall distinguish the sort of a term (which is the codomain of its dominating operator) and the type in which it is T-reducible (i.e. the type in which its dominating operator is defined).

A type T is said consistent if its specification does not modify the pre-existing types of the library. More precisely, T is consistent if and only if :

$\forall$ u, $\forall$ v $\in P_T$, $u \underset{R_T}{=} v \Longrightarrow u \underset{R-R_T}{=} v$, where R is the whole union of the axioms in the types library and $u \underset{R-R_T}{=} v$ is a theorem that may be proved without R_T (this definition is equivalent to the one given by [Gu-Ho 78]).

Now, if we suppose that the type B is defined by $S_B = \{B\}$; $\Sigma_B = \{vv, ff, \wedge\}$ with the profiles : $vv, ff \to B$, $B \wedge B \to B$ and the rules $R_B = \{vv \wedge x \to x,\ ff \wedge x \to ff\}$, then the type A is consistent. In fact, only one A-reducible term-g(b)-rewrites into two different A-primitive terms-ff and vv $\wedge$ ff - which are equal by R_B.
The consistency is related to the confluence of the rewriting system R.
A type T has the strong consistency if and only if every T-reducible term with a primitive sort is rewritten by R_T in a unique T-primitive term :

Let $u,v \in P_T$ then $u \underset{R_T}{\equiv} v \Longrightarrow \exists\, w \in P_T$ such that $u \overset{*}{\underset{R_T}{\rightarrow}} w$ and $v \overset{*}{\underset{R_T}{\rightarrow}} w$.

The type A given before is not strongly consistent, whereas the type "sequence" given in section III has the property.

As immediate consequences of the definitions, we now give some properties.

Property 1 : R_T confluent ==> T is strongly consistent ==> T is consistent.

This proceeds from the definitions.
For certain types it is possible to verify the consistency by proof confluence of their rewriting systems.

Property 2 : If R_T is noetherian, then if every T-reducible term of primitive sort has a T-primitive normal form then T has the property of sufficient completeness.

By definition, a type T has the sufficient completeness property if and only if for every T-reducible term "t" with a primitive sort there is a theorem in the equational theory that states $t \underset{R_T}{\equiv} t'$ and $t' \in P_T$.
The hypothesis of the property 2 are sufficient : indeed the finite termination (noetherian) property of R_T implies that there is a reduct (on which no rule ($\in R_T$) can be further applied) and then the second hypothesis corresponds with the sufficient completeness.

This last property ensures that a new type T does not introduce new terms (in their reduced form) in the T-primitive types.

In fact, the introduction of a new type described with a rewriting system implies two sorts of control : sufficient completeness and consistency.

Usually their proof takes into account the whole library of axioms R (at least those which are transitively T-primitive). The properties 1 and 2 make it possible to consider local conditions which can be studied with only the axioms of R_T.

Now we are concerned with the property of canonicity. A rewriting system is said to be canonical if it is noetherian and confluent.

Definition : a type T is canonical if and only if every T-reducible term has a unique normal form of its sort and T has the sufficient completeness property.

Of course, the normal form is computed with the rewriting system R of the library.

It is clear that if R is canonical and T is sufficiently complete then T is canonical
The previous properties (1 & 2) allow a sufficient condition to be given for the hierarchical construction of types by using only local properties of the type T :

Theorem 1 : Let T be a type ;
If : - the T-primitive types are canonical
- R_T is canonical
- every T-reducible term with a T-primitive sort reduces through R_T to a

T-primitive term,

then T is canonical.

These conditions ensure the sufficient completeness of T, the finite termination and confluence of the set of axioms used for the rewriting of T-reducible terms.

In other words, if T is canonical, then every T-**reducible** term can be rewritten alternatively in its unique normal form in the type T or as a term reducible in one of the primitive types of T.

In other words, if T is canonical, then every T-reducible term can be rewritten alternatively in its unique normal form in the type T or as a term reducible in one of the primitive types of T.

II - PROGRAMMING IN PROLOG OF ABSTRACT DATA TYPES DEFINED WITH A REWRITING SYSTEM

The goal of the types programming is to reduce a given term at most : this means that a ground (closed) term has to be reduced to its canonical form in its type and a term with free variable (s) must be reduced as much as possible.

First, we shall present the programming of one-sorted types then the general case of many-sorted ones and finally we shall indicate how to program the conditional.

Brief recall on PROLOG :

The programming language PROLOG is mathematically based on the first order logic, namely on the subset of Horn-clauses. In PROLOG, a predicate P is specified as a disjunction of clauses which can have 2 forms :

(1) P <== Q.R....S

or (2) P <==

where P,Q,R,...S are predicate names.

The clause (1) must be interpreted as a short hand for : P if Q and R and...S in order that Q.R....S is the conditional part and the clause is a conditional assertion.

The second form (2) means an assértion which is an axiom.

Then the different clauses of a same predicate P describe the different way in order that P would be true.

Now the operational semantics of PROLOG can be presented by considering assertion (clauses) as rewrite rules.

The form (1) means : P rewrites in the right part of the clause and Q.R....S becomes a list of goals.

The form (2) means : P desappears.

For a given list of goals (resolvant) PROLOG tries to get a rewritten empty list of goals.

If a goal cannot be rubbed, the demonstration fails.

The variables appearing in a clause are bounded in that clause ; they may be instanciated by any arbitrary term : this occurs when a clause is selected by unification both on its leftmost predicate name and the arguments of the same (the unification proceeds from the leftmost litteral of the resolvant). This implies that the semantics are dependent one the order of the clauses of a predicate and the order of the litterals of the conditional part of the clauses.

Programming of one sorted types :

The strategy consists in considering the AAT as an interpretor the semantical actions of which are the rewrite rules. In order to do that, the analysis of the term t is performed by selection of its dominating operator the arguments of which being reduced with calls by value.

After that an attempt is made to apply a rewrite <u>rule</u> the left term of which is dominated by the same operator as t. If the rewriting succeeds then the analysis continues with the reduced term or else the term is <u>normalized</u>.

This methodology results in the three PROLOG predicates :

$$\text{ANALYSE}\ (f(x),y) \Longleftrightarrow f(x) \overset{*}{\underset{R_T}{\rightarrow}} y \quad \forall f \in \Sigma_T$$

$$\text{NORMALIZE}\ (x,y) \Longleftrightarrow x \overset{+}{\underset{R_T}{\rightarrow}} y \text{ or } x = y$$

$$\text{RULE}\ (x,y) \Longleftrightarrow \langle x,y \rangle \in R_T$$

One interest of the programming in PROLOG is to make it possible that the signature Σ_T and the given functionalities are strictly followed and the form of the rules are preserved.

For instance, for the type Integer defined by :

<u>Type</u> : Integer ;

<u>syntax</u> :

$$\Sigma_{Int} \begin{cases} 0 \rightarrow Int ; \\ s(Int) \rightarrow Int ; \\ +(Int \times Int) \rightarrow Int. \end{cases}$$

<u>semantics</u> :

$$R_{Int} \begin{cases} 0+i \rightarrow i ; \\ s(i)+j \rightarrow s(i+j). \end{cases}$$

We shall program :

$$\begin{cases} \text{Analyse}_{Int}(0,0) \Longleftarrow \\ \text{Analyse}_{Int}(s(i),t) \Longleftarrow \text{Analyse}_{Int}(i,i').\text{Normalize}_{Int}(s(i'),t) \\ \text{Analyse}_{Int}(i+j,t) \Longleftarrow \text{Analyse}_{Int}(i,i').\text{Analyse}_{Int}(j,j').\text{Normalize}_{Int}(i'+j',t) \end{cases}$$

$$\begin{cases} \text{Normalize}_{Int}(x,y) \Longleftarrow \text{Rule}_{Int}(x,x').\text{Analyse}_{Int}(x',y) \\ \text{Normalize}_{Int}(x,x) \Longleftarrow \end{cases}$$

$$\begin{cases} \text{Rule}_{Int}(0+i,i) \Longleftarrow \\ \text{Rule}_{Int}(s(i)+j,s(i+j)) \Longleftarrow \end{cases}$$

The strategy consists in searching the deepest redex which maches a left rule member, then rewrites it by "Normalize" and continues the redex search on such a term that no rule can further be applied.

The three predicate names are indexed by the name of the type which they specify : this enables a modular programming.

If no rule may be applied by the first clause of "Normalize", the program backtracs on the second one which delivers the input as output (case of term already normalized).

<u>Programming of many-sorted types</u> :

In the general case, the strategy of programming is the same but the signature contains other types than the type of interest : this implies that the reduction has to be made by invocation of the predicate "Analyse" corresponding to the right type of the argument according to the functionalities.

Now we give the abstract specification of the classical example of Stack of Integer and its programming.

<u>Type</u> : Stack <u>of</u> Integer ;

<u>Syntax</u> :

$$\begin{cases} \text{Newstack} \rightarrow \text{Stack} ; \\ \text{Push(Stack,Int)} \rightarrow \text{Stack} ; \\ \text{Pop(Stack)} \rightarrow \text{Stack} ; \end{cases}$$

<u>Semantics</u>

$$\begin{cases} \text{Pop(Newstack)} \rightarrow \text{Newstack} ; \\ \text{Pop(Push(s,i))} \rightarrow \text{s} . \end{cases}$$

The PROLOG program will be :

$$\begin{cases} \text{Analyse}_{St}\text{(Newstack ,Newstack)} \Leftarrow \\ \text{Analyse}_{St}\text{(Push(s,i),t)} \Leftarrow \text{Analyse}_{St}\text{(s,s').Analyse}_{Int}\text{(i,i')} \\ \qquad\qquad\qquad\qquad\qquad\qquad \text{.Normalize}_{St}\text{(Push(s',i'),t)} \\ \text{Analyse}_{St}\text{(Pop(s),t)} \Leftarrow \text{Analyse}_{St}\text{(s,s').Normalize}_{St}\text{(Pop(s),t)} \end{cases}$$

$$\begin{cases} \text{Normalize}_{St}\text{(x,y)} \Leftarrow \text{Rule}_{St}\text{(x,x').Analyse}_{St}\text{(x',y)} \\ \text{Normalize}_{St}\text{(x,x)} \Leftarrow \end{cases}$$

$$\begin{cases} \text{Rule}_{St}\text{ (Pop(Newstack),Newstack)} \Leftarrow \\ \text{Rule}_{St}\text{(Pop(Push(s,i)),s)} \Leftarrow \end{cases}$$

Here we have only In-operators, i.e. for which the range is "Stack". But if we add to the signature the functionality of the Top function :

Top(Stack) → Int

and a rewrite rule :

Top(Push(s,i)) → i

then we have to add to the Analyse_{St} predicate the clause :

$$\text{Analyse}_{St}\text{(Top(s),t)} \Leftarrow \text{Analyse}_{St}\text{(s,s').Normalize}_{St}\text{(Top(s'),t').Analyse}_{Int}\text{(t',t)}$$

and we must add to the Rule_{St} predicate the clause :

$$\text{Rule}_{St}\text{(Top(Push(s,i),i)} \Leftarrow$$

Here the call to normalization is always made in the type "Stack" but the result t' has to be analysed in its type : namely "Int" ; the condition of sufficient completeness ensures that Analyse_{Int}(t',t) holds.

Now a problem may arise : let us suppose that s is a Stack of Integer and we want to reduce the term :

0 + Top(s).

We shall invoque

Analyse_{Int}(0 + Top(s),u)

as a goal, but the reduction fails because in the type Integer the operator Top is unknown. To avoid this "misapplication" we have to add a clause of "$Analyse_{Int}$" :

$Analyse_{Int}(Top(x),t) \Leftarrow Analyse_{St}(Top(x),t)$

We must note that this clause doesn't modify the type Integer because its rewriting system R_{Int} is not changed.

<u>The conditional case</u> :

Now we are concerned with polymorphic conditional "If".

For each type T its functionality is :

$If(BOOL,T,T) \to T$

where BOOL is the boolean type and its semantics is given by the rewrite rules :

$If(True,t,t') \to t$

$If(False,t,t') \to t'$

Its implementation in the program doesn't work with a sort of call by value as for the other operators but with a call by name. This point is important because the normalization of the term :

$if(b,t,t')$

where b is the boolean, does not need the normalization of both t and t'. Moreover one of the terms t and t' may be unnormalizable.

Then the preceding way of programming cannot be applied because the normalization of "If" term may fail.

In order to avoid this, we program with the normalization of the boolean followed immediately by the use of the rewrite rules devoted to the "If" operator in the $Rule_T$ predicate.

We add the clauses

$Analyse_T(If(b,t,t'),u) \Leftarrow Analyse_{Bool}(b,b').Normalize_T(If(b',t,t'),u).$

⋮

and

$Rule_T(If(True,t,t'),t) \Leftarrow$

$Rule_T(If(False,t,t'),t') \Leftarrow$

So the first clause of $Normalize_T$ will perform the call by name. Of course these clauses may be automatically generated.

We must note that there is no possible ambiguity because the ranges of the IF operators for different types are distinct. We can now summarize the advantages of programming in PROLOG :

- The program is close to the abstract specification
- programming is rapid, clear and fully automatisable
- the programming is most modular and the canonicity of types is respected.
- even if the control (with the "/" PROLOG predicate) is not explicited here, it is also automatisable and gives efficiency by cutting unuseful tries.

III - HELP TO WRITING SPECIFICATIONS

We show here that it is possible to prove equational properties in the canonical types programmed as before.

Many authors have already realized systems allowing such facilities as in DTVS [Mus 77], AFFIRM [Mus 78] or OBJ [Go-Ta 79]. The proofs can be with or without induction and using the Knuth-Bendix algorithm [Hu-Hu 80]. All are interested in properties in the initial models of the types.

It seems to us that the loss of "expression power" due to the limitation of axioms as canonical rewriting systems may be balanced by references to non-initial models for the types of the library.

We outline here two strategies allowing proof in final models.

The first method is based on the "visible equality" in the type T, noted $\underset{T}{=}$; given t_1 and t_2 two terms of T, then $t_1 \underset{T}{=} t_2$ iff $\forall f \in O_T$, $f(C(t_1)) \underset{T'}{=} f(C(t_2))$ where T' is the visible equality in the primitive type T' after reduction in T and where C represents all possible contexts with the operators of I_T. (In order to simplify the definitions, we use only unary operators. The definitions may be straightforwardly extended with any operators). For instance, in the type stack the functional f = Top and C is any combination of Push and Pop functionals.

A proof is possible with this definition for the equality if we can find an induction on the contexts ; in particular, if the type verifies the property : (Context Induction Property)

$$\forall u, \forall v \in T \ \ [\forall f \in O_T, f(u) = f(v)] \Longrightarrow [\forall f \in O_T, \forall g \in I_T, f(g(u)) = f(g(v))]$$

In this case, the proof comes to :

$$t_1 \underset{T}{=} t_2 \text{ iff } \forall f \in O_T, f(t_1) \underset{T'}{=} f(t_2)$$

Such a property is a syntactical one. It is not verified by the presentation of the type Stack in section 2 but it is true for the presentation of the type "sequence" defined with :

- I_T = {emptyseq, add, sub} ;
- O_T = {i^{th}, lgt},

where "i^{th}" gives the access to the ith element and "lgt" is the length of the sequence ;

- $O_I = \{\underset{Int}{=}, >, +, -\}$

where Int is for Integer and >,+,- have the usual signification.

The rules are given by (s is a variable of type seq and e of type element)

$$R_{seq} = i^{th}(\text{emptyseq}, i) = \text{undef}.$$
$$i^{th}(\text{add}(s,e),i) = \underline{\text{if}}\ \text{lgt}(s) + 1 \underset{Int}{=} i\ \underline{\text{then}}\ e\ \underline{\text{else}}\ i^{th}(s,i)$$
$$i^{th}(\text{sub}(s),i) = \underline{\text{if}}\ \text{lgt}(s) > i\ \underline{\text{then}}\ i^{th}(s,i)\ \underline{\text{else}}\ \text{undef}.$$

$lgt(emptyseq) = 0$

$lgt(add(s,e)) = lgt(s) + 1$

$lgt(sub(s)) = lgt(s) - 1.$

It is easy to show that the rewriting system R_{seq} is finite terminating and confluent. Thus the type "sequence" is canonical if the type "integer" is.

The property we want to prove is the following : $\forall$ s $\in$ sequence, $\forall$ e $\in$ element :

$sub(add(s,e)) = s.$

As this representation verifies the Context Induction Property, it is sufficient to prove the two equations :

$$i^{th}(sub(add(s,e)),i) \underset{Int}{=} i^{th}(s,i)$$

$$lgt(sub(add(s,e))) \underset{Int}{=} lgt(s).$$

This can be done by constructor induction, trying to rewrite each member in the context of the encountered "if" condition taken alternatively true or false. This gives a proof by cases. This method may be illustrated by the rewriting of the left member of the first equation :

$$i^{th}(sub(add(s,e)),i)$$

$$\overset{*}{\underset{R_{seq}}{\rightarrow}} \ \underline{if}\ lgt(s) + 1 > i\ \underline{then}\ i^{th}(s,i)\ \underline{else}\ undef$$

(using the fact that $lgt(s) + 1 > i \Longrightarrow lgt(s) + 1 \neq i$).

<u>case</u> : $lgt(s) + 1 > i = true$

The identity holds.

<u>case</u> : $lgt(s) + 1 > i = false$

We have to prove the property :

$$lgt(s) + 1 > i = false \Longrightarrow i^{th}(s,i) = undef.$$

This proof can be achieved by constructor induction on s (cases : s = emptyseq, add(s',e') and supp(s',e')).

The Context Inference Property is a very useful one by avoiding the study of infinite context cases. This property may be immediately inferred from types without axiom on internal operators (Operators of I_T).

<u>The second method</u> : here we give some indication on a second strategy that can be used. It is based on the following theorem :

<u>Theorem</u> 2 : If a type T is consistent and sufficiently complete, then if the algorithm of Knuth-Bendix halts with success for a set $R \cup \{t_1 = t_2\}$($t_1 = t_2$ equation on terms of T), then $t_1 = t_2$ holds at least in the final model of T.

<u>Sketch of the proof</u> : the Knuth-Bendix algorithm tries to build an equivalent and canonical set of rewrite rules from $R \cup \{t_1 = t_2\}$. If it succeeds, the new rewriting system defines a new type T' such that all models of the new type are models of the old one. As the new type is consistent and sufficiently complete, there is at least

one model of this type [Pai 80], the final one. Thus, the equation $t_1 = t_2$ is valid in the final model of type T.

This theorem gives a semi-decision procedure, to decide properties validating final models of a type. It is in fact sufficient to verify that during the completion process,rules are not introduced, which reduce the class of models of a type to the unique trivial one (all terms of the type are interpreted as equal).

This can be avoided by introducing in some primitive types a complete equality and by verifying that "true" is always different to "false" in the "boolean" type as in [Mus 78].

Programming in PROLOG :

As the considered types are canonical ones, the equality of two ground terms is decidable and can be programmed in PROLOG by a predicate representing the polymorphic function : equal(T,T) → Boolean.

$t_1, t_2 \in T$: equal $(t_1,t_2) \Longleftarrow \text{Analyse}_T(t_1,z).\text{Analyse}_T(t_2,z)$

For non-ground terms, the presented methods are beeing implemented in PROLOG. The first method applies if the type has the Context Induction Property but gives a complete approach to the problem. The second method applies with any set of axioms (not only for canonical types) [Ho 80], but has strong practical limitations. One possibility may be to combine the methods. Both require human control.

CONCLUSION

Implementing Algebraic Abstract Data Types needs to know their proper formal semantics and their accurate operational semantics.

The rewriting systems seem to us to be an adequate and natural way of expressing an abstract specification and the hypothesis of finite termination and Church-Rosser property cannot be avoided in order to implement one and many-sorted types.

The notion of T-reducibility (closed to that of "based type" in [Pai 80]) has the vertue of clarifying the class of the abstract data types for which we offer an explicit way of implementation which is automatizable.

Intrinsically PROLOG works as a rewriting system so that it is an adequate language for programming data types. The strategy results in a clear, rapid and short programming close to the abstract definition.

Even they are defined as an initial or final object in a category, it is possible to use them for proving property of the form M = N where both M and N belong the free Σ_T-algebra on which = is a congruence. This can be used as an aid to the mathematical specification.

Even if it is not shown in this paper, the error treatment as well as the genericity (with the sense of overloading operators) are possible.

Finally must to remark that the call by value can perform useless computing. Of course a lazy evaluation would be better but this needs a more sophisticated programming : this work is under development but we are convinced that this can be done after a more elaborated formal construction for abstract data types as a complete theory of types.

ACKNOWLEDGEMENTS

We are indebted to Professor Claude Pair for his contribution to the prof methods, especially the proof of the theorem 2 and to Marie-Claude Gaudel and Martin Wirsing for their helpful comments on the first version of this draft.

Références

Fu-Ve 80 A.L. FURTADO, P.A.S. VELOSO
Procedural Verification and Implementations for Abstract Data Types,
Pontificia Universidade Catolica do Rio de Janeiro.
Private communication.

Gau 80 M.C. GAUDEL
Compiler Proof and Generation based on semantics of programming Languages (French)
Thèse de Doctorat - 10 th March 1980 - Nancy -

Gog 80 J.A. GOGUEN.
How to prove algebraic inductive hypotheses without induction with applications to the correctness of data type implementation. 5th Conference on Automated Deduction. Les Arcs, France, Juillet 1980. (LNCS 87, pp. 356-373).

Go-Ta 79 J.A. GOGUEN, J.J. TARDO,
"An introduction to OBJ, A Language for writing and testing Formal Algebraic Program Specification". Specification of Reliable Software. Boston Apr. 1979. pp. 170-189.

Gu-ho 78 J.V. GUTTAG, J.J. HORNING,
The Algebraic Specification of Abstract Data Types. Acta Informatica 10 (1978), 27-52.

Gut 78 J. GUTTAG
Abstract Data Types and the Development of Data Structures.
CACM - V20 - N6 - June 1977 - pp 396-404.

Hu-Hu 80 G. HUET, J.M. HULLOT,
Proofs by Induction in Equational Theories with Constructors.
Rapport INRIA n°28.

Hu-Op 80 G. HUET, D. OPPEN
Equations and Rewrite Rules : a Survey.
"Formal Languages : Perspectives and Open Problems".
Ed. Book R, Academic Press 1980. Also TR-CSL-111, SRI International, Jan. 1980.

Ma-Se 80 E. MADELAINE, S. SENESI,
On using LCF and PROLOG to prove Correctness of Abstract Data Type Representation (French).
Ecole Polytechnique - INRIA - July 1980.

Mus 77 D.R. MUSSER
"A Data Type Verification System Based on Rewrite Rules".
Proceedings of the Sixth Texas Conference

Mus 78 D.R. MUSSER,
Convergent Set of Rewrite Rules for Abstract Data Types. USC - ISI. Californie. Déc. 78 - Nov. 79.

Pai 80 C. PAIR
On the Models of Algebraic Abstract Data Types (French)
Centre de Recherche en Informatique de Nancy - Report 80 P 052, May 1980.

Rou 75 P. ROUSSEL
PROLOG Reference Manual (French)
Université d'Aix - Marseille II - Septembre 1975.

TOP-DOWN TREE-TRANSDUCERS FOR INFINITE TREES I

J. Bilstein and W. Damm

Lehrstuhl für Informatik II , RWTH Aachen
Büchel 29-31 , D-5100 Aachen

Introduction

Consider a denotational semantic specification for an imperative programming language. Essentially, such a definition can be viewed as a translation, which describes the control-structure of the programming language in terms of certain elementary operations (as e.g. updating the store, reserving storage, composing two continuations) using the concepts of the meta language used in the semantic-specification. Clearly the "complexity" of the meta-language depends on the expressive power inherent in the control-structures of the programming language. As an example, to describe the procedure concept of ALGOL 68, the full λ-calculus is needed as meta language, while the restriction to finite modes can be handled by a typed λ-calculus with fixedpoint operators [Da 2, DF 1, DF 2] . Hence, by viewing the elementary operations in the meta language as uninterpreted operation symbols, a programming language canonically induces a class of program schemes. In the deterministic case, it is well known how to associate with a scheme an infinite tree over the operation symbols which completely describes its semantics (over all possible interpretations of the operation symbols; see [Sco] , [Niv] , [Da 1] , [Fe] for the case of flowchart-, recursive-, level-n-, and λ-schemes, respectively). In particular, when starting with some program P , the infinite tree of the associated scheme S_P completely specifies the semantics of P for all possible inputs. Hence the process of compiling a higher level language into some target language PL can be viewed as a semantic preserving translation of the infinite trees of the higher level language to the class of infinite trees associated with PL This paper should be viewed as a starting point in investigating the power of transducers needed to describe such transformations on infinite trees. We investigate the extension of the simplest kind of tree transducers, the deterministic (finite-state) top-down tree transducers,to infinite trees. It is shwon, that the class n-$\mathcal{R}$ of level-n trees (corresponding to ALGOL 68 programs with finite modes), is closed under deterministic top-down translations. The proof is based on the operational characterization of level-n trees as join of the schematic language generated by the corresponding scheme using outermost-innermost-derivations given in [Da 2] . As an application, we give a full proof of the result stated in [ES], that the OI-hierarchy of string languages (c.f. [Da 2, ES]) starts with the regular, context-free and OI-macro languages. In fact, this paper was motivated by the wish to give a precise meaning to the ideas suggested in the proof of this result in [ES].

1 BASIC NOTIONS

Let I be a set of *base types*. $w \in I^*$ of *length* k is viewed as a mapping $w : [k] \to I$, where $[k] := \{1,\dots,k\}$, hence $w(j)$ is the j-th letter of w. We denote the empty string by e.

An *I-set* A is a family of sets $(A^i \mid i \in I)$. Define A^w inductively by $A^e := \{()\}$, $A^{wi} := A^w \times A^i$. For I-sets A,B we write $A \subseteq B$ iff $A^i \subseteq B^i$ for all $i \in I$ and $A \cup B$ to denote the I-set $(A^i \cup B^i \mid i \in I)$.

The set of *derived types over* I, $D^*(I)$, is defined by

$$D^o(I) := I \ , \quad D^{n+1}(I) := D^n(I)^* \times D^n(I) \ , \quad D^*(I) := \bigcup_n D^n(I) \ .$$

For $\alpha \in D^n(I)^*$ of length l, we define $Y_\alpha := \{y_{1,\alpha(1)},\dots,y_{l,\alpha(l)}\}$, and $y_\alpha := (y_{1,\alpha(1)},\dots,y_{l,\alpha(l)})$ - in particular $Y_e = \emptyset$ and $y_e = ()$. We extend this notation to $\tau = (\alpha_m,\dots,(\alpha_o,i)\dots) \in D^*(I)$ by setting $Y^\tau := \bigcup_{o \leq j \leq m} Y_{\alpha_j}$.

The basic objects of this paper are typed *applicative terms* over sets Y of *parameters*, X of *procedurenames* and Σ of *operationsymbols*. Formally, let $Y \subseteq \{y_{j,\tau} \mid j \in \omega, \tau \in D^*(I)\}$, X a $D^*(I)$-set, Σ a $D(I)$-set, then the $D^*(I)$-set $T_{\Sigma,X,Y}$ is the smallest $D^*(I)$-set T satisfying

(i) $\Sigma \subseteq T$, $X \subseteq T$, $y_{j,\tau} \in T^\tau$

(ii) $t \in T^{(\alpha,\tau)}$, $\underline{t} \in T^\alpha \leadsto t\underline{t} \in T^\tau$.

Intuitively, the parameters denote positions, where terms of suitable type should be subtituted (as specified by the rewriting rules of the scheme), based on the elementary operation of substituting (in parallel) $\underline{t} \in T^\alpha_{\Sigma,X,Y}$ for y_α in t, $t[y_\alpha/\underline{t}]$.
Note, that $T_\Sigma{}^i$ ($= T_{\Sigma,\emptyset,\emptyset}{}^i$) coincides with the set of Σ-*trees* of sort i. If Σ is *monadic* ($\Sigma^{(e,i)} = \{e\}$, $\Sigma^\tau = \emptyset$ for $\tau \neq (i,i)$), $T_\Sigma{}^i$ is isomorphic to $(\Sigma^{(i,i)})^*$.
If Y happens to be an I-set, we write $T_\Sigma(Y)$ for $T_{\Sigma,Y}{}^i$, the set of Σ-*trees generated by* Y.
We use the following auxiliary functions:

- $\tau \in D^n(I)$ has *level* n

- $t \in T^{\tau}_{\Sigma,X,Y}$ has *type* τ and *level* *level*(τ) .

  We often write $t : \tau$ for *type*$(t) = \tau$.

- For $t : (\alpha_m,\dots,(\alpha_o,i)\dots)$, $t\!\downarrow := ty_{\alpha_m}\dots y_{\alpha_o}$.

Let $\Sigma_\perp$ denote the $D(I)$-set $\Sigma \cup \{\perp_i \mid (e,i) \in D(I)\}$. Note that $FT_{\Sigma,X,Y} := T_{\Sigma_\perp,X,Y}$ is partially ordered by $t \leq t'$ iff $t = \perp_{type(t')}$ or there exist t_j, t'_j s.t. $t_j \leq t'_j$ and $t = t_o(t_1,\dots,t_r)$, $t' = t'_o(t'_1,\dots,t'_r)$.

A subset Δ of FT_Σ is *directed* iff $\forall t_1,t_2 \in \Delta\ \exists t \in \Delta\ \{t_1,t_2\} \leq t$. We denote by CT_Σ the set of *infinite trees* over Σ (with possible $\perp$-leaves). Recall that CT_Σ is Δ-*complete*, i.e. each directed $\Delta \subseteq CT_\Sigma$ has a *least upper bound* $\sqcup\Delta$ in CT_Σ . An infinite tree t can be represented by *ideal*$(t) = \{s \in FT_\Sigma \mid s \leq t\}$.

The notation of substitution is extended

- to infinite trees by setting $t_o[y_w/(t_1,\dots,t_r)] = \sqcup\{s_o[y_w/(s_1,\dots,s_r)] \mid s_j \in ideal(t_j)\}$
- to *tree languages* $L \subseteq T_\Sigma(Y)$: $L_o[y_w/(L_1,\dots,L_r)] = \{ s_o[y_w/(s_1,\dots,s_r)] \mid s_j \in L_j\}$
- as *OI-substitution* on tree languages: $L \overset{\leftarrow}{_{OI}} (L_1,\dots,L_r) = \bigcup_{t \in L} t \overset{\leftarrow}{_{OI}} (L_1,\dots,L_r)$, where

$$a \overset{\leftarrow}{_{OI}} (L_1,\dots,L_r) := \{a\} \qquad \text{for } a \in \Sigma^{(e,i)}$$

$$f(t_1,\dots,t_k) \overset{\leftarrow}{_{OI}} (L_1,\dots,L_r) := \{f(s_1,\dots,s_k) \mid s_j \in t_j \overset{\leftarrow}{_{OI}} (L_1,\dots,L_r)\}$$

We assume, that the reader is familiar with the notion of Δ- ($\sqcup$-) *continuous* Σ-*algebra* as treated in [GTWW, ES] . In particular we will make use of the following facts:

- the Δ-continuous *algebra of infinite trees over* Σ, CT_Σ , is initial in the category of Δ-continuous Σ-algebras with strict Δ-continuous Σ-homomorphisms .
- the $\sqcup$-continuous *algebra of tree languages over* Σ , PT_Σ , is initial in the category of $\sqcup$-continuous Σ-algebras with $\sqcup$-continuous Σ-homomorphisms .

The following definitions, which make use of initiality, are basic for this paper.

- Let $\Sigma_+ := \Sigma \cup \{+_i \mid (ii,i) \in D(I)\}$. We denote by $(PT_\Sigma)_+$ the Δ-continuous extension of PT_Σ with + denoting *finite union* . The unique strict Δ-continuous Σ_+-homomorphism $CT_{\Sigma_+} \to (PT_\Sigma)_+$ will be denoted *set*. It is easy to see, that *set* commutes with (OI-)substitution:

$$set(t[y_w/(t_1,\dots,t_r)]) = set(t) \overset{\leftarrow}{_{OI}} (set(t_1),\dots,set(t_r))$$

- Let $D(\Sigma)$ denote the *derived alphabet of* Σ (c.f. [Mai, ES, Da 1]) . The Δ-continuous $D(\Sigma)$-algebra *der*-CT_Σ has carrier $(CT_\Sigma(Y_w)^i \mid (w,i) \in D(I))$ and assignmentfunction φ_{der} given by

$f' \quad : (e,(w,i)) \mapsto (() \mapsto f(y_w)) \quad \text{for } f \in \Sigma^{(w,i)}$

$pr_j^w \quad : (e,(w,w(j)) \mapsto (() \mapsto y_{j,w(j)})$

$abs_{(w,i)} \quad : ((e,i),(w,i)) \mapsto (t \mapsto t)$

$sub^v_{(w,i)} \quad : ((v,i)(w,v(1))\dots(w,v(r)),(w,i)) \mapsto ((t,t_1,\dots,t_r) \mapsto t[y_v/(t_1,\dots,t_r)])$.

The unique strict Δ-continuous $D(\Sigma)$-homomorphism $CT_{D(\Sigma)} \to der\text{-}CT_\Sigma$ will be denoted *yield* .

2 DETERMINISTIC TOP-DOWN TRANSDUCERS AND LEVEL-N TREES

Let us start by extending the top-down tree transducers to infinite trees. The simple underlying idea is to consider a conventional tree transducer working in a top-down fashion on approximations of the input tree. By restriction ourselves to deterministic transducers which map $\perp$ to $\perp$, the translation of finite approximations is order-preserving, hence we can define the output-tree as the join over all initial pieces of output obtained from finite approximations of the input.

Definition 1

A deterministic top-down-transducer on infinite trees (in short a DT^∞) is a family of $D(\{i\})$-mappings

$$M = (M_q : \Sigma \to FT_\Omega([k] \times Z) \mid q \in [k])$$

for some $k \in \omega$. We call $[k]$ the set of *states* , 1 the *initial state* , Σ the *input alphabet* , Ω the *output alphabet* , and Z the $(\{i\}\text{-})$ set of *input-parameters*. □

The definition of the translation induced by M on finite trees is standard. It is straightforward to show, that this mapping is monotonic.

Defininition 2

Let $M \in DT^\infty$ with k states, input alphabet Σ and output alphabet Ω . Define

$$\overline{M}_q : FT_\Sigma \to FT_\Omega$$

by

$$\overline{M}_q(\perp) = \perp \qquad \overline{M}_q(a) = M_q(a)$$

$$\overline{M}_q(ft_1\dots t_m) = M_q(f)[([1,z_1],\dots,[k,z_m]) / (\overline{M}_1(t_1),\dots,\overline{M}_k(t_m))] .$$

The *tree translation realized by* M , $\overline{M} : CT_\Sigma \to CT_\Omega$ is defined by $\overline{M}(t) := \bigsqcup\{\overline{M}_1(s) \mid s \in ideal(t)\}$. The class of tree translations realized by DT^∞'s will be denoted $\mathcal{DT}^{\infty}$. □

For an example, we refer to section 3 . It should be clear to the reader, that the above definition coincides on trees in T_{Σ} with translations realized by (total) deterministic top-down transducers.

We will now prove, that the class of *level-n trees* (corresponding to ALGOL 68 programs with mode depth restricted by n) is closed under DT^{∞}'s . To this end, we will use the operational characterization of level-n trees as join over all approximations generated by the corresponding schematic grammar as definition. For motivation, proofs, and the connection to ALGOL 68 we refer the reader to [Da 2] and [DF 2] .

Definition 3

A *level-n scheme* over a D({i})-set Ω of operation symbols is a mapping

$$S : X_N \to FT^{\,i}_{\Omega,X_N,Y}$$

s.t. (1) $X_N = \{x_o,\dots,x_N\}$ is a $D^*(\{i\})$-set of *procedure names*

(2) $\mathit{level}(x_o) = o \;\wedge\; \forall j \in [N] \quad \mathit{level}(x_j) \leq n$

(3) $\forall j \in \{o,\dots,N\} \quad \mathit{type}(x_j) = \tau \rightsquigarrow S(x) \in FT^{\tau}_{\Omega,X_N,Y}$

The class of level-n schemes over Ω will be denoted $n\text{-}\lambda(\Omega)$. □

As usual, we will write such a scheme as a system of equations

$$x_j{\downarrow} = S(x_j) \quad \text{for} \quad j = o,\dots,N \;.$$

The procedure name x_o corresponds to the main-program. To view such a scheme as a schematic grammar $S_{\perp}$ simply apply the ALGOL 60 copy rule to procedure-calls, where all actual parameters (down to *level* o) are supplied and allow in addition the replacement of such "complete" procedure calls by $\perp$. The *schematic language generated by* S , $L(S_{\perp})$, is the set of all trees derivable from the axiom. For examples, we refer to [Da 1] .

In [Da 2] it is shown, that $L(S_{\perp})$ is directed, hence we can define the *infinite tree generated by* S to be $T(S) := \bigsqcup L(S_{\perp})$. We denote by $n\text{-}\mathcal{R}$ the class of infinite trees generated by level-n schemes. It should be clear to the reader, that $o\text{-}\mathcal{R}$ and $1\text{-}\mathcal{R}$ coincide with the class of *regular* and *context-free infinite trees* , respectively. In [Da 2] it is shown, that level-n trees form a strict hierarchy with increasing n .

In the definition of translations realized by DT^{∞}'s , an infinite tree was represented by its ideal. The following *Chomsky-Normalform* of a level-n scheme S generates all approximations of $T(S)$.

Theorem 1 [Da 2]

Let $S \in n\text{-}\lambda(\Omega)$, then there exists $S' \in n\text{-}\lambda(\Omega)$ s.t. $T(S') = T(S)$ and all equations of S' are of one of the following forms (1) to (5) :

(1) $x{\downarrow} = f y_{\alpha_o}$ with $f \in \Omega$

(2) $x{\downarrow} = \perp$

(3) $x{\downarrow} = x_1(x_2(y_{\alpha_p}),\dots,x_r(y_{\alpha_p}))(y_{\alpha_{p-1}})\dots(y_{\alpha_o})$

(4) $x{\downarrow} = y_{j,\alpha_p(j)}(y_{\alpha_{p-1}})\dots(y_{\alpha_o})$

(5) $x{\downarrow} = y_{1,\alpha_p(1)}(y_{2,\alpha_p(2)}(y_{\alpha_{p-1}}),\dots,y_{r,\alpha_p(r)}(y_{\alpha_{p-1}}))(y_{\alpha_{p-2}})\dots(y_{\alpha_o})$

where *type* $(x) = (\alpha_m,\dots,(\alpha_o,i)\dots)$, $p \leq m$, $r = l(\alpha_p)$ □

For the proof of the closure result it is convenient to consider only outermost-innermost derivations.

Notation

(1) Let N denote the set of *level*-o applicative terms over procedurenames. For $A \in N$, we denote by *head*(A) the leftmost-outermost procedure-identifier in A .

(2) A *linearisation* of $t \in FT_\Omega(N)$ is a list $(s,A_1,\dots,A_p)$, where $s \in FT_\Omega(Y_w)$, $l(w) = p$, each parameter $y_{1,w(1)},\dots,y_{p,w(p)}$ occurs exactly once in s in this order, and $t = s[A_1,\dots,A_p] := s[y_w/(A_1,\dots,A_p)]$. □

Definition 4

Let $S \in n\text{-}\lambda(\Omega)$ be in Chomsky Normalform, and $t,t' \in FT_\Omega(N)$ with $t = s[A_1,\dots,A_p]$.
t is OI-*derivable* to t' *in* $S_\perp$ ($t \underset{OI,S_\perp}{\Rightarrow} t'$) in one step iff

$$t' = s[A_1,\dots,A_{j-1},\perp,A_{j+1},\dots,A_p]$$

or

$$t' = s[A_1,\dots,A_{j-1},S(x)[y_{\alpha_m}/s_m]\dots[y_{\alpha_o}/s_o],A_{j+1},\dots,A_p]$$

for some j s.t. $A_j = x(s_m)\dots(s_o) \in N$.

The *schematic* OI-*language generated by* S is defined by

$$L_{OI}(S_\perp) = \{t \in FT_\Omega \mid x_o \overset{*}{\underset{OI,S_\perp}{\Rightarrow}} t\} \quad .$$ □

Lemma 1

Let $S \in n\text{-}\lambda(\Omega)$ be in Chomsky Normalform, then $\mathit{ideal}(T(S)) = L_{OI}(S_\perp)$.

Proof:

"$\supseteq$" by definition. For "$\subseteq$" it suffices to show

$x_o \overset{*}{\underset{OI,S_\perp}{\Rightarrow}} t \in FT_\Omega(N) \rightsquigarrow \forall t' \leq t \; x_o \overset{*}{\underset{OI,S_\perp}{\Rightarrow}} t'$. This can be done by induction on the length of the derivation; the induction step follows by a straightforward caseanalysis according to the right hand side of the replaced procedure identifier. □

Now consider some level-n scheme S in Chomsky Normalform and some $M \in DT^\infty$ working on approximations generated by $S_\perp$. The scheme S_M simulating S and M will memorize the states reached by M after processing the allready OI-derived terminal part in the *head*-procedurnames of the leaves. When ever a production $x\downarrow = fy_{\alpha_o}$ is applied, the current state will be attached to x and the corresponding translation of f is chosen, with the proper states stored in the *heads* of the leaves to be substituted for y_{α_o} . While simulating a production $x\downarrow = x_1(x_2(y_\alpha),\ldots,x_r(y_\alpha))\downarrow$, we do not know in advance, which state the automaton will have reached when working on trees generated from $x_2,\ldots,x_r$. To cope with this, we allow all possible states by encoding them in additional parameter positions and choose the right alternative by projections as soon as the correct state information is available.

Construction

(i) Types: $\overline{}: D^*(\{i\}) \to D^*(\{i\})$

$$\overline{i} := i \; , \quad \overline{(\alpha(1)\cdot\ldots\cdot\alpha(r),\nu)} := (\overline{\alpha(1)}^k\cdot\ldots\cdot\overline{\alpha(r)}^k,\overline{\nu})$$

We identify $[k] \times Y_\alpha$ and $Y_{\overline{\alpha}^k}$ via

$$(q,y_{j,\alpha(j)}) \mapsto y_{l,\overline{\alpha}^k(l)} \quad \text{with} \quad l = (j-1)\cdot k+q$$

(ii) Terms:

Let $t \in T^\tau_{X,Y}$ be an applicative term.

We define inductively $\overline{t} \in T^{\overline{\tau}^k}_{[k]\times X,[k]\times Y}$ by

$\overline{x} := ([1,x],\ldots,[k,x])$ with $\mathit{type}\,[q,x] := \overline{\mathit{type}(x)}$, $\overline{y} := ([1,y],\ldots,[k,y])$

$$\overline{t(t_1,\dots,t_r)} := (pr_1(\bar{t})(\bar{t}_1,\dots,\bar{t}_r),\dots,pr_k(\bar{t})(\bar{t}_1,\dots,\bar{t}_r))$$

(iii) Let $S \in n\text{-}\lambda(\Sigma)$ in Chomsky normalform and $M \in DT^{\infty}$ with states $[k]$, input alphabet Σ, and output alphabet Ω. If $X = \{x_o,\dots,x_N\}$ is the set of procedurenames of S then S_M uses the set $[k] \times X$ with $[1,x_o]$ as main procedure. For each equation of the form (1) ... (5) in S, S_M contains the equations

(1) $[q,x]\downarrow = M_q(f)[([1,z_1],\dots,[k,z_m])/([1,y_{1,\alpha_o(1)}],\dots,[k,y_{m,\alpha_o(m)}])]$

(2) $[q,x]\downarrow = \perp$

(3) $[q,x]\downarrow = [q,x_1](\overline{x_2(y_{\alpha_p})},\dots,\overline{x_r(y_{\alpha_p})})\downarrow$

(4) $[q,x]\downarrow = [q,y_{j,\alpha_p(j)}]\downarrow$

(5) $[q,x]\downarrow = [q,y_{1,\alpha_p(1)}](\overline{y_{2,\alpha_p(2)}(y_{\alpha_{p-1}})},\dots,\overline{y_{r,\alpha_p(r)}(y_{\alpha_{p-1}})})\downarrow$ □

Since the infinite tree generated by S_M is just the translation of $T(S)$ under $\bar{M}$, the class of level-n trees is closed under deterministic top-down translations.

Theorem 2

$\mathcal{DT}^{\infty}(n\text{-}\mathcal{R}) = n\text{-}\mathcal{R}$

Proof: Clearly it suffices to prove $\subseteq$.

Let $M \in DT^{\infty}$ with k states, input alphabet Σ, and output alphabet Ω, and let $S \in n\text{-}\lambda(\Sigma)$ be in Chomsky normalform. We will show

(1) $\bar{M}_1(L_{OI}(S_\perp)) \subseteq L_{OI}(S_{M_\perp})$

(2) $\forall t \in L_{OI}(S_{M_\perp})\ \exists\ s \in L_{OI}(S_\perp)\quad t \leq \bar{M}_1(s)$.

This implies

$$\begin{aligned}\bar{M}(T(S)) &= \bigsqcup \bar{M}_1(\mathit{ideal}(T(S))\\ &= \bigsqcup \bar{M}_1(L_{OI}(S_\perp)) && \text{by lemma 1}\\ &= \bigsqcup L_{OI}(S_{M_\perp}) && \text{by (1), (2)}\\ &= T(S_M)\end{aligned}$$

and thus proves the theorem.

For the prove of the assertions, we extend $\bar{M}_q$ to trees in $FT_\Sigma(N)$ by setting $\bar{M}_q(A) = pr_q(\bar{A})$. To establish (1), we prove

$$x_o \underset{OI,S_\perp}{\overset{*}{\Rightarrow}} t \sim [1,x_o] \underset{OI,S_M}{\overset{*}{\Rightarrow}} \overline{M}_1(t)$$

by induction on the length of the derivation. The induction step follows by a straightforward case analysis.

For the proof of (2), we have to introduce some notation. Let $\nu \in \{o,\dots,N\}^*$, where N+1 is the number of equations of S, then each string $o\nu$ determines uniquely a tree $\nu(x_o) \in FT_\Sigma(N)$ by

$$e(x_o) = x_o \ , \quad \nu j(x_o) \text{ is the tree obtained from } \nu(x_o)$$

by OI-deriving all *head*-occurences of x_j in $\nu(x_o)$ from left to right using $S(x_j$

Similary, $\nu([1,x_o])$ uniquely defines a tree in $FT_\Omega(\underline{N})$, where $\underline{N} = \{pr_q(\overline{A}) \mid A \in N$ and j denotes all procedurenames $[q,x_j]$.

We then prove by induction on the length of ν

$$\overline{M}(\nu(x_o)) = \nu([1,x_o]) \quad .$$

This implies (2), since any tree OI-derivable from $[1,x_o]$ is majorized by some $\nu([1,x_o])$. □

In [Da 3] we prove, that extending DT^∞'s by *level-n stacks* (c.f. [DG]) leads to a strict hierarchy of tree translations.

3 A CHARACTERIZATION OF THE OI-HIERARCHY

There are (at least) two canonical ways of embedding string-language into tree-languag theory, which essentially depend on the algebraic structure one has in mind when talking about "the algebra of formal languages". In one case, left-concatenation with the elements of the alphabet V is taken as the basic operation, while e.g. the structure underlying the fixedpoint-characterization of context-free languages views concatenation of (arbitrary) languages as the basic operation. In both cases, the algebraic structure induces suitable ranked (i.e. D({i})-sorted) alphabets V^m and V^c : V^m contains a o-ary symbol e (denoting the language consisting of the empty string), and monadic operation symbols $a \in V$ (corresponding to left-concatenation with a), while V^c has all symbols in V and e as o-ary symbols (denoting the obvious one element languages) and a binary symbol c for concatenation The corresponding algebras of formal languages will be denoted $m\text{-}PV^*$ and $c\text{-}PV^*$, respectively. (Note, that essentially the same relation holds between CT_Ω and $der\text{-}CT_\Omega$). Clearly the algebras PT_{V^m} and $m\text{-}PV^*$ are isomorphic (and can thus be identified in the sequel). On the other hand, PT_{V^c} and $c\text{-}PV^*$ are *not* isomorphic since the (unique) V^c-homomorphism *front* : $T_{V^c} \to c\text{-}V^*$ is not one-one.

From the previous section and the above it should be clear, that two hierarchies of string-languages can be defined using level-n schemes over V^m_+ and V^c_+, which we denote by $n\text{-}L^m_{OI}(V)$ and $n\text{-}L^c_{OI}(V)$, respectively. The hierarchy $n\text{-}L^m_{OI}$ is the OI-*hierarchy* discussed in detail in [Da 2]. The original definition of the above hierarchies as "OI(n)-equational subsets" of $m\text{-}PV*$ and $c\text{-}PV*$ is given in [ES]. In this section we prove the result conjectured in [ES], that the n-th language class in the c-hierarchy coincides with the n+1-st language family in the OI-hierarchy. In particular, this gives a full proof of the result stated in [ES], that the OI-string language hierarchy starts with the regular, context-free, and OI-macro-languages.

We will now formally define the m - and c - hierarchies as homomorphic images of level-n trees under *set* and *front* ∘ *set*, respectively. This is equivalent to the definition used in [ES], and [Da 2].

Definition 5

$$n\text{-}L^m_{OI}(V) := set(n\text{-}R(V^m_+)) \qquad n\text{-}L^c_{OI}(V) := front(set(n\text{-}R(V^c_+))) \qquad \square$$

In order to apply the closure result of the previous section, we are going to represent a *level*-n+1 tree over V^m_+ by a *level*-n tree over $D(V^m)_+$ and define a suitable "language preserving" translation in $\mathcal{DT}^\infty$ from infinite trees over $D(V^m_+)$ to infinite trees over V^c_+. Before motivating this translation, we recall the following characterization of *level*-(n+1) trees proved in [Da 2].

Theorem 3

For any D(I)-set Ω $\qquad (n+1)\text{-}R(\Omega) = yield(n\text{-}R(D(\Omega))^{(e,i)})$ $\square$

It can easily be shown for the case $\Omega = V^m_+$, that + can be kept at the top-level, i.e. any *level*-(n+1) tree t over V^m_+ is in fact the image under $yield_+$ of a *level*-n tree t' over $D(V^m)_+$, where (informally) $yield_+(+_{(i,i)}(t_1,t_2)) = +_i(yield_+(t_1), yield_+(t_2))$.

Now consider some tree $t \in CT^{(i^k,i)}_{D(V^m)_+}$, then in general t will contain many superfluous subtrees with respect to *set* ∘ $yield_+$, since the resulting trees in $T_{V^m}(Y)$ can at most contain one parameter. As a simple example, take

t = 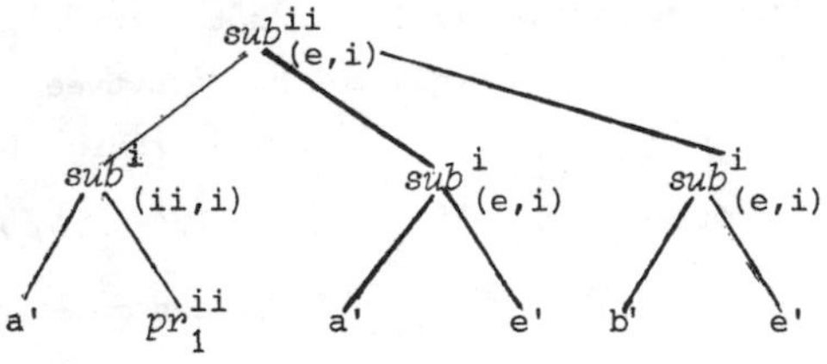 $\in T_{D(\{a,b\}^m)_+}$.

Clearly the translation under $set \circ yield$ gives the language $\{aa\}$; in particular, the subtree containing b' is redundant and could be deleted, "leading" to $c(c(a,e),c(a,e)) \in T_{\{a,b\}_+^c}$.

We will now construct $M \in DT^{\infty}$, which at each subtree $sub(t_o,t_1,\ldots,t_r)$ will guess, which translation $yield_+(t_j)$ will be substituted in $yield_+(t_o)$ and evaluate all guesses (combined by +) in parallel. Subtrees, which in the course of the computation turn out to be "redundant" will eventually receive $\perp$-leaves and thus be mapped to $\emptyset$ under *set*. The guess made will be memorized in the state; hence if the computation proceeds in state q , M will expect $yield_+(t_q)$ to be substituted into $yield_+(t_o)$.

In the construction below we allow all natural numbers as states; however, in all applications, M will be allowed to use only a fixed initial segment [k] of ω .

Construction

We define $M \in DT^{\infty}$ with input alphabet $D(V_+^m)$ and output alphabet V_+^c by

$M_o(e') = e \qquad M_q(e') = \perp \qquad$ for $q \neq 0$

$M_1(a') = a \qquad M_q(a') = \perp \qquad$ for $q \neq 1$

$M_j(pr_j^{i^k}) = e \qquad M_q(pr_j^{i^k}) = \perp \qquad$ for $q \neq j$

$M_q(+) = +([q,z_1],[q,z_2]) \qquad$ for all q

$M_o(abs_{(i^k,i)}) = [o,z_1] \qquad M_q(abs_{(i^k,i)}) = \perp \qquad$ for $q \neq 0$

$M_o(sub^{i^k}_{(i^r,i)}) = +([o,z_1],+(c([1,z_1],[o,z_2]),\ldots c([k,z_1],[o,z_{k+1}])\ldots))$

$M_q(sub^{i^k}_{(i_r,i)}) = +(c([1,z_1],[q,z_2]),\ldots c([k,z_1],[q,z_{k+1}])\ldots) \qquad$ for $q \neq o$

Remark

We slightly deviated from the notation in definition 1 by including 0 as initial state. Moreover, we need in fact the concept of a DT^{∞} for the many sorted case, where the z_j are taken from some $D(\{i\})$-set $\{z_{j,\nu} \mid j \in \omega\ ,\ \nu \in D(\{i\})\}$. We leave the (easy) generalization of the results of section 2 to the many sorted case to reader and view the z_j in the above construction as abreviation for the obvious $z_{j,\nu}$. □

Before proving the correctness of the construction, we illustrate the construction by computing $\overline{M}(t)$ for the previously considered example tree t .

Example

Consider again the sample-tree $t \in T_{D(\{a,b\}^m)}$. Then $\overline{M}(t) = \overline{M}_o(t)$

=

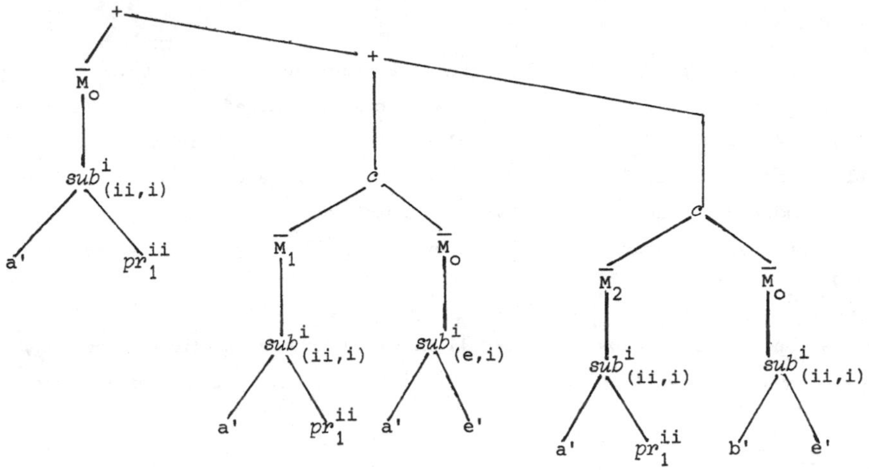

=

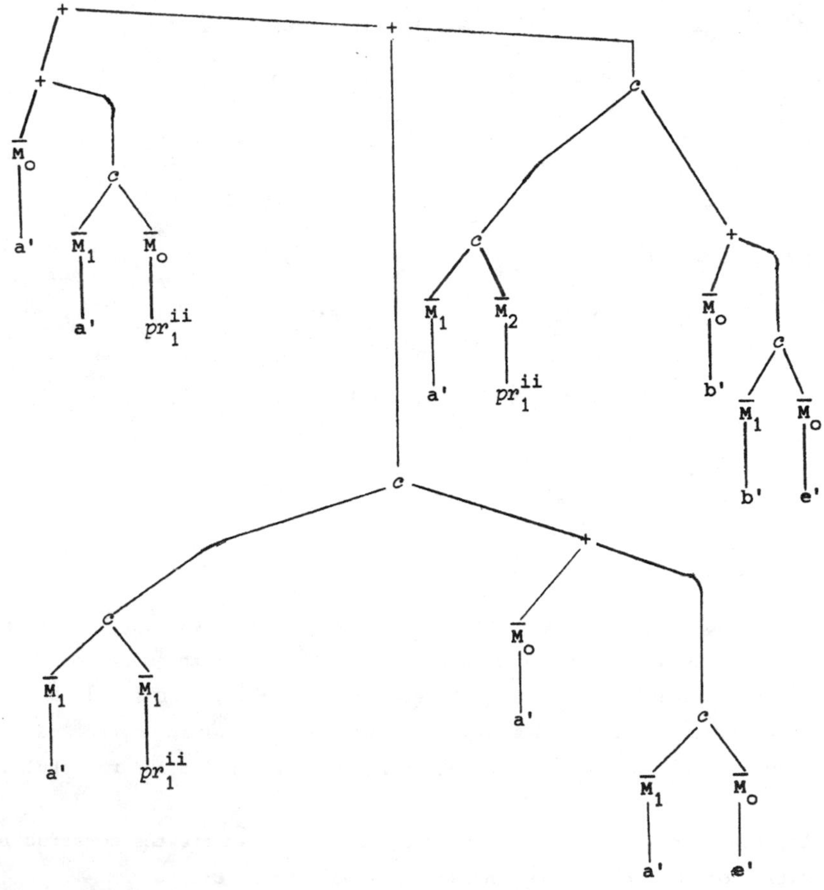

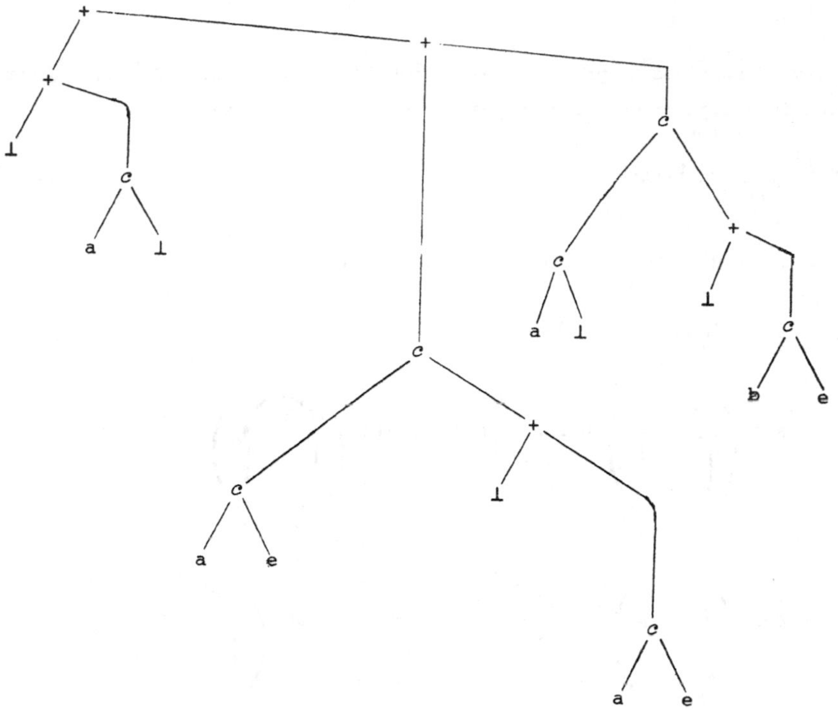

Thus

$$front\,(set\,(\overline{M}(t))) \;=\; front\;(\{\;\;\;\;\;\}) \;\;=\;\{aa\}$$

□

Showing that $\overline{M}$ is "language-preserving" requires a proof of commutativity of the following diagram, whose induction invariant is established in the next lemma.

$$
\begin{array}{ccc}
CT^{(e,i)}_{D(V^m)_+} & \xrightarrow{\quad\overline{M}\quad} & CT^{\;\;V^c_+} \\[4pt]
\downarrow{\scriptstyle yield_+} & & \downarrow{\scriptstyle set} \\[4pt]
CT^{\;\;V^m_+} & \xrightarrow{\;set\;} P(V^*) \xleftarrow{\;front\;} & PT^{\;V^c}
\end{array}
$$

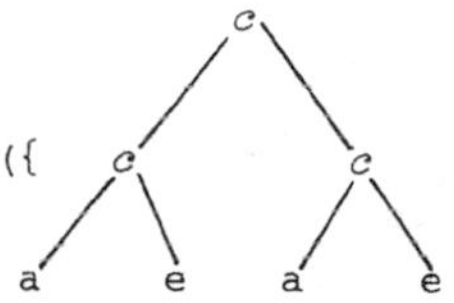

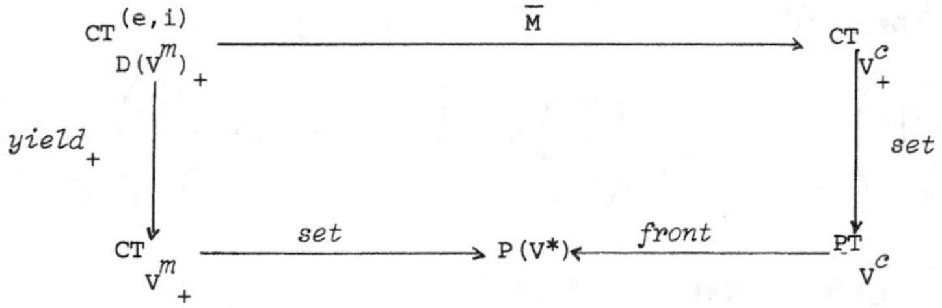

Lemma 3

$$\forall t \in CT^{(i^k,i)}_{D(V^m)_+}$$

$$set(yield_+(t)) \;=\; (front \circ set \circ \overline{M}_o)(t) \;\cup\; \bigcup_{q\,\in\,[\,k]}\;(front \circ set \circ \overline{M}_q)(t)\cdot y_{q,i}$$

Proof:

by continuity if suffices to prove the assertion for finite trees. The only nontrival steps in the induction proof arise in the treatment of substitution.

Let $t = sub^{i^r}_{(i^k,i)}\ t_o\, t_1 \ldots t_r$.

$$(front \circ set \circ \overline{M}_o)(t) \cup \bigcup_{q \in [k]} (front \circ set \circ \overline{M}_q)(t) \cdot y_{q,i}$$

$$= \quad (front \circ set \circ \overline{M}_o)(t_o)$$

$$\cup\ front \circ set \left(\begin{array}{ccc} & c & \\ \overline{M}_1 & & \overline{M}_o \\ | & & | \\ t_o & & t_1 \end{array}\right) \cup \ldots \cup\ front \circ set \left(\begin{array}{ccc} & c & \\ \overline{M}_r & & \overline{M}_o \\ | & & | \\ t_o & & t_r \end{array}\right)$$

$$\vdots$$

$$\cup\ front \circ set \left(\begin{array}{ccc} & c & \\ \overline{M}_1 & & \overline{M}_k \\ | & & | \\ t_o & & t_1 \end{array}\right) \cdot y_{k,i} \cup \ldots \cup\ front \circ set \left(\begin{array}{ccc} & c & \\ \overline{M}_r & & \overline{M}_k \\ | & & | \\ t_o & & t_r \end{array}\right) \cdot y_{k,i}$$

=: E

Now consider the j-th column in the above "matrix-like" expression. By induction hyphothesis this evaluates to

$$front \circ set \left(\begin{array}{ccc} & c & \\ \overline{M}_j & & \overline{M}_o \\ | & & | \\ t_o & & t_j \end{array}\right) \cup \bigcup_{q \in [k]} front \circ set \left(\begin{array}{ccc} & c & \\ \overline{M}_j & & \overline{M}_q \\ | & & | \\ t_o & & t_j \end{array}\right) \cdot y_{q,i}$$

$$= front \circ set \left(\begin{array}{c} \overline{M}_j \\ | \\ t_o \end{array}\right) \cdot \left(front \circ set \left(\begin{array}{c} \overline{M}_o \\ | \\ t_j \end{array}\right) \cup \bigcup_{q \in [k]} front \circ set \left(\begin{array}{c} \overline{M}_q \\ | \\ t_j \end{array}\right) \cdot y_{q,i}\right.$$

$$= \quad (front \circ set \circ \overline{M}_j)(t_o) \cdot (set \circ yield_+)(t_j)$$

Hence exchanging rows and colums gives E

$$= \quad (front \circ set \circ \overline{M}_o)(t_o) \cup \bigcup_{q \in [r]} (front \circ set \circ \overline{M}_q)(t_o) \cdot (set \circ yield_+)(t_q)$$

$= \langle (front \circ set \circ \overline{M}_o)(t_o) \cup \bigcup_{q \in [r]} (front \circ set \circ \overline{M}_q)(t_o) \cdot y_{q,i} \rangle \overset{\leftarrow}{_{OI}} \underline{L}$

where $\underline{L} = (set(yield_+(t_1)),\ldots,set(yield_+(t_r)))$

$= set(yield_+(t_o)) \overset{\leftarrow}{_{OI}} \underline{L}$ by induction hypothesis

$= set(yield_+(t_o)[y_{i_k}/(yield_+(t_1),\ldots,yield_+(t_r))])$ c.f. section 1

$= set(yield_+(t))$ □

We now define a language preserving translation from trees over V_+^c to trees over $D(V^m)_+$ realized by $M' \in DT^\infty$. The definition of this mapping is straigthforward: we use $sub^i_{(i,i)}$ to represent c and the projection symbol pr^i_1 instead of e. The resulting tree of *type* (i,i) is then called with e'.

Construction

We define $M' \in DT^\infty$ with input alphabet V_+^c, output alphabet $D(V_+^m)$, and states $\{0,1\}$ by

$M'_o(e) = e'$ $M'_1(e) = pr^i_1$

$M'_o(a) =$ 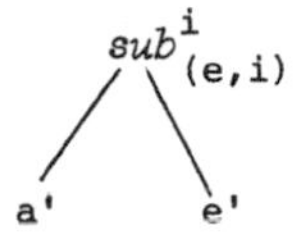

$M'_1(a) = a'$

$M'_o(c) =$ 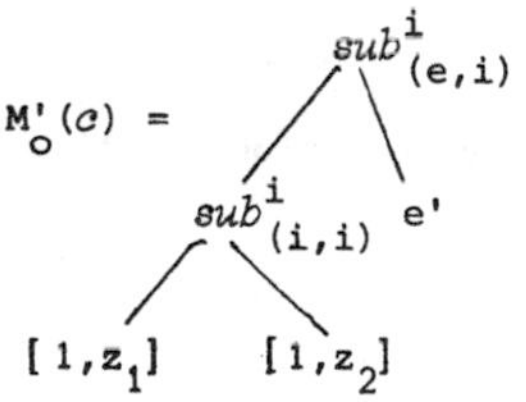

$M'_1(c) =$ 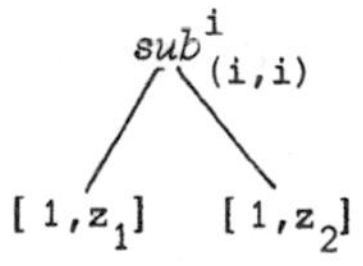

$M'_o(+) =$ 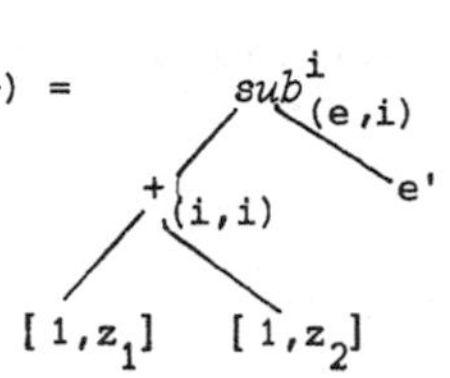

$M'_1(+) =$ 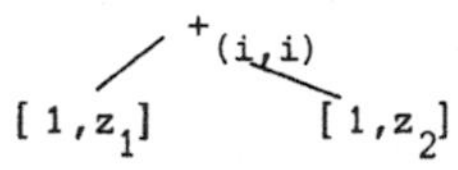

□

The correctness of the construction follows immediately from the following lemma.

Lemma 3

$\forall t \in CT_{V_+^c} \quad \forall w \in V^*$

$set(yield_+(\overline{M}_1'(t))) \overleftarrow{_{OI}} \{w\} = front(set(\overline{M}_1(t))) \cdot w$

Proof:

Again it suffices to prove the assertion for finite trees. The only interesting case is $t = c(t_1,t_2)$.

$$set(yield_+(\overline{M}_1'(c(t_1,t_2)))) \overleftarrow{_{OI}} \{w\}$$

$$= set(yield_+(\overline{M}_1'(t_1))[y_{1,i}/yield_+(\overline{M}_1'(t_2))]) \overleftarrow{_{OI}} \{w\}$$

$$= [set(yield_+(\overline{M}_1'(t_1))) \overleftarrow{_{OI}} set(yield_+(\overline{M}_1'(t_2)))] \overleftarrow{_{OI}} \{w\}$$

c.f. section 1

$$= set(yield_+(\overline{M}_1'(t_1))) \overleftarrow{_{OI}} [set(yield_+(\overline{M}_1(t_2))) \overleftarrow{_{OI}} \{w\}]$$

by associativity of $\overleftarrow{_{OI}}$

$$= set(yield_+(\overline{M}_1'(t_1))) \overleftarrow{_{OI}} front(set(t_2)) \cdot w$$

by induction hypothesis

$$= \bigcup \{set(yield_+(\overline{M}_1'(t_1))) \overleftarrow{_{OI}} \{w_2 \cdot w\} \mid w_2 \in front(set(t_2))\}$$

since all trees in $set(yield_+(\overline{M}_1'(t_1)))$ contain at most one occurence of $y_{1,i}$

$$= \bigcup \{front(set(t_1)) \cdot w_2 \cdot w \mid w_2 \in front(set(t_2))\}$$

by induction hypothesis

$$= front(set(t_1)) \cdot front(set(t_2)) \cdot w$$

$$= front(set(c(t_1,t_2))) \cdot w$$

□

Since the class of level-n trees is closed under DT^{∞}'s , we have established the characterization of *level*-n+1 OI string languages as the n-th language family in the *c*-hierarchy.

Theorem 4

$n{+}1 - L_{OI}^{m}(V) = n{-}L_{OI}^{c}(V)$

Proof:

By theorem 3 (and the remark following it) it remains to prove

$set(yield_+(n\text{-}R(D(V^m)_+)^{(e,i)})) = front(set(n\text{-}R(V^c_+)))$

"$\subseteq$" Let $t \in n\text{-}R(D(V^m)_+)^{(e,i)}$, then by theorem 2 $\overline{M}(t) \in n\text{-}R(V^c_+)$ and by lemma 2 $set(yield_+(t)) = front(set(\overline{M}(t)))$. Clearly M needs only the states from O up to the maximal rank of the substitution symbols occuring in t .

"$\supseteq$" Let $t \in n\text{-}R(V^c_+)$. A straigthforward case analysis using lemma 2 shows $set(yield_+(\overline{M}(t))) = front(set(t))$ (for the case $t = c(t_1,t_2)$ apply the same argumentation as in the proof of lemma 2), hence the assertion follows from theorem 2 . □

In particular, *level*-1/*level*-2 OI languages coincide with images under *front* ∘ *set* of regular/context-free infinite trees and thus (see e.g. [Da 2, ES, Niv]) with front-languages of regular/context-free tree languages over V^c. This shows, that the OI-hierarchy starts with the regular, context-free and OI-macro languages.

Corollary 1

$0\text{-}L^m_{OI}(V) = REG(V)$, $1\text{-}L^m_{OI}(V) = CF(V)$, $2\text{-}L^m_{OI}(V) = MAC(V)$ □

4 REFERENCES

[Da 1] DAMM, W. *Languages defined by higher type program schemes*, 4 th international colloquium on Automata, Languages and Programming, Lecture Notes in Computer Science 52 (1977), 164-179, Springer Verlag

[Da 2] DAMM, W. *The IO- and OI-Hierarchies*, Schriften zur Informatik und Angewandten Mathematik, Bericht Nr. 41 (1980), RWTH Aachen

[Da 3] DAMM, W. *Top-down tree transducers for infinite trees*, in preparation

[DF 1] DAMM, W. / FEHR, E. *On the power of self application and higher type recursion*, Proc. 5th international colloquium on Automata, Languages and Programming, Lecture Notes in Computer Science Science 62 (1978), 177-191, Springer Verlag

[DF 2] DAMM, W. / FEHR, E. *A schematological approach to the analysis of the procedure concept in ALGOL-languages*, Proc. 5 ième colloque sur les Arbres en Algebre et en Programmation, Lille, (1980), 130-134 (abstract)

[DG] DAMM, W. / GUESSARIAN, I. *Combining T and level-N* , LITP-report, Université Paris VII, (1981), to appear

[ES] ENGELFRIET, J. / SCHMIDT, E.M. *IO and OI*, JCSS Vol. 15, 3 (1977), 328-353 and Vol. 16, 1 (1978), 67-99

[Fe] FEHR, E. *Lambda calculus as control structure of programming languages*, Schriften zur Informatik und Angewandten Mathematik, Bericht Nr. 57 (1980), RWTH Aachen

[GTWW] GOGUEN, J.A. / THATCHER, J.W. / WAGNER, E.G. / WRIGHT, J.B., *Initial Algebra Semantics and Continuous Algebras* JACM Vol. 24, 1 (1977), 68-95

[Mai] MAIBAUM, T.S.E. *A generalized approach to formal languages*, JCSS Vol. 8, (1974), 409-439

[Niv] NIVAT, M. *On the interpretation of recursive program schemes*, Atti del convegno d'Informatica theoretica (1972), Rome

[Sco] SCOTT, D. *The Lattice of Flow Diagrams*, Symp. on Semantics of Algorithmic Languages, (ed. E. Engeler), Lecture Notes in Math. Vol. 188 (1971), 311-366, Springer Verlag

Easy Solutions are Hard To Find

Stephen L. Bloom*

Mathematical Sciences Department
IBM T.J. Watson Research Center
Yorktown Heights, N.Y. 10598

on leave from

Department of Pure and Applied Mathematics
Stevens Institute of Technology
Hoboken, N.J. 07030

David B. Patterson
Division of Mathematics and Science
St. John's University
Staten Island, N.Y. 10301

Abstract

There are two main results in this paper.

A new NP-complete problem is found: given a system (S) of recursion equations, determine whether (S) has a solution in a nontrivial "contraction algebra" in which one of the components is a projection. This problem, which arose in [3] where all solutions of a system of recursion equations in a contraction algebra A were found, is related to the equivalence problem for deterministic pushdown automata.

Secondly, for signatures Σ with a finite number of function symbols of positive rank, the free complete contraction Σ-algebras are shown to be isomorphic to algebras of "Σ-trees". When Σ has an infinite number of function symbols of positive rank, it is shown that there are no free complete contraction Σ-algebras.

* Partially supported by NSF Grant MCS 78-00882

1. Contraction Σ-algebras.

In this section we review some definitions in order to motivate the problem discussed in Section 2. We also use this opportunity to present a new fact about free complete Σ-algebras.

1.1 Σ-algebras. We assume Σ is a "ranked set", i.e. the disjoint union of sets Σ_n, $n \geq 0$. The elements of Σ_n, $n \geq 0$, are "function constants of rank n". A Σ-algebra A is a set (also denoted A) equipped with functions $\sigma: A^n \to A$ for each $\sigma \in \Sigma_n$, $n \geq 0$. (If necessary we will write σ_A to avoid confusion. The empty set is a Σ-algebra iff Σ_0 is empty.

We will always assume that our Σ-algebras A are equipped with a metric d such that $d(x,y) \leq 1$ for all $x,y \in A$. (The reader will recall that a metric on A is a function from A^2 into the nonnegative real numbers satisfying three well-known properties.)

The metric on A is extended to each power A^k of A, $k \geq 2$ by defining for $\bar{x},\bar{y} \in A^k$:

$$d(\bar{x},\bar{y}) = \max(d(x_i,y_i): i=1,2,\ldots,k) \tag{1}$$

where $\bar{x} = (x_1,\ldots,x_k)$, $\bar{y} = (y_1,\ldots,y_k)$.

Definition 1.2. The Σ-algebra A (equipped with the metric d) is a contraction algebra if for each $\sigma \in \Sigma_n$, $n \geq 0$, the function $\sigma: A^n \to A$ is a proper contraction; i.e. there is a real number $0 \leq c < 1$ such that for all $\bar{x},\bar{y}$ in A^n,

$$d(\sigma(\bar{x}), \sigma(\bar{y})) \leq c\, d(\bar{x},\bar{y}). \tag{2}$$

(A number c with the property (2) is called a "contraction constant" for σ. We assume that for each $\sigma \in \Sigma_n$, $n > 0$, a fixed contraction constant for $\sigma: A^n \to A$ in each contraction algebra A is selected, so that we may speak of "the" contraction constant for σ in A.) If the metric d on the contraction algebra A is complete (i.e. every Cauchy sequence converges) then A is a complete contraction algebra.

The reasons for considering complete contraction algebras in connection with solving systems of recursion equations will be discussed below. First we find the free complete contraction algebras.

Let X be a set and let $\Sigma_X \mathrm{Tr}$ be the set of all (finite or infinite) rooted, labeled, locally-ordered trees t whose leaves are labeled either by elements of Σ_0 or elements of X; any vertex v of t of outdegree $k, k > 0$, is labeled by an element of Σ_k and v has a first, second,...,k-th successor (see [1] or [7] for a more formal definition). $\Sigma_X\ \mathrm{Tr}$ is a complete metric space when the metric is defined by:

$$d(t,t') = {}^1/2^n$$

where n is the first depth at which some vertex of t is labeled differently from one in t'. If no such vertex exists we identify t and t', so that $d(t,t')=0$ iff $t = t'$.

We make $\Sigma_X\ \mathrm{Tr}$ into a Σ-algebra by defining, (for $\sigma \in \Sigma_n$, $n \geq 0$,) $\sigma(t_1,\dots,t_n)$ to be the tree obtained from $t_1,\dots,t_n$ by adding a new root, labeled σ having n successors; the i-th successor of the root is the root of t_i, $i=1,2,\dots,n$. For $x \in X$, let $\hat{x}$ be the tree having a root labeled x with no successors.

<u>Proposition 1.3 ([9],[4])</u>. For any set X, $\Sigma_X\mathrm{Tr}$ is a complete contraction Σ-algebra.

For any ranked set Σ, let $K(\Sigma)$ be the category whose objects are complete contraction Σ-algebras and whose morphisms are continuous homomorphisms. For any set X, let Σ_X be the ranked set obtained from Σ by adding X to Σ_0; i.e. $(\Sigma_X)_0$ is the disjoint union of Σ_0 and X; $(\Sigma_X)_n = \Sigma_n$, $n \geq 1$. A Σ-algebra A becomes a Σ_X-algebra by assigning an element of A to each element of X.

An algebra F in $K(\Sigma_X)$ is <u>initial</u> if for every algebra A in $K(\Sigma_X)$ there is a unique continuous homomorphism $F \to A$. (Using other terminology, F is "freely generated by X in $K(\Sigma)$".) Clearly, any two initial algebras in $K(\Sigma_X)$ are isomorphic.

We may now state one of the new results.

<u>Theorem 1.4</u>. If $\bigcup_{n=1}^{\infty} \Sigma_n = \Sigma^+$ is finite, then for every set X, $\Sigma_X\ \mathrm{Tr}$ is an initial algebra in $K(\Sigma_X)$. If Σ^+ is infinite, there is no initial algebra in $K(\Sigma_X)$, for any set X (unless both Σ_0 and X are empty).

Proof. If Σ_0 and X are both empty, the initial Σ_X-algebra (and hence the initial algebra in $K(\Sigma_X)$) is empty. Thus we may assume not both Σ_0 and X are empty. First we assume Σ^+ is finite and let A be a complete contraction Σ_X-algebra. Let a_x be the element of A corresponding to $x \in X$. We will show there is a unique continuous homomorphism $h:\Sigma_X\ \mathrm{Tr} \to A$ such that for $x \in X$, $h(\hat{x}) = a_x$. Note that the value of h on finite trees (i.e. Σ_X-terms) is determined by the requirement that h be a homomorphism. Any infinite tree t is the limit of a Cauchy sequence of finite trees (t_n); we will show that the sequence $h(t_n)$ converges in A. Then we will define $h(t) = \lim_{n\to\infty} h(t_n)$.

Lemma 1.5. Suppose t and t' are finite trees in $\Sigma_X Tr$ with $d(t,t') \le {}^1/2^k$. Then $d(h(t), h(t')) \le c^k$ where c is the maximum of the contraction constants of the finite set of functions σ_A, $\sigma \in \Sigma^+$.

Proof of the Lemma, by induction on k. When k=0, there is nothing to prove since the metric on A is bounded by 1. Assume the Lemma holds for k. If $d(t,t') = {}^1/2^{k+1}$, we may write

$$t = \sigma(t_1,\ldots,t_n) \ , \ t' = \sigma(t_1',\ldots,t_n')$$

for some $\sigma \in \Sigma_n$, $n > 0$, where $d(t_i,t_i') \le 1/2^k$. But then $h(t) = \sigma_A(h(t_1),\ldots,h(t_n))$, $h(t') = \sigma_A(h(t_1'),\ldots,h(t_n'))$ since h is a homomorphism on the finite trees and thus

$$d(h(t), h(t')) \le c_\sigma \max\{d(h(t_i), h(t_i')) : i=1,\ldots,n\}$$

where c_σ is the contraction constant for σ_A. From the induction hypothesis, it follows $d(h(t), h(t')) \le c^{k+1}$, completing the proof of the Lemma.

From the Lemma it easily follows that we may uniformly define h on all trees, finite or infinite, by

$$h(t) = \lim_{n\to\infty} h(t_n) \qquad (3)$$

where (t_n) is any Cauchy sequence of finite trees with limit t. It is easy to check that h is a well-defined continuous homomorphism. Clearly any continuous homomorphism must satisfy (3), so that $\Sigma_X Tr$ is initial.

Now assume that Σ^+ is infinite. To obtain a contradiction, suppose that an initial algebra F in $K(\Sigma_X)$ exists. If the rank of $\sigma \in \Sigma^+$ is n, let $h_\sigma : F \to F$ be the function defined by: $u \mapsto \sigma_F(u,\ldots,u)$. Then one may find a positive odd integer N_σ such that the function $g_\sigma : F \to F$ defined by:

$$g_\sigma(u) = h_\sigma(h_\sigma(\ldots(h_\sigma(u)\ldots)) \quad (N_\sigma \text{ times})$$

has a contraction constant $\le 1/2$ (since σ_F is a proper contraction). Let $\Sigma^+ = \{\sigma 1, \sigma 2, \ldots\}$ and let α be some element in $\Sigma_0 \cup X$, with $u_0 = \alpha_F$. Then the sequence $t_n = g_{\sigma 1}(g_{\sigma 2}(\ldots g_{\sigma n}(u_0)\ldots))$ is a Cauchy sequence (since $d(t_n,t_{n+k}) \leq {}^1/2^n$) and converges in F, say to t.

Now consider the following algebra A in $K(\Sigma_X)$. The underlying set of A is the closed interval of real numbers $[-1,1]$ equipped with the metric $d(x,y) = 1/2\cdot|x-y|$. (The factor 1/2 is used to keep d below 1.) Clearly, this metric is complete. The elements of A corresponding to symbols in $\Sigma_o \cup X$ may be chosen arbitrarily except the element $\bar{u}$ corresponding to α (see preceding paragraph) must be nonzero. For $\sigma = \sigma n$ in Σ^+, if the rank of σ is $s > 0$, define $\sigma_A(x_1,\ldots,x_s) = -c_n x_1$, where the constants c_n are chosen as follows, where $0 < \lambda < 1$:

$$c_n = \lambda^{N_\sigma/2^n}$$

Then in A the function $(g_{\sigma n})_A$ is given by:

$$(g_{\sigma n})_A(x) = (\lambda^m)x, \; x \in [-1,1]$$

where $m = 1/2^n$. But then $(t_n)_A = (-1)^n\beta\bar{u}$, where $\beta=\lambda^{(1/2+\ldots+1/2^n)}$, so that as $n\to\infty$, $(t_{2n})_A$ converges to $\lambda\bar{u}$ and $(t_{2n+1})_A$ converges to $-\lambda\bar{u}$. But then there can be no continuous homomorphism $h:F \to A$ since if h is continuous, $h(t) = \lim h(t_{2n}) = \lim h(t_{2n+1})$, but for each n, $h(t_n) = (t_n)_A$. This completes the proof.

<u>1.6 Systems of Recursion Equations</u>. Let $\Phi = \{\varphi_1,\ldots,\varphi_n\}$ be a finite ranked set disjoint from the ranked set Σ. The elements of Φ will be called function variables, and we let ki denote the rank of φ_i, $i=1,2,\ldots,n$. Let $V = \{x_1,x_2,\ldots\}$ be a countably infinite set disjoint from $\Sigma \cup \Phi$ of "individual variables". The set of all terms (i.e. finite trees) built from $\Sigma \cup \Phi \cup \{x_1,\ldots,x_k\}$ is denoted $(\Sigma \cup \Phi)T_k$, $k \geq 0$.

A "system of recursion equations" (or "recursive program scheme") over $\Sigma \cup \Phi$ is a system of equations

$$(S):(\varphi_i(x_1,\ldots,x_{ki}) = t_i,\; i=1,\ldots,n)$$

where for each i, t_i is a term in $(\Sigma \cup \Phi)T_{ki}$.

As an example, the three equations

$$(4) \qquad \begin{aligned} \varphi_1(x_1,x_2) &= \sigma_2(\varphi_2(x_1,x_2),\; \varphi_1(\sigma_1(x_1),x_2)) \\ \varphi_2(x_1,x_2) &= \varphi_3(\sigma_1(\varphi_1(x_1,x_2))) \\ \varphi_3(x_1) &= \varphi_3(\sigma_1(x_1)) \end{aligned}$$

form such a system, where the function variables φ_1 and φ_2 have rank 2, and φ_3 has rank 1.

A solution of a system (S) of recursion equations in a Σ-algebra A is an n-tuple of total functions $\overline{\varphi}_i : A^{k_i} \to A$ such that for each i, $\overline{\varphi}_i$ is the function defined by the term t_i, the righthand side of the ith equation in (S). The functions $\overline{\varphi}_i$ are the components of the solution. Not every system of recursion equations has a solution in every Σ-algebra. In order to insure the existence of solutions, it has been customary to solve such systems in Σ-algebras A equipped with a partial ordering which is at least ω-complete and such that every function $\sigma : A^m \to A$, $\sigma \in \Sigma_m$, is order preserving and ω-continuous (see [10] or [8]). Recursion equations will have solutions in such "ω-complete ordered Σ-algebras" and the least solution is easily described as the least fixed point of a certain functional.

In [2] and [3] another way of guaranteeing that solutions of systems of recursion equations will exist is studied. Namely, one solves the system in complete contraction Σ-algebras. In such algebras, solutions exist and not just one solution but all solutions are easily described. In [3] it is shown how every system of recursion equations is equivalent to one (like the example (4)) such that for each i, if the i-th equation has the form $\varphi_i(x_1,\dots,x_{ki}) = \varphi_j(u_1,\dots,u_{kj})$, where φ_j is a function variable, then the j-th equation must be "reflexive", i.e. have the form $\varphi_j(x_1,\dots,x_{kj}) = \varphi_j(t_1,\dots,t_{kj})$ for some terms $t_1,\dots,t_{kj}$. Furthermore it is shown that once one assumes that none of the reflexive variables φ_j will be "trivial", i.e. projection functions, one can solve these equations one at a time, independently of the other equations in (S). The remaining nonreflexive equations can be solved uniquely, once one chooses a solution to each reflexive equation. It remained to determine whether there were any trivial solutions of the reflexive equations. A method for doing so which takes an exponential number of steps was given in [3]. It was not known whether a more efficient method existed. The main result of Section 2 shows that this problem is NP-complete and it is therefore highly probable that an exponential number of steps is necessary.

Section 2. Trivial Solutions of Reflexive Equations.

Let A be a contraction Σ-algebra (not necessarily complete). Let $\underline{C}(A)$ be the class of functions $A^k \to A$, $k \geq 0$, consisting of all the projection functions, and all proper contractions. The class $\underline{C}(A)$ is closed under composition: i.e. if $f : A^k \to A$ and $g_i : A^n \to A$, $i=1,\dots,k$ are in $\underline{C}(A)$, then $h : A^n \to A$ is in $\underline{C}(A)$, where for $\bar{a} \in A^n$,

$$h(\bar{a}) = f(g_1(\bar{a}),\dots,g_k(\bar{a})).$$

In order to insure that the projection functions are distinct we henceforth assume that all contractions algebras are nontrivial-i.e. contain at least two elements.

In this section we will consider the following question. Given a system $(S) = \{\varphi_i(x_1 \ldots x_{ki}) = t_i,\ i=1,\ldots,n\}$ of reflexive recursion equations (in the function variables $\Phi = \{\varphi_1,\ldots,\varphi_n\}$, functions constants Σ and individual variables $V = \{x_1,x_2,\ldots\}$ does there exist any contraction Σ-algebra A (containing at least two elements) and a solution $\overline{\varphi}_1,\ldots,\overline{\varphi}_n$ of (S) in A such that each $\overline{\varphi}_i \in \underline{C}(A)$ and such that at least one component of the solution is a projection?

We will abbreviate this problem as "EPS" for the "Existence of Projections as Solutions".

In the rest of this section we will show that this problem is NP complete, by showing that the "Satisfiability Problem" [5] is reducible to EPS.

An <u>instance</u> I of the satisfiability problem (from now on, denoted "SAT") is given by a finite set $B = \{b_1,\ldots,b_k\}$ of "Boolean variables" and a finite set $C = \{c_1,\ldots,c_n\}$ of <u>clauses</u>. (A clause is a subset of elements of $B \cup \overline{B}$, where $\overline{B} = \{\overline{b}_1,\ldots,\overline{b}_k\}$ is disjoint from B. Elements of $B \cup \overline{B}$ are called <u>literals</u>.) A truth function is a mapping h of $B \cup \overline{B}$ into $\{1,2\}$ such that $h(\overline{b}_i) = 3-h(b_i)$; i.e. $h(\overline{b}_i)=2$ iff $h(b_i) = 1$, $i=1,\ldots,k$. A truth function h satisfies a clause c if $h(u)=1$ for at least one literal $u \in c$; h satisfies the set C of clauses if h satisfies each clause in C. The satisfiability problem is: given an instance $I = (B,C)$, does there exist a truth function satisfying C?

For each instance $I = (B,C)$ of SAT, we will construct an instance $(\hat{I})$ of EPS such that there is a truth function satisfying C iff the system $(\hat{I})$ has a solution in $\underline{C}(A)$ (for some contraction Σ-algebra A) for which at least one function is a projection.

Let $I = (B=\{b_1,\ldots,b_k\},\ C = \{c_1,\ldots,c_n\})$ be a given instance of the satisfiability problem. The instance $(\hat{I})$ of EPS will have function variables $\varphi_b, \varphi_{\hat{b}}$, $b \in B$, φ_c, $c \in C$ all of rank 2.

First, for every subset c of $B \cup \overline{B}$, we define a <u>term</u> t_c of rank 2, by induction.

<u>Definition 2.1.</u> (a) If $c = \{b\}$, $t_c(x_1,x_2)$ is $\varphi_b(x_1,x_2)$; (b) If $c = \{\overline{b}\}$, $t_c(x_1,x_2)$ is $\varphi_b(x_2,x_1)$; (c) If $c = \{u_1,\ldots,u_n,u\}$, $n > 0$ then $t_c(x_1,x_2) = t_{c1}(x_1,\ t_u(x_1,x_2))$, where $c1$ is $\{u_1,\ldots,u_n\}$. (We assume that $B \cup \overline{B}$ is linearly ordered and that if $c = \{u_1,\ldots,u_n\}$ is

is a clause, then $u_1 < u_2 < \ldots < u_n$. Thus Definition 2.1 is unambiguous.)

Remark 2.2. In order to see how many steps will be needed to construct $(\hat{I})$ note that, considered as a finite tree, t_c has $2m+1$ vertices when c has m elements.

We let $\Pi_1(a_1,a_2) = a_1$, $\Pi_2(a_1,a_2) = a_2$ denote the two projection functions $A^2 \to A$, on any set A.

The following simple Lemma will be used repeatedly.

Lemma 2.3. Suppose each function variable φ_b is interpreted as a projection function (on some Σ-algebra A). For any clause c in C, let $\bar{t}_c$ be the resulting interpretation of t_c as a function $A^2 \to A$. Obviously $\bar{t}_c$ is a projection function. Then: $\bar{t}_c = \Pi_1$ iff the truth function h satisfies the clause c, where

$$h(b_i) = 1 \text{ iff } \varphi_{bi} = \Pi_1. \tag{2.4}$$

Proof by induction on the number of literals in c. If $c = \{b\}$ or $c = \{\bar{b}\}$, then the Lemma is obvious. So assume $c = c_1 \cup \{u\}$, so that for $a_1, a_2 \in A$

$$\bar{t}_c(a_1,a_2) = \bar{t}_{c1}(a_1, \bar{t}_u(a_1,a_2)).$$

First suppose $\bar{t}_c = \Pi_1$. Then either $\bar{t}_{c1} = \Pi_1$ or $[\bar{t}_{c1} = \Pi_2$ and thus $\bar{t}_u = \Pi_1]$. In either case the truth function h defined by (2.4) satisfies either c_1 or u, by the induction hypothesis, and hence h satisfies c. Conversely if h satisfies c_1 then $\bar{t}_{c1} = \Pi_1$, by the induction hypothesis and hence $\bar{t}_c = \Pi_1$. Otherwise h satisfies u and $\bar{t}_c(a_1,a_2) = \Pi_2(a_1, \bar{t}_u(a_1,a_2)) = \bar{t}_u(a_1,a_2) = a_1$. Thus $\bar{t}_c = \Pi_1$, completing the induction step.

For any clause c in C, consider the equation

$$\varphi_c(x_1,x_2) = \varphi_c(t_c(x_1,x_2), x_1) \tag{2.5}$$

in the function variables φ_c, φ_b, $b \in B$.

Lemma 2.6. There is a solution $\bar{\varphi}_c, \bar{\varphi}_b$, $b \in B$, of (2.5) in any contraction algebra A for which each of the components is a projection iff the truth function h of (2.4) satisfies c.

Proof. Notice that if $\bar{\varphi}_c$ is a projection and $\bar{\varphi}_c(a_1,a_2) = \bar{\varphi}_c(\bar{t}_c(a_1,a_2), a_1)$, for all a_1,a_2 in A, then $\bar{\varphi}_c$ must be Π_1 and hence $\bar{t}_c = \Pi_1$. Now we may apply Lemma 2.3. For the converse, if h satisfies c, then $\bar{t}_c = \Pi_1$ by Lemma 2.3 and we may define $\bar{\varphi}_c = \Pi_1$, getting a solution of (2.5).

Remark. It follows from Lemma 2.6 that the following problem is NP-complete. Given a system (S) of reflexive recursion equations determine whether there is a solution of (S) in any nontrivial contraction Σ-algebra for which all of the components of the solution are projections.

Indeed, given an instance $I = (B,C)$ of SAT, let $(\hat{I})$ be the system consisting of an equation (2.5) for each clause $c \in C$, as well as a dummy equation $\varphi_b(x_1,x_2) = \varphi_b(x_1,x_2)$, for each $b \in B$. Then Lemma 2.6 implies that there is a truth function satisfying C iff there is a solution of $(\hat{I})$ in which each component is a projection.

We now define 3 auxiliarly terms τ_C, τ_B and τ of rank 2, corresponding to the given instance $I = (B,C)$ of the satisfiability problem. The function variables φ_c, $c \in C$, φ_b, $b \in B$ and new variables $\varphi_{\hat{b}}$, $b \in B$, will be used.

Definition 2.7. Define the terms τ_C, τ_B and τ by:

$$\begin{aligned} \tau_C(x_1,x_2) &= \varphi_{c1}(\varphi_{c2}(\ldots(\varphi_{cn}(x_1,x_2),x_2),\ldots,x_2) \\ \tau_B(x_1,x_2) &= \varphi_{\hat{b}1}(\varphi_{\hat{b}2}(\ldots(\varphi_{\hat{b}k}(x_1,x_2),x_2),\ldots,x_2) \\ \tau(x_1,x_2) &= \tau_C(\tau_B(x_1,x_2),x_2). \end{aligned}$$

For example, if $B = \{b_1,b_2\}$ and $C = \{c_1,c_2\}$ then

$$\tau(x_1x_2) = \varphi_{c1}(\varphi_{c2}(\varphi_{\hat{b}1}(\varphi_{\hat{b}2}(x_1,x_2),x_2),x_2),x_2).$$

Remark 2.8. As finite trees, τ_C has $2n+1$ vertices, τ_B has $2k+1$ vertices and τ has $2(n+k)+1$ vertices.

Now we can define the instance $(\hat{I})$ of EPS corresponding to the given instance $I = (B,C)$ of SAT.

Definition 2.9. $(\hat{I})$ consists of the following three groups of reflexive equations in the function variables φ_c, $c \in C$, $\varphi_b, \varphi_{\hat{b}}$, $b \in B$. There are no function constants.

(2.9.1) $\qquad \varphi_b(x_1,x_2) = \varphi_b(\tau(x_1,x_2), \tau(x_2,x_1))$, $b \in B$

(2.9.2) $\qquad \varphi_{\hat{b}}(x_1,x_2) = \varphi_{\hat{b}}(\varphi_b(x_1,x_1),x_1)$, $b \in B$

(2.9.3) $\qquad \varphi_c(x_1,x_2) = \varphi_c(t_c(x_1,x_2),x_1)$, $c \in C$.

Remark. The purpose of the equations (2.9.1) and (2.9.2) is to force all components of the solution to be projections if any one of the components is.

The proof of the next theorem depends on the fact that the solutions of systems of reflexive equations are assumed to be either proper contractions or projections. Note that if $f_0,\ldots,f_k$ are functions in $\underline{C}(A)$ and the composition $f = f_0(f_1,\ldots,f_k)$ is defined then:

(2.10) $\qquad f_0$ is a projection if f is.

<u>Theorem 2.11</u>. The following statements are equivalent, for any nontrivial contraction algebra A.

(i) There is a solution of $(\hat{I})$ in $\underline{C}(A)$ for which all of the components are projections.

(ii) There is a solution of $(\hat{I})$ in $\underline{C}(A)$ in which at least one component is a projection.

<u>Proof</u>. We need only show (ii) implies (i). Thus suppose $\bar{\varphi}_b$, $\bar{\varphi}_{\hat{b}}$, $\bar{\varphi}_c$, $b \in B$, $c \in C$ is a solution of $(\hat{I})$ in $\underline{C}(A)$. Let $\bar{\tau}_B$, $\bar{\tau}_C$ and $\bar{\tau}$ be the resulting functions $A^2 \to A$ defined by 2.7.

First we show that if one function φ_b is a projection, then all of the component functions are. Indeed, by using the equation (2.9.1), we will show that if $\bar{\varphi}_b$ is either Π_1 or Π_2, $\bar{\tau}$ must be the projection Π_1. This fact is obvious if $\bar{\varphi}_b$ is Π_1; if $\bar{\varphi}_b$ is Π_2, then $\Pi_2(a_1,a_2) = \bar{\tau}(a_2,a_1)$, which implies $\bar{\tau}(a_1,a_2) = a_1$, all a_1,a_2 in A. The crucial fact is that

(2.12) $\qquad$ If $\bar{\tau}$ is Π_1, then each of the functions $\bar{\varphi}_c$, $\bar{\varphi}_{\hat{b}}$, $c \in C$, $b \in B$ is also the projection Π_1.

[<u>Proof</u>. By (2.10), each function $\bar{\varphi}_c$, $\bar{\varphi}_{\hat{b}}$ is a projection. From Definition 2.7 it is easily seen that these projections must be Π_1.]

Since each $\bar{\varphi}_{\hat{b}}$ is the projection Π_1, then by (2.10) each $\bar{\varphi}_b$ is also a projection (not necessarily Π_1).

Now suppose that one function $\bar{\varphi}_{\hat{b}}$ is a projection. It follows as above that $\bar{\varphi}_{\hat{b}}$ must be Π_1 and hence $\bar{\varphi}_b$ is a projection. By the above argument, all of the components of the solution are projections.

Lastly, if one function $\bar{\varphi}_c$ is a projection, it too must be Π_1 and so $\bar{t}_c$ is the projection Π_1. But then at least one φ_b is a projection by (2.10) and hence all components are projections. This completes the proof.

<u>Corollary 2.12</u>. For any instance $I = (B,C)$ of SAT, there is a

truth function satisfying C iff for any nontrivial contraction algebra A, there is a solution of $(\hat{I})$ in $\underline{C}(A)$ such that at least one component is a projection.

Proof. First suppose there is a truth function h satisfying C. Define the functions $\bar{\varphi}_b$ (in any contraction algebra A) to be projections such that (2.4) holds. Let $\bar{\varphi}_c = \bar{\varphi}_{\hat{b}} = \Pi_1$, all $c \in C$, $b \in B$. Then using Lemma 2.6 it is clear that $\bar{\varphi}_c, \bar{\varphi}_b, \bar{\varphi}_{\hat{b}}$, $c \in C$, $b \in B$ is a solution of $(\hat{I})$ for which each component is a projection.

Conversely, if for some contraction algebra A there is a solution $\bar{\varphi}_c, \bar{\varphi}_b, \bar{\varphi}_{\hat{b}}$, $c \in C$, $b \in B$ of (I) (in $\underline{C}(A)$) for which one component is a projection, by Theorem 2.12, all of the components are projections and then necessarily each $\bar{\varphi}_c = \Pi_1$. Again by 2.6, the truth function h defined by (2.4) will satisfy C.

Corollary 2.13. The problem EPS is NP-complete.

Proof. It is not difficult to see that EPS is in NP (this is proved in [3]). From remarks 2.2 and 2.8 and Definition 2.9 it is easy to see that the instance $(\hat{I})$ of EPS is constructible in a polynomial amount of steps from the instance (I) of SAT. The proof is completed by Corollary 2.12.

Remark 2.14. The construction of the instance $(\hat{I})$ is independent of the signature Σ. The only property of contraction algebras used in the proof of 2.13 is the property of the collection of functions $\underline{C}(A)$ mentioned in (2.10).

Remark 2.15. The problem EPS is somewhat analogous to a problem discussed in [6]. According to Courcelle, the equivalence problem for deterministic pushdown automata (as yet unsolved, as far as we know) is in turn equivalent to the following problem. Given a system $(S):(\varphi_i(x_1,\dots,x_{ki}) = t_i,\ i=1,\dots,n,\ n \geq 2)$ of recursion equations over $\Sigma \cup \Phi$ such that no term t_i has the form $\varphi_j(u_1,\dots,u_{kj})$, determine whether $\bar{\varphi}_1 = \bar{\varphi}_2$, where $\bar{\varphi}_1,\dots,\bar{\varphi}_n$ is the unique solution of (S) in the complete contraction algebra $\Sigma_V\mathrm{Tr}$.

References

[1] E.G.Wagner, J.B.Wright, J.B.Thatcher, "Rational algebraic theories and fixed point solutions", Proc. 17th IEEE Symposium on Foundations of computing, Houston (1976), 147-158.

[2] S.Arnold, M.Nivat, "Metric interpretation of infinite trees and semantics of nondeterministic recursive programs", Universite de Lille, No. I-T-3-78, 1978.

[3] S.L.Bloom, "All solutions of a system of recursion equations in infinite trees and other contraction theories", to appear in J. Comp. Syst. Sci.

[4] S.L.Bloom, C.C.Elgot, J.B.Wright, "Vector iteration in pointed iterative theories", SIAM J. Computing, Vol. 9, No. 3(1980), 525-540.

[5] S.A.Cook, "The complexity of theorem-proving procedures", Proc. 3rd Ann. ACM Symp. Theory of Computing, ACM New York, 151-158.

[6] B.Courcelle, "A representation of trees by languages, I,II", Theoretical Comp. Sci 6 (1978), 255-279; 7 (1978), 25-55.

[7] C.C.Elgot, S.L.Bloom, R.Tindell, "On the algebraic structure of rooted trees", J. Comp. Sys. Sci, 16, No. 3(1978), 362-399.

[8] Z.Manna, Mathematical Theory of Computation, McGraw-Hill,(1974).

[9] J.Mycielski, W.Taylor, "A compactification of the algebra of terms", Algebra Universalis 6 (1976), 159-163.

[10] M.Nivat, "On the interpreation of recursive polyadic program schemes", Symp. Math. 15 (1975), 255-281.

UNE SEMANTIQUE POUR LES ARBRES NON DETERMINISTES

G. BOUDOL

Université Paris 7 - LITP

1-Introduction

Le rôle central de la notion d'arbre en théorie de la programmation ,et tout particulièrement dans le domaine de la sémantique des programmes,est maintenant bien établi. Il est bien connu par exemple (D. Scott [23],Z. Manna [16],M. Nivat [19],G. Cousineau [10]) que les exécutions possibles d'un schéma de charte ("flow diagram"), telle que:

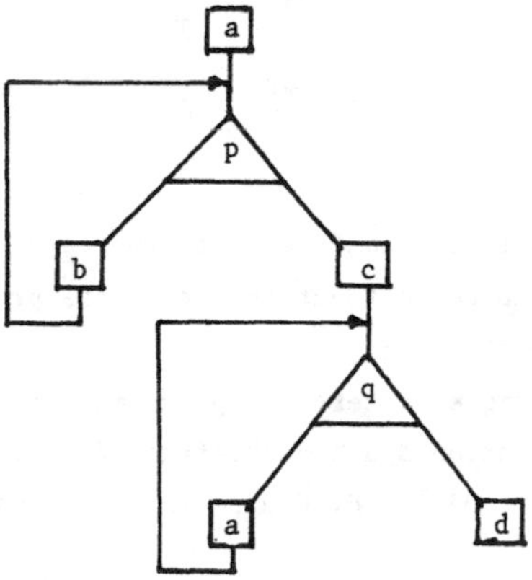

peuvent se représenter par un arbre, ici:

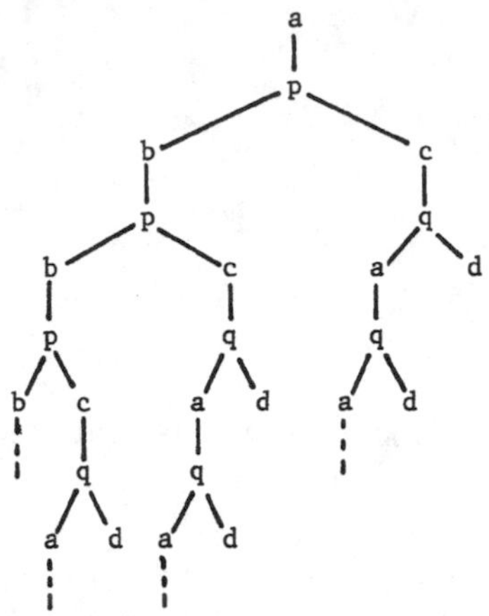

De même, à un schéma de définition récursive (Mc Carthy [15], J.-M. Cadiou [8],M.Nivat [17]) tel que:

F(x) = f(x,g(x,F(a(x))))

on peut associer un arbre (B.K. Rosen [21], M. Nivat & B. Courcelle [9]):

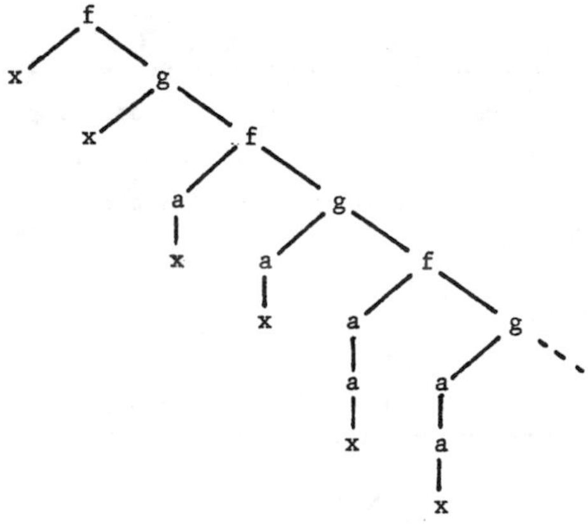

qui, pour toute interprétation des fonctions f, g et a, représente la fonction définie par le schéma en ce sens que cette fonction est l'interprétation de l'arbre ainsi associé au programme récursif.

On peut donc définir la sémantique de certains programmes par l'intermédiaire d'un arbre, qui est leur "interprétation sybolique", façon de faire que l'on peut qualifier de *sémantique algébrique* (ADJ [1],B. Courcelle & M. Nivat [9], M. Nivat [3],[18]):

programme P ⟼ arbre A(P) (sémantique symbolique de P)

de telle sorte que:

$$\mathrm{Val}_I(P) = A(P)_I \qquad \text{pour toute interprétation } I$$

Le problème est posé depuis longtemps de définir en particulierla sémantique des programmes comportant une structure de contrôle *non déterministe*, que l'on représente en général par un opérateur de choix binaire <u>or</u>. Diverses propositions ont été faites pour résoudre ce problème, soit en définissant un comportement opérationnel des structures de contrôle non déterministes (R.W. Floyd [12], Z. Manna [16],E.W. Dijkstra [11], J.W. de Bakker [4],M. Hennessy & E.A. Ashcroft [13]), soit en prolongeant les méthodes de D. Scott ([22]) de sémantique par point fixe (G. Plotkin [20], M. Smyth [24], M. Hennessy & E.A. Ashcroft [13]). Plus récement, A. Arnold & M. Nivat ([2],[3]) ont donné une définition algébrique de cette sémantique, proposant une notion d'ensemble d'arbres engendrés par un programme récursif non-déterministe. Il est bien clair en effet que l'on peut définir une sémantique (algébrique) des programmes non-déterministes si l'on est capable de leur associer des ensembles d'arbres "déterministes" (ie ne contenant pas d'occurrence de choix <u>or</u>), que l'on sait interpréter: la relation calculée par un tel programme sera alors l'union des fonctions interprétant les arbres déterministes qui lui sont associés.

programme $P \longmapsto L(P)$ ensemble d'arbres déterministes

$$\mathrm{Val}_I(P) = \bigcup \{ A_I \ / \ A \in L(P) \}$$

Or on sait associer à un programme non-déterministe un arbre "non-déterministe" (contenant éventuellement des choix): son interprétation symbolique, en ignorant la signification possible du <u>or</u>.

<u>Exemple 1</u>

si l'on considère le systême de définitions récursives

$$\begin{cases} F(x) = \underline{or}(x, F(a(x))) \\ G = a(G) \\ H(x) = \underline{or}(G, F(x)) \end{cases}$$

alors à F,G et H sont resp. associés les arbres suggérés par

Exemple 2

l'arbre des exécutions de la charte:

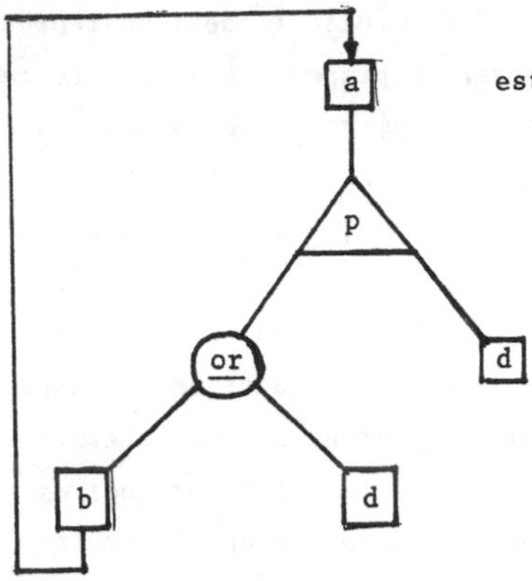

est représenté par:

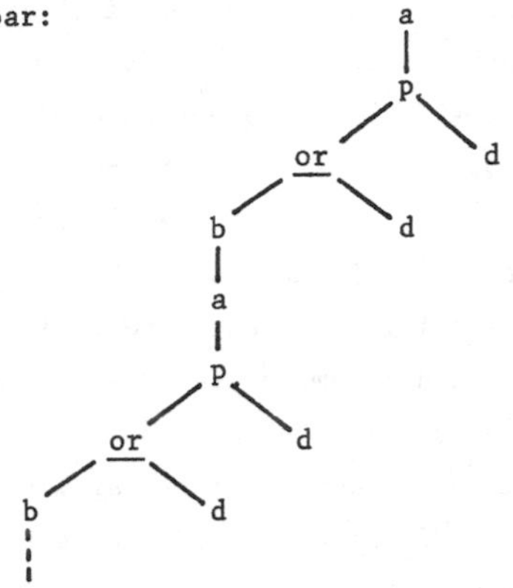

Ce que nous proposons ici est de définir l'ensemble d'arbres (déterministes) L(P) associé à un programme non-déterministe P comme "interprétation" (tenant compte de la signification intuitive du or) de l'arbre (non-déterministe) A(P) (mais nous ne définirons pas formellement ici la fonction $P \longmapsto A(P)$, voir par exemple [5], [6]). L'idée que nous développons ci-dessous est que L(P) est l'ensemble des arbres déterminés par des choix effectifs dans l'arbre A(P), choix qui à chaque occurrence de or indiquent quel est l'argument choisi (ie quel est l'argument à calculer, ou comment est éxécutée cette occurrence particulière de choix au cours d'un calcul de P).

Mentionnons le fait que, dans le cas où P est un programme récursif, cette sémantique algébrique a une exacte contrepartie opérationnelle: l'ensemble L(P) des arbres ainsi défini est exactement l'ensemble des arbres "calculés" par le programme, si l'on donne à la structure de contrôle non-déterministe sa signification intuitive: un or s'éxécute en choisissant l'un ou l'autre alternant; ce résultat d'adéquation est établi par ailleurs ([5],[6]). Nous indiquerons également pour conclure comment situer la sémantique proposée ici par rapport à la théorie de la sémantique par point fixe, et en particulier par rapport aux constructions de "power-domain" de G. Plotkin ([20]) et M. Smyth ([24]).

2-Arbres

Les arbres (finis ou infinis) que nous considérons sont constitués d'un ensemble de noeuds, qui forme un domaine d'arbre, chaque noeud étant étiqueté par un symbôle pris dans un alphabet (fini) S. Pour plus de commodité nous supposerons que chaque fois qu'un symbôle de S apparaît dans un arbre, c'est avec un nombre fixé de fils. Cela revient à dire que l'on suppose que S est gradué:

$$S = S_o + S_1 + \ldots + S_m \qquad \text{(union disjointe)}$$

où S_n est l'ensemble des symbôles de S qui ont n successeurs dans un arbre.(on peut encore considérer que chaque $s \in S_n$ est un symbôle de fonction à n arguments).

Dans le cas où l'on considère des arbres d'éxécutions de chartes, on peut supposer par exemple que S_o contient des symbôles de fin, S_1 est un ensemble d'actions élémentaires (affectations par exemple), S_2 contient des tests et l'opérateur de choix <u>or</u>.

Pour présenter les arbres infinis comme limites de suites croissantes d'arbres finis -c'est à dire pour définir un ordre syntaxique sur les arbres- on suppose en général que S_o contient un élément particulier Ω. L'ordre syntaxique étant en fait "est un sous-arbre initial de", il est plus commode pour le définir de supposer, de façon équivalente, que les arbres peuvent contenir des occurrences de l'arbre vide. Pour définir maintenant formellement les arbres, nous avons besoin des notations suivantes:

notations: N est l'ensemble des entiers, muni de son ordre naturel (strict) $<$
N^* est l'ensemble des mots sur N (suites finies d'entiers), la concaténation de u et de v étant notée uv et le mot vide 1.

<u>définition 1</u>: <u>arbres (bien formés) sur S</u>.

Soit $S=S_o+S_1+\ldots+S_m$ *un alphabet gradué (tel que* <u>or</u> $\in S_2$ *). Un arbre (bien formé) sur* S *est une application partielle* t *de* N^* *dans* S *dont le domaine* dom(t) *est un domaine d'arbre c'est à dire:*

$$u \in \mathrm{dom}(t) \;\&\; u=vw \Rightarrow v \in \mathrm{dom}(t)$$

et telle que:

$$u \in \mathrm{dom}(t) \;\&\; t(u) \in S_n \Rightarrow \{i /\ i \in N \;\&\; ui \in \mathrm{dom}(t)\} \subseteq \{j /\ j \in N \;\&\; j<n\}$$

On désignera par $A^\infty(S)$ l'ensemble de ces arbres, et par $A(S)$ le sous-ensemble des arbres finis (ie de domaine fini).

<u>définition 2</u>: <u>ordre syntaxique sur $A^\infty(S)$</u>.

pour t et t' dans $A^\infty(S)$:

$$t \subseteq t' \Leftrightarrow_{def} \forall u \in N^* \;\; u \in \mathrm{dom}(t) \Rightarrow u \in \mathrm{dom}(t') \;\&\; t'(u)=t(u)$$

Cet ordre n'est donc rien d'autre que l'ordre par prolongement sur les applications partielles, d'où la notation $\subseteq$. Pour cet ordre l'arbre vide (de domaine vide), que nous noterons $\perp$, est le plus petit élément. Il est bien connu (cf [1],[9]) que $\langle A^\infty(S), \subseteq, \perp \rangle$ est un *ensemble ordonné complet,* c'est à dire que toute partie dirigée D de $A^\infty(S)$ admet une borne supérieure pour l'ordre $\subseteq$, que nous notons $\bigcup D$ (rappelons que D est dirigé ssi $t_1 \in D \;\&\; t_2 \in D \Rightarrow \exists t \in D \;\; t_1 \subseteq t \;\&\; t_2 \subseteq t$). L'ensemble $A^\infty(S)$ est même une complétion par ideaux de $A(S)$ (cf [9]), c'est à dire que:

$$\forall t \in A^\infty(S) : \quad t = \bigcup \{t' /\ t' \in A(S) \;\&\; t' \subseteq t\}$$

On désignera par App(t) l'ensemble (dirigé) $\{t' / t' \in A(S) \ \& \ t' \subseteq t\}$ des approximannts (finis) de t. Nous noterons toujours: $S'=S-\{\underline{or}\}$, et nous convien-drons de dire qu'un arbre est déterministe s'il est dans $A^{\infty}(S')$.

Puisque nous avons supposé que S est un alphabet gradué, $A^{\infty}(S)$ est aussi une S-*algèbre*, chaque symbôle s de S définissant canonique ment une fonction $\hat{s}$ sur $A^{\infty}(S)$:

si $s \in S_n$ alors pour $(t_1,\ldots,t_n) \in A^{\infty}(S)^n$, $\hat{s}(t_1,\ldots,t_n)$ est l'arbre t défini par:

$$\mathrm{dom}(t) = \{1\} \cup \bigcup_{1 \leq i \leq n} \{i\}\mathrm{dom}(t_i)$$

$$t(1)=s \ \ \& \ \ \text{pour} \ \ 1 \leq i \leq n \ \ \text{et} \ \ u \in \mathrm{dom}(t):\ t(iu)=t_i(u)$$

(implicitement: si n=o alors t est l'arbre réduit au symbôle s)

On peut mentonner le fait que A(S) est la S-*algèbre ordonnée libre engendrée par* $\{\perp\}$,c'est à dire l'ensemble des *termes* construits sur $\perp$ avec les symbôles de S. Pour simplifier l'écriture, nous identifierons arbres finis et termes dans les notations, confondant l'arbre $\hat{s}(\hat{t}_1,\ldots,\hat{t}_n)$ et le terme $s(t_1,\ldots,t_n)$, où les arbres $\hat{t}_i$ sont représentés par les termes t_i . Quant à $A^{\infty}(S)$, c'est la S-*algèbre ordonnée complète libre engendrée par* $\{\perp\}$ (cf [1],[9]). En particulier les applications $\hat{s}$ sont continues, et nous nommerons *congruence* toute équivalence $\equiv$ sur $A^{\infty}(S)$ telle que:

$$\forall s \in S_n \ \ (n>0) \ \text{si pour} \ 1 \leq i \leq n \ :\ t_i \equiv t'_i \ \ \text{alors} \ \ \hat{s}(t_1,\ldots,t_n) \equiv \hat{s}(t'_1,\ldots,t'_n)$$

Nous avons dit que nous voulions associer à un arbre $t \in A^{\infty}(S)$ l'ensemble des arbres déterministes -ensemble que nous noterons rep(t)- qu'il représente lorsque l'on interprète le <u>or</u> , de façon à ce que cette sémantique reflète les propriétés attendues pour le <u>or</u> que l'on conçoit comme une union. Un premier pas dans ce sens, si nous notons par $P(A^{\infty}(S'))$ l'ensemble des parties non vides de $A^{\infty}(S')$, est de définir $\mathrm{rep}(t) \in P(A(S'))$,lorsque t est un arbre fini,par induction structurelle -ie par morphisme d'algèbre:

$$\mathrm{rep}(x) = \{x\} \quad \text{si} \quad x \in S_o \cup \{\perp\}$$

$$\mathrm{rep}(s(t_1,\ldots,t_n)) = \{s(t'_1,\ldots,t'_n) / \ t'_i \in \mathrm{rep}(t_i)\} \quad \text{si} \quad s \in S_n \ (n>0) \ \text{et} \ s \neq \underline{or}$$

$$\mathrm{rep}(\underline{or}(t_1,t_2)) = \mathrm{rep}(t_1) \cup \mathrm{rep}(t_2)$$

En fait, rep est l'unique morphisme qui prolonge $\perp \mapsto \{\perp\}$ de A(S) dans P(A(S')) muni de la stucture de S-algèbre décrite par:

$$f(s)(T_1,\ldots,T_n) = \{s(t_1,\ldots t_n) / \ t_i \in T_i\} \ \text{pour} \ s \in S_n - \{\underline{or}\} \ \ (\text{si } n=0 \ f(s)=\{s\})$$

$$f(\underline{or})(T_1,T_2) = T_1 \cup T_2$$

<u>Exemple 3</u>

avec $a,b \in S_o$, $s \in S_2$, si t est l'arbre $t = \underline{or}(a,s(b,\underline{or}(a,b)))$ alors

$\mathrm{rep}(t) = \{a \ ,\ s(b,a) \ ,\ s(b,b)\}$

Puisque les arbres de programmes sont en général des arbres infinis, le problème est d'étendre rep à $A^{\infty}(S)$. Nous verrons plus loin qu'il n'est pas possible de présenter le prolongement que nous proposons ci-dessous comme "prolongement par continuité" ($A^{\infty}(S)$ étant une complétion de $A(S)$.) de rep, supposant $\mathcal{P}(A^{\infty}(S'))$ muni d'une structure d'ensemble ordonné complet, pour laquelle rep serait croissante.

3-Sémantique des arbres non-déterministes

Notre solution au problème précédent repose sur la remarque suivante: reprenant l'exemple 3, on peut observer que chaque élément de rep(t) correspond à un "choix" (éventuellement plusieurs) dans l'arbre t qui à chaque occurrence de <u>or</u> (noeud étiqueté <u>or</u>) dans t indique quel est l'alternant choisi. Ainsi par exemple:

s(b,a)$\in$ rep(t) correspond au choix figuré par:

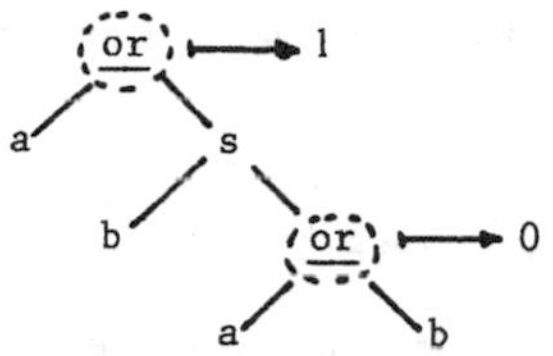

où 0 et 1 indiquent resp. que l'on choisit le premier ou le second alternant.

On peut ainsi définir la notion de *choix* sur un arbre non-déterministe quelconque (éventuellement infini):

<u>définition 3</u>: <u>choix sur les arbres non-déterministes</u>

un choix h *sur un arbre* $t \in A^{\infty}(S)$ *est une application*

$$h: \{u/\ u \in dom(t)\ \&\ t(u)=\underline{or}\} \longrightarrow \{0\ ,\ 1\}$$

On pourrait donner une définition un peu plus restrictive, dans la mesure où certains choix sont inutiles (par exemple pour a$\in$rep(t) de l'exemple 3), mais cela n'est pas vraiment nécéssaire.

On notera par choix(t) l'ensemble des choix sur t (on peut remarquer que choix(t) n'est jamais vide: si t est un arbre déterministe, choix(t) est réduit à l'application vide). On conçoit aisément ce que peut être le résultat de l'application d'un choix h à un arbre fini t: c'est un arbre fini déterministe que nous noterons $\hat{h}(t)$. Pour le définir formellement, il nous faut un peu de notations:

pour $t \in A^{\infty}(S)$ et $u \in dom(t)$, on désigne habituellement par (t/u) le sous-arbre de t au noeud u , donné par:

$dom(t/u) = \{ v/\ v \in N^{*}\ \&\ uv \in dom(t)\}$

$(t/u)(v) = t(uv)$ pour $v \in dom(t/u)$

Avec les mêmes hypothèses, si $h \in choix(t)$, on désignera par (h/u) le choix sur (t/u) défini par:

$(h/u)(v) = h(uv)$ pour $uv \in dom(t)\ \&\ t(uv) = \underline{or}$

<u>définition 4</u>: <u>résultat de l'application d'un choix à un arbre fini</u>

Soit $t \in A(S)$ et $h \in choix(t)$. L'arbre $\hat{h}(t)$ est défini par:

$\hat{h}(t) = t$ si $t \in S_o \cup \{\bot\}$

$\hat{h}(s(t_1,\ldots,t_n)) = s(\widehat{(h/0)}(t_1),\ldots,\widehat{(h/n-1)}(t_n))$ si $s \in S_n - \{\underline{or}\}$

$$\hat{h}(\underline{or}(t_1,t_2)) = \begin{cases} \widehat{(h/0)}(t_1) & \text{si } h(\mathit{1})=0 \\ \widehat{(h/1)}(t_2) & \text{si } h(\mathit{1})=1 \end{cases}$$

ou plus brièvement dans ce dernier cas:

$\hat{h}(\underline{or}(t_o,t_1) = \widehat{(h/h(\mathit{1}))}(t_{h(\mathit{1})})$

Il est clair que l'on a toujours:

$\hat{h}(t) \in A(S')$

etl'on peut alors prouver l'exactitude de la remarque faite au début de ce paragraphe:

<u>lemme 1</u>

pour tout $t \in A(S)$: $rep(t) = \{\hat{h}(t) / h \in choix(t)\}$

<u>preuve</u>: par induction structurelle sur t

-si $t \in S_o \cup \{\bot\}$ alors $choix(t) = \{\emptyset\}$ et $\hat{\emptyset}(t) = t$

donc $\{\hat{h}(t) / h \in choix(t)\} = \{t\} = rep(t)$

-si $t=s(t_1,\ldots,t_n)$ avec $s \in S_n - \{\underline{or}\}$ $(n>0)$ alors l'application

$$h \longmapsto ((h/0),\ldots,(h/n-1))$$

établit une bijection de $choix(t)$ sur $choix(t_1) \times \ldots \times choix(t_n)$ de telle sorte que:

$$\{\hat{h}(t) / h \in choix(t)\} = \{s(\hat{h}_1(t_1),\ldots,\hat{h}_n(t_n)) / h_i \in choix(t_i)\}$$

Par hypothèse d'induction:

$$rep(t_i) = \{\hat{h}_i(t_i) / h_i \in choix(t_i)\}$$

d'où, par définition de la structure de S-algèbre sur $\mathcal{P}(A(S'))$:

$$rep(t) = \{\hat{h}(t) / h \in choix(t)\}$$

-si $t = \underline{or}(t_1,t_2)$ alors l'application

$$h \longmapsto j(h)=(h(\mathit{1}),(h/h(\mathit{1})))$$

est une surjection de $choix(t)$ dans $\{0\} \times choix(t_1) \cup \{1\} \times choix(t_2)$ telle que $j(h)=j(h') \Rightarrow \hat{h}(t)=\hat{h}'(t)$,et comme:

$$\begin{aligned}\{\hat{h}(t) / h \in choix(t)\} &= \{\hat{h}(t) / h \in choix(t) \;\&\; h(\mathit{1})=0\} \cup \{\hat{h}(t) / h \in choix(t) \& h(\mathit{1})=1\} \\ &= \{\hat{h}_1(t_1) / h_1 \in choix(t_1)\} \cup \{\hat{h}_2(t_2) / h_2 \in choix(t_2)\}\end{aligned}$$

on utilise l'hypothèse d'induction pour obtenir:

$$\{\hat{h}(t) / h \in choix(t)\} = rep(t) \qquad \blacksquare$$

Pour définir maintenant le résultat de l'application d'un choix h à un arbre *infini* t, résultat que nous noterons encore $\hat{h}(t)$, il suffit d'observer, en notant, pour $t \subseteq t'$., par $h|t'$ le choix sur t' qui est restriction de h au sous-arbre initial t' de t , que:

lemme 2

pour $t \in A(S)$, $t' \subseteq t$ et $h \in \text{choix}(t)$: $\widehat{h|t'}(t') \subseteq \hat{h}(t)$

preuve: par induction structurelle sur t'

-si $t' \in S_o \cup \{\perp\}$ alors

--soit $t' = \perp$, $h|t' = \emptyset$ et $\widehat{h|t'}(t') = t' = \perp$ donc

$\widehat{h|t'}(t') \subseteq \hat{h}(t)$ pour tout $t \in A(S)$ et tout $h \in \text{choix}(t)$

--soit $t' \in S_o$ et alors $t' \subseteq t \Rightarrow t' = t$ d'où

$\widehat{h|t'}(t') = \hat{h}(t)$ pour tout $h \in \text{choix}(t)$ (en fait ici $\text{choix}(t) = \{\emptyset\}$)

-si $t' = s(t'_1, \ldots, t'_n)$ où $s \in S_n - \{\underline{or}\}$ $(n > 0)$ alors

$t' \subseteq t \Rightarrow t = s(t_1, \ldots, t_n)$ avec $\forall i$ $(1 \leq i \leq n)$ $t'_i \subseteq t_i$

pour $h \in \text{choix}(t)$, en posant $h_i = (h/i-1)$ $(1 \leq i \leq n)$ on a

$\hat{h}(t) = s(\hat{h}_1(t_1), \ldots, \hat{h}_n(t_n))$

et si $h' = h|t'$ et $h'_i = (h'/i-1)$ (pour $1 \leq i \leq n$), on a

$\hat{h}'(t') = s(\hat{h}'_1(t'_1), \ldots, \hat{h}'_n(t'_n))$

Mais il est clair que pour tout i $(1 \leq i \leq n)$ $h'_i = h_i|t'_i$

donc par hypothèse d'induction $\hat{h}'_i(t'_i) \subseteq \hat{h}_i(t_i)$

d'où $\hat{h}'(t') \subseteq \hat{h}(t)$

-si $t' = \underline{or}(t'_o, t'_1)$ alors $t' \subseteq t \Rightarrow t = \underline{or}(t_o, t_1)$ avec $t'_i \subseteq t_i$ $(i=0,1)$.

Soit $h \in \text{choix}(t)$ tel que $h(\mathit{1}) = 0$. Alors si $h' = h|t'$ on a $h'(\mathit{1}) = 0$ et

$\hat{h}'(t') = \widehat{(h'/0)}(t'_o)$ et $\hat{h}(t) = \widehat{(h/0)}(t_o)$

Mais il est bien clair là encore que $(h'/0) = (h/0)|t'_o$, et l'hypothèse d'induction entraine trivialement $\hat{h}'(t') \subseteq \hat{h}(t)$

L'argument est exactement le même lorsque $h(\mathit{1}) = 1$ ∎

Ce lemme à pour conséquence évidente que, pour $t \in A^{\infty}(S)$ et $h \in \text{choix}(t)$ l'ensemble

$$\{ \widehat{h|t'}(t') \ / \ t' \in \text{App}(t) \}$$

est une partie dirigée de $A(S')$. On est donc conduit à la

<u>définition 5: sémantique symbôlique des arbres non-déterministes</u>

- pour $t \in A^{\infty}(S)$ et $h \in choix(t)$ le résultat de l'application de h à t est l'arbre $\hat{h}(t)$ de $A^{\infty}(S')$:

  $\hat{h}(t) = \bigcup \{ \widehat{h|t'}(t') / \ t' \in App(t) \}$
- pour $t \in A^{\infty}(S)$ le sous ensemble $rep^{\infty}(t)$ de $A^{\infty}(S')$ des arbres (déterministes) représentés par t est

  $rep^{\infty}(t) = \{ \hat{h}(t) / \ h \in choix(t) \}$

Il est clair que lorsque t est un arbre fini, la définition 5 pour $\hat{h}(t)$ coïncide avec la définition 4 déjà donnée, donc dans ce cas $rep^{\infty}(t)=rep(t)$. L'application rep^{∞} ainsi définie est bien un prolongement de rep.

<u>Exemple 4</u> (suite de l'exemple 1)
reprenant les arbres t , t" , t' resp. associés à F,G,H, avec $x \in S_0$ et $a \in S_1$, nous laissons le lecteur se convaincre de ce que

$$rep^{\infty}(t) = \{\bot\} \cup \{ a^n(x) / \ n \in N \}$$
$$rep^{\infty}(t'') = \{a^{\omega}\}$$
$$rep^{\infty}(t') = \{\bot\} \cup \{a^n(x) / \ n \in N\} \cup \{a^{\omega}\}$$

où l'on note:

$$a^0(x) = x$$
$$a^{n+1}(x) = a(a^n(x))$$
$$a^{\omega} = \bigcup \{a^n(\bot) / \ n \in N \}$$

dans t et t' l'élément $\bot$ des ensembles $rep^{\infty}(t)$ et $rep^{\infty}(t')$ est obtenu comme $\hat{h}(t)$ et $\hat{h}(t')$ pour h tel que h(u)=1 pour tout noeud u étiqueté <u>or</u>. On obtient de même l'arbre a^{ω} dans $rep^{\infty}(t_1)$ pour

$t_1 =$

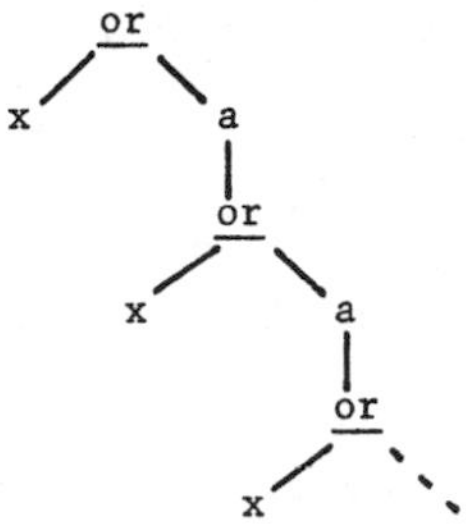

4-Résultats et conclusion

On peut d'abord montrer, ce qui intuitivement est à peu près évident, que la sémantique symbôlique des arbres non-déterministes ci-dessus définie donne du <u>or</u> une interprétation en terme d'union, c'est à dire que l'application rep^{∞} est un morphisme de S-algèbre de $A^{\infty}(S)$ dans $P(A^{\infty}(S'))$, muni de la structure analogue à celle qui a été définie pour $P(A(S'))$;on tire de là la conclusion que l'équivalence sémantique des arbres non-déterministes est une congruence:

<u>théorème 1</u>: <u>adéquation</u>

- l'application $rep^{\infty} : A^{\infty}(S) \longrightarrow P(A^{\infty}(S'))$ est un morphisme d'algèbre:

  $rep^{\infty}(s) = \{s\}$ pour $s \in S_o$

  $rep^{\infty}(s(t_1,\ldots,t_n)) = \{s(t'_1,\ldots,t'_n) / t'_i \in rep^{\infty}(t_i)\}$ pour $s \in S_n - \{\underline{or}\}$ $(n>0)$

  $rep^{\infty}(\underline{or}(t_1,t_2)) = rep^{\infty}(t_1) \cup rep^{\infty}(t_2)$

- l'équivalence sémantique sur $A^{\infty}(S)$ définie par:

$$t \equiv t' \iff_{def} rep^{\infty}(t) = rep^{\infty}(t')$$

  est une congruence telle que, pour tout $t,t_1,\ldots,t_i,t'_i,\ldots,t_n$ dans $A^{\infty}(S)$:

  $\underline{or}(t_1,t_2) \equiv \underline{or}(t_2,t_1)$ (commutativité)

  $\underline{or}(t_1,\underline{or}(t_2,t_3)) \equiv \underline{or}(\underline{or}(t_1,t_2),t_3)$ (associativité)

  $\underline{or}(t,t) \equiv t$ (idempotence)

  $s(t_1,\ldots,\underline{or}(t_i,t'_i),\ldots,t_n) \equiv \underline{or}(s(t_1,\ldots,t_i,\ldots,t_n),s(t_1,\ldots,t'_i,\ldots,t_n))$

  (distributivité par rapport aux fonctions $s \in S_n$, $n>0$)

<u>preuve</u>

1-Pour $s \in S_o$ on a $rep^{\infty}(s) = \{s\} = rep(s)$

Soit $s \in S_n - \{\underline{or}\}$ $(n>0)$ et $t_1,\ldots,t_n$ dans $A^{\infty}(S)$ tels que $t=s(t_1,\ldots,t_n)$ et soit $h \in$ choix(t). Montrons d'abord que, si l'on pose (pour $1 \leq i \leq n$) $h_i=(h/i-1)$ on a: $\hat{h}(t) = s(\hat{h}_1(t_1),\ldots,\hat{h}_n(t_n))$

Il est clair que, pour $(t'_1,\ldots,t'_n) \in App(t_1) \times \ldots \times App(t_n)$, on a, en posant $t' = s(t'_1,\ldots,t'_n)$: $((h|t')/i-1) = h_i|t'_i$

et que $App(t) = \{\bot\} \cup \{s(t'_1,\ldots t'_n) / t'_i \in App(t_i)\}$

Donc, comme $\widehat{h|\bot}(\bot) = \bot$:

$$\hat{h}(t) = \bigcup \{ s(\widehat{h_1|t'_1}(t'_1),\ldots,\widehat{h_n|t'_n}(t'_n) / t'_i \in App(t_i)\} \quad \text{(def 4 et 5)}$$

$$= \hat{s}(\bigcup_{t'_1 \in App(t_1)} \widehat{h_1|t'_1}(t'_1),\ldots, \bigcup_{t'_n \in App(t_n)} \widehat{h_n|t'_n}(t'_n)) \quad \text{(continuité de } \hat{s})$$

$$= s(\hat{h}_1(t_1),\ldots\hat{h}_n(t_n)) \quad \text{(def 5)}$$

Comme $h \longmapsto (h_1,\ldots,h_n)$ établit une bijection de choix(t) dans choix$(t_1) \times \ldots \times$choix$(t_n)$ on a

$$rep^{\infty}(t) = \{s(t'_1,\ldots,t'_n) / t'_i \in rep^{\infty}(t_i)\}$$

$$= f(s)(rep^{\infty}(t_1),\ldots,rep^{\infty}(t_n))$$

Soient t_1 et t_2 dans $A^{\infty}(S)$ et $t=\underline{or}(t_1,t_2)$, et soit $h \in$ choix(t).

Posons $i_h=h(1)$ et $h'=(h/i_h)$. Alors pour $t'=\underline{or}(t'_1,t'_2)$, $t'_i \in App(t_i)$, on a

$h'|t'_{i_h} = ((h|t')/i_h)$

et puisque $\mathrm{App}(t) = \{\bot\} \cup \{\, \underline{or}(t'_1,t'_2)/\ t'_i \in \mathrm{App}(t_i)\}$ on a

$$\begin{aligned}\hat{h}(t) &= \bigcup \{\, \widehat{h|t'}(t')/\ t' \in \mathrm{App}(t)\} && \text{(def 5)}\\ &= \bigcup \{\, \widehat{h'|t'_{i_h}}(t'_{i_h})/\ t'_{i_h} \in \mathrm{App}(t_{i_h})\} && \text{(def 4)}\\ &= \hat{h}'(t_{i_h}) && \text{(def 5)}\end{aligned}$$

Donc, $h \longmapsto j(h)=(i_h,(h/i_h))$ étant une surjection de choix(t) dans

$\{0\} \times \mathrm{choix}(t_1) \cup \{1\} \times \mathrm{choix}(t_2)$ telle que $j(h)=j(h') \Rightarrow \hat{h}(t)=\hat{h}'(t)$,

on a par conséquent:

$$\mathrm{rep}^{\infty}(\underline{or}(t_1,t_2)) = \mathrm{rep}^{\infty}(t_1) \cup \mathrm{rep}^{\infty}(t_2)$$

Puisque rep^{∞} est un morphisme, l'équivalence sémantique associée $\equiv$ est une congruence; la vérification de ses propriétés annoncées dans le théorème est triviale. ∎

On peut donc dire que notre sémantique constitue une formalisation adéquate du non-déterminisme. Par contre, on peut montrer, comme on l'a déjà signalé, que cette sémantique ne peut être continue, relativement à l'ordre syntaxique sur $A^{\infty}(S)$, et c'est le moins que l'on puisse demander à une sémantique continue que de l'être par rapport à cet ordre (cf [1]):

<u>théorème 2</u>: <u>non-continuité de rep^{∞}</u>
Il n'existe pas de relation d'ordre faisant de $P(A^{\infty}(S'))$ un ensemble ordonné complet et pour lequel rep^{∞} soit continue ($A^{\infty}(S)$ étant ordonné par $\subseteq$)

<u>preuve</u>: ce résultat suppose en fait que: $\exists n > 0 \quad S'_n \neq \emptyset$. Pour montrer un exemple simple, on supposera que S' contient un symbôle monadique a et (mais cela n'est pas nécéssaire) un symbôle de constante x.
On considère les suites $\subseteq$-croissantes $(t_n)_{n\in N}$ et $(t'_n)_{n\in N}$ données par:

$t_n =$ 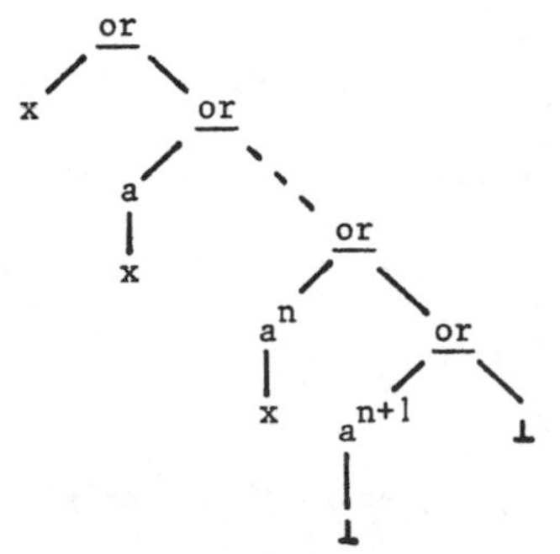

$t'_n =$ 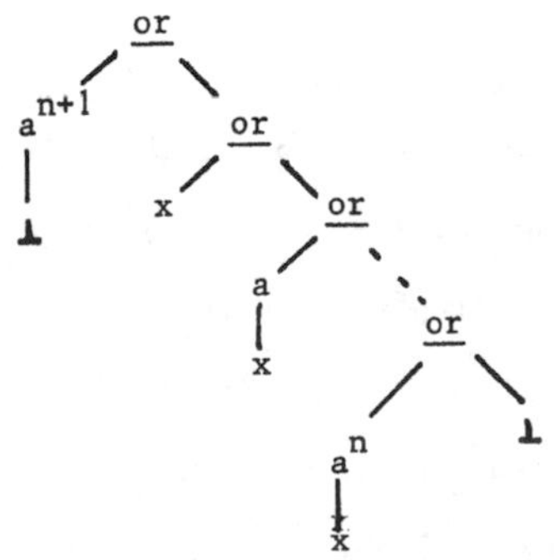

On peut remarquer que ces suites convergent resp. vers les arbres t et t de l'exemple 4 (et qui sont les arbres associés aux procédures récursives F,G e H de l'exemple 1). On a:

$$\underline{or}(x_o,\underline{or}(x_1,x_2) \equiv \underline{or}(\underline{or}(x_1,x_2),x_o) \quad \text{(commutativité)}$$
$$\equiv \underline{or}(x_1,\underline{or}(x_2,x_o)) \quad \text{(associativité)}$$
$$\equiv \underline{or}(x_1,\underline{or}(x_o,x_2)) \quad \text{(commutativité)}$$

et l'on voit facilement (par récurrence par exemple) que, pour $n > 0$:

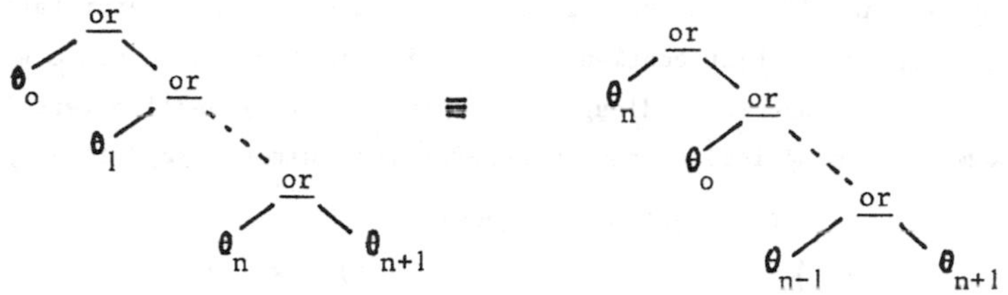

On a donc pour tout $n \in N$ $t_n \equiv t'_n$ (ce que l'on constate d'ailleurs directement sans difficulté).

Supposons $P(A^{\infty}(S'))$ muni d'un ordre $\sqsubseteq$ pour lequel cet ensemble soit complet, et supposons que rep^{∞} soit continue, on devrait donc avoir:

$$rep^{\infty}(t) = rep^{\infty}(\bigcup_{n \in N} t_n) = \bigsqcup_{n \in N} rep^{\infty}(t_n) = \bigsqcup rep^{\infty}(t'_n)$$
$$= rep^{\infty}(\bigcup t'_n) = rep^{\infty}(t')$$

Or: $rep^{\infty}(t') = \{a^{\omega}\} \cup rep^{\infty}(t)$ et $a^{\omega} \notin rep^{\infty}(t')$ (voir exemple 4) ■

Ce résultat a pour conséquence qu'on ne peut faire de $P(A^{\infty}(S'))$ un "power domain" (cf [20] & [24]) pour lequel rep^{∞} soit continue. Plotkin et Smyth prennent le parti de considérer en fait un quotient de $P(A^{\infty}(S'))$, c'est à dire d'identifier certains sous-ensembles de $A^{\infty}(S')$; Smyth ([24]) note cependant que cette façon de faire introduit quelques difficultés, et il ne nous parait pa justifié de procéder ainsi ici, dans la mesure où t et t' sont "arbres de programmes", et qu'il est possible de montrer ([5],[6]) que les arbres de rep^{∞}(t lorsque t est l'arbre d'un programme récursif non-déterministe, sont les résultats obtenus par sémantique opérationnelle pour ce programme.

Pour voir quels peuvent être de "bons" (par rapport à notre sémantique) modèles du non-déterminisme, reprenons l'exemple du théorème 2: on voit que si l'on se donne un choix h' sur t' tel que $h'(\mathit{1})=0$ (ie tel que $\hat{h}'(t')= a^{\omega}$) alors à chaque $h'_n=h'|t'_n$ correspond un choix $h_n \in$ choix(t_n) tel que

$$\hat{h}_n(t_n) = \hat{h}'_n(t'_n)$$

Mais la suite h_n n'est pas croissante, c'est à dire qu'il n'existe pas de choi h sur t tel que $h_n=h|t_n$ pour tout n. Lorsqu'on change d'approximant pour il faut changer de choix sur l'approximant correspondant de t.

Si nous voulons présenter les suites d'ensembles (d'approximations de résultats) $(rep^{\infty}(t_n))_{n\in N}$ et $(rep^{\infty}(t'_n))_{n\in N}$ comme "convergeant" vers $rep^{\infty}(t)$ et $rep^{\infty}(t')$, il n'est pas suffisant de dire que ces suites sont croissantes pour un certain ordre qui, selon l'idée de D. Scott ([22]) représente l'accroissement d'information. Il faut aussi pouvoir dire quelque chose sur "comment augmente l'information": ici il faut indiquer qu'un élément de $rep(t_n)$ est relié à un élément de $rep(t_{n+1})$ dans la mesure où ils sont obtenus par l'application de la restriction d'un même choix sur t. C'est ce qu'il est possible de faire, comme le propose Lehmann ([14]), en supposant que les domaines d'interprétations sont des catégories, où un objet contient moins d'information qu'un autre s'il y a une flèche (qui indique une façon d'augmenter l'information) de l'un vers l'autre. Dans l'exemple examiné ici, on peut présenter les suites $(rep^{\infty}(t_n))_{n\in N}$ et $(rep^{\infty}(t'_n))_{n\in N}$ de la façon suivante: $(n>0)$

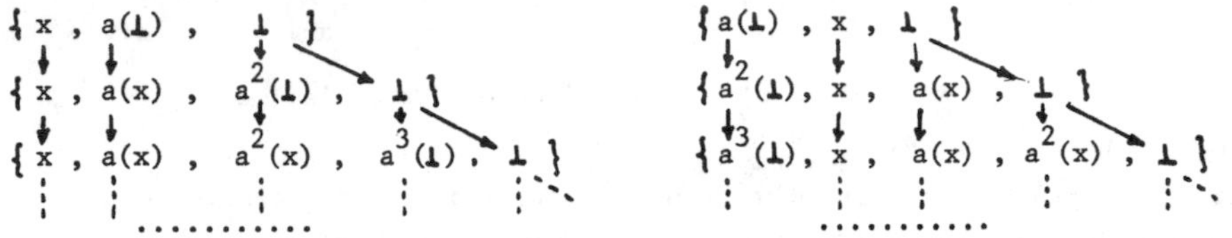

qui auront une "limite" différente, puisque les systèmes de flèches sont distincts. On voit ici que chaque suite de flèches correspond à un choix sur l'arbre infini; notre travail sur ce sujet est maintenat consacré à montrer que l'on obtient bien là un "bon" modèle du non-déterminisme ([7]).

Références

[1] ADJ : "Initial algebra semantics and continuous algebras", JACM 24 (1977) 68-95

[2] A. Arnold & M. Nivat: "Formal computations of non-deterministic recursive schemes", MST 13 (1980) 219-236

[3] A. Arnold &M. Nivat: "Algebraic semantics of non-deterministic recursive program schemes", Rapport du LITP n° 78-4, Univ. Paris 7 (1978)

[4] J.W. de Bakker: "Semantics and termination of non-deterministic recursive programs", Proc. of the 3rd ICALP, Edimburgh (1976) 435-477

[5] G. Boudol: "Sémantique opérationnelle et algébrique des programmes récursifs non-déterministes", Thèse, Univ. Paris 7 (1980)

[6] G. Boudol: "On the semantics of non-deterministic recursive programs", à paraître

[7] G. Boudol: "Category-theoretic models of non-determinism", à paraître

[8] J.-M. Cadiou: "Recursive definitions of partial functions and their computations", Ph. D. Thesis, Stanford (1972)

[9] B. Courcelle & M. Nivat: "Algebraic families of interpretations", 17th FOCS (1976) 137-146

[10] G. Cousineau: "An algebraic definition for control structures", TCS 12 (1980) 175-192

[11] E.W. Dijkstra: "Guarded commands, non-determinacy and formal derivation of programs", CACM 18 (1975) 453-457

[12] R.W. Floyd: "Non-deterministic algorithms", JACM 14 (1967) 636-644

[13] M. Hennessy & E.A. Ashcroft: "A mathematical semantics for non-deterministic typed λ-calculus", TCS 11 (1980) 227-246

[14] D.J. Lehmann: "Categories for fixpoint semantics", 17th FOCS (1976) 122-126

[15] J. Mc Carthy: "A basis for a mathematical theory of computation", in "Computer programming and formal systems" (Braffort & Hirschberg, Eds) (1963) 33-70

[16] Z. Manna: "The correctness of non-deterministic programs", Artificial Intelligence 1 (1970) 1-26

[17] M. Nivat: "On the interpretation of recursive polyadic program schemes", Symposia Mathematica 15, Bologna (1975) 225-281

[18] M. Nivat: "Non-deterministic programs: an algebraic overview", IFIP Congress 1980

[19] M. Nivat: "Chartes, arbres, programmes itératifs", Rapport du LITP n° 78-28, Univ. Paris 7 (1978)

[20] G. Plotkin: "A power-domain construction", SIAM J. on Computing 5 (1976) 452-487

[21] B.K. Rosen: "Program equivalence and context-free grammars", 13th SWAT (1972) 7-18

[22] D. Scott: "Outline of a mathematical theory of computation", Technical Monograph PRG 2, Oxford (1970)

[23] D. Scott: "The lattice of flow-diagrams", Symp. on Semantics of Algorithmic Languages, Lecture Notes in Mathematics n° 182 (1971) 311-366

[24] M. Smyth: "Power-domains", JCSS 16 (1978) 23-36

ON THE ALGEBRAIC SPECIFICATION OF NONDETERMINISTIC PROGRAMMING LANGUAGES *)

M. Broy, M. Wirsing **)
Technische Universität München, Institut für Informatik
Arcisstraße 21, D-8000 München 2

Abstract

Different semantic models for a nondeterministic programming language are defined, analysed, and compared in the formal framework of algebraic specifications of programming languages by abstract types. Four abstract types are given representing *choice ("erratic") nondeterminism*, *backtrack ("demonic") nondeterminism*, *unbounded ("angelic") nondeterminism* and *loose nondeterminism*. The classes of algebras of these types represent classes of semantic models. A comparison of these classes of semantic models shows the connections and differences between the four different concepts of nondeterminism as found in programming languages.

1. Introduction

The concepts of nondeterminism and nondeterminacy have found their way into programming languages only during the middle of the last decade, although McCarthy in his pioneering paper /McCarthy 63/ already introduced an "ambiguity operator" and Floyd in /Floyd 67/ suggested nondeterministic programs for the implicit formulation of backtrack programs.

Recently the growing interest in rigorous methods for formal specification and program development and numerous attempts to define a formal semantics for concurrent programming languages has led to intensive investigations in the theory and formal foundations of nondeterminism. However, a careful study of the different approaches indicates, that not only the formal description methods are different, but there are actually different concepts described, although the differences are often rather

*) This work was carried out within the Sonderforschungsbereich 49 - Programmiertechnik - Munich

**) Present address: Department of Computer Science, University of Edinburgh, Edinburgh EH9 3JZ

sophisticated but nevertheless of great importance. Strictly speaking essentially ("extensionally") different semantic models can be given for nondeterministic programming languages reflecting the different concepts of nondeterminism.

Recent studies have shown, that algebraic methods allow the specification of programming languages by abstract (data) types in a short, flexible way (cf./Broy, Wirsing 80a/). There the *context free syntax* corresponds to the signature (the term-algebra represents the set of syntactically correct programs), the *context conditions* (sometimes called "static semantics") are expressed by particular definedness predicates (restricting the term algebra), and the *semantics* is specified by a number of (conditional) equations. Then each model of that type can be considered as a particular semantic model of the programming language. Due to the termination problem of partial recursive functions such an algebraic specification generally includes semantic models where optimal or even maximal fixed points are associated with recursive definitions. The minimality property of least fixed points, however, can be conveniently expressed by *weakly terminal models*, the existence of which is guaranteed under certain (syntactic) conditions (cf. /Broy, Wirsing 80b/).

The class of *extensionally equivalent* models of the type containing the weakly terminal models comprises all possible semantic models which specify the semantics of least fixed points (syntactic, operational, algorithmic and mathematical models). In particular the initial model of the type lies in this class which forms a complete lattice of models (in the usual sense, cf. /Wirsing, Broy 80/).

In this formal framework it is also possible to discuss the semantic models of nondeterministic (applicative or procedural) programming languages. The various concepts of nondeterminism such as *backtrack* nondeterminism versus *choice* nondeterminism (cf. /Broy et al. 80/, /Kennaway, Hoare 80/) as well as *loose* versus *tight* nondeterminism (cf. /Park 80/) may be discussed conveniently in the algebraic approach by the particular classes of models of a nondeterministic programming language characterized by the resp. semantic equations.

We show that *backtrack* nondeterminism, *unbounded* nondeterminism and *choice* nondeterminism admit terminal semantics. The weakly terminal models of *backtrack* nondeterminism as well as of *unbounded* nondeterminism are properly weaker than those of *choice* nondeterminism. In the *(partial) initial semantics* of both forms of nondeterminism nondeterministic statements differ only in their evaluation, while the induced equalities between them are the same.

Loose nondeterminism does not allow terminal or initial semantics, but only minimal models which correspond to all possible deterministic and nondeterministic least fixed point semantics which implement nondeterministic statements. The weakly terminal model of *backtrack* nondeterminism is one of these minimal models. By introducing an "implementation" relation $\subseteq_I$ we can structurize these minimal models in such a way that the $\subseteq_I$ - minimal models are exactly the deterministic least fixed point implementations. The weakly terminal models of *choice* nondeterminism are optimal in the following sense: They are the weakest models which are $\subseteq_I$ - greater than all $\subseteq_I$ - minimal models.

Finally we show that the so-called Egli-Milner Ordering is a consequence of the specification using weak homomorphisms and thus is "natural" in the weakly terminal models.

In fact, the goal of this case study is twofold: First, we want to demonstrate how algebraic methods can be used as a powerful, flexible tool for the formulation and analysis of semantic specifications. Second, we give an attempt to clarify, unify, and compare several notions of nondeterminism with rather sophisticated differences as found in the literature.

We demonstrate our approach by means of abstract data types specifying the sort <u>sta</u> of nondeterministic statements. The types define procedural programming languages very similar to Dijkstra's language of guarded commands. We investigate several closely related versions:

- a type AN the weakly terminal model of which corresponds to *unbounded ("angelic")* nondeterminism (this type resembles to the wlp-calculus definition of Dijkstra).
- a type BN the weakly terminal model of which corresponds to *backtrack ("demonic")* nondeterminism (this type resembles to the wp-calculus definition of Dijkstra).
- a type CN the weakly terminal model of which corresponds to *choice ("erratic")* nondeterminism. Every model of CN implements a model of BN in a "natural" way*)
- a type LN corresponding to *loose* nondeterminism. For this type there does not exist a weakly terminal model. However all models of AN, BN as well as all models of CN are models of LN, too. Each minimal model of LN represents the mathematical semantics of a particular (possibly deterministic) programming language.

---

*)This type resembles to the wp/wlp-calculus definition of Dijkstra.

2. Basic Definitions

Before we define one type we briefly give the most important definitions (for a complete definition see /Broy, Wirsing 80a/). We consider *hierarchical abstract types* with *primitive subtypes* and *finitely generated partial heterogeneous Σ-algebras* as models; i.e. partial heterogeneous Σ-algebras without proper subalgebras. Between two Σ-algebras A und B a family φ of total mappings is called *(partial)* Σ *-homomorphism* (cf./Grätzer 68/), if for all operations f

$$\varphi(f^A(x_1,\dots,x_n)) = \begin{cases} f^B(\varphi(x_1),\dots,\varphi(x_n)) & \text{if } f^A(x_1,\dots,x_n) \text{ is defined} \\ \text{undefined} & \text{otherwise} \end{cases}$$

and if

$$f^A(x_1,\dots,x_n) \text{ defined} \Rightarrow f^B(\varphi(x_1),\dots,\varphi(x_n)) \text{ defined}$$

A model I of T is called *initial* , if for all models A of T there exists a unique homomorphism $\varphi : I \to A$. An initial model I is *minimally defined,* i.e. every term t which is undefined in some model of T is undefined in I, too.

The properties of homomorphisms for total algebras are generalized by the following notion (cf. /Broy, Wirsing 80b/).

A family φ of partial mappings is called *weak* Σ *-homomorphism,* if for all operations f

$$\varphi(f^A(x_1,\dots,x_n)) = \begin{cases} f^B(\varphi(x_1),\dots,\varphi(x_n)) & \text{if } f^B(\varphi(x_1)\dots,\varphi(x_n)) \text{ is defined} \\ \text{undefined} & \text{otherwise} \end{cases}$$

If such a weak Σ-homomorphism exists, then B is called *weaker* than A . A mapping which is both a partial Σ-homomorphism and a weak Σ-homomorphism is called a *strong Σ-homomorphism.*

In order to describe observable equivalence we need a notion of terminality for partial algebras. Let I be an initial model of T and consider the class $W =_{def} \{A \mid$ there exists a strong Σ-homomorphism $\varphi : I \to A\}$. Then a model Z of T is said to be *weakly terminal* if Z is strongly terminal in W, i.e. for all $A \in W$ there exists a strong Σ-homomorphism $\varphi : A \to Z$. The weakly terminal models as well as all elements of W are minimally defined.

Let us fix a single model P' of the primitive subtype P of T and consider only the models of T which are extensions of P'. Then every two models A and B for which a strong Σ-homomorphism $\varphi : A \to B$ or $\varphi : B \to A$ exists are *extensionally equivalent*, i.e. for every function f with range in P and every nonprimitive term t we have $f(\ldots,t,\ldots)^A = f(\ldots,t,\ldots)^B$. In particular, W forms a class of extensionally equivalent models. Every Σ-homomorphism between two extensionally equivalent models is a strong Σ-homomorphism. If C is a class of extensionally equivalent models then a strongly initial (terminal) model $A \in C$ is called *(C-)extensionally* initial (terminal). For example, the initial models of T are W-extensionally initial and the weakly terminal models are W-extensionally terminal (cf. figure 1).

The extensional equivalence leads to another definition of terminality. A model R of T is called *reachable* , if for all models A of T there exists an extensionally equivalent model B of T such that there is a weak Σ-homomorphism $\varphi : B \to R$. Every reachable model is minimally defined. T is *reachably terminal* if it is strongly terminal in the class of all reachable models.

If an initial model exists, then every reachably terminal model is weakly terminal (but in general not vice versa).

A model A of a hierarchical abstract type is called *fully abstract.* (cf. /Milner 77/) if for every pair of terms t1, t2 of nonprimitive sort $t1^A = t2^A$ iff for every primitive context K[x]: $K[t1]^A = K[t2]$; a *primitive context* K[x] for terms of sort <u>s</u> is a term K[x] with the only free variable x such that for every term t of sort <u>s</u> , K[t] is a term of primitive sort.

Obviously (cf. /Broy, Wirsing 80b/) a fully abstract model is minimal with respect to strong homomorphisms. Furthermore, if there exists a fully abstract, minimally defined model of a type and a weakly terminal model, then both are isomorphic. Both notions of minimal full abstractness and weak terminality therefore capture the notion of observable equality or functional equivalence. This means that in a fully abstract model two terms are considered to be equal, iff all observable results of applications of this term (the result of this term in all primitive contexts) are equal. Then the two terms are called *visibly equivalent.*

3. The Abstract Type of Choice Nondeterminism

We define an abstract type comprising the following primitive sorts:

- <u>dom</u> , the sort of a semantic objects (including the truth values tt and ff and their characteristic operations) with an equality operation ~ ,
- <u>var</u> , the sort of identifiers for programming variables,
- <u>proc</u> , the sort of identifiers for procedures,
- <u>exp</u> , the sort of arithmetic expressions over <u>var</u> together with a total evaluation function eval : <u>exp</u> → <u>dom</u> , which yields error for free identifiers (where error is a defined constant of <u>dom</u>). We denote by e1[e2/v] the

substitution of v in e1 by e2.

bexp, the set of boolean expressions (also with evaluation function eval).

For simplicity we may assume that these sorts are given by abstract types, which are *monomorphic*, i.e. for which up to isomorphic only one model exists. Equivalently we might assume to take always initial (or terminal) models of the primitive subtype (cf. /Broy, Wirsing 80b/).

As the only nonprimitive sort we specify the sort sta of nondeterministic statements with the *constructor functions:*

```
nop, abort    : → sta,
assign        : var × exp → sta,
if            : bexp × sta × sta → sta,
semi, choice  : sta × sta → sta,
letrec        : proc × sta → sta,
call          : proc → sta,
```

As *semantic functions* we use

```
loops : sta → {tt,ff}
elem  : sta × exp × dom → {tt,ff}
```

with the meaning

loops(S) = ff iff the execution of S cannot lead to a non-terminating computation

elem(S,e,x) = tt iff after the execution of S the expression e may be evaluated to x .

First we specify a number of semantic equalities for statements:

```
        semi(abort,S) = abort = semi(S,abort),
        semi(nop,S)   = S = semi(S,nop),
        semi(semi(S1,S2),S3) = semi(S1, semi(S2,S3)),
        choice(S1,choice(S2,S3)) = choice(choice(S1,S2),S3),
        letrec(p,S) = S[letrec(p,S)/call(p)],
(STA)   semi(if(b,S1,S2),S3) = if(b,semi(S1,S3), semi(S2,S3)),
        semi(assign(v,e), if(b,S1,S2)) = if(b[e/v], semi(assign(v,e),S1),
                                                    semi(assign(v,e),S2)),
        semi(choice(S1,S2),S3) = choice(semi(S1,S3), semi(S2,S3)),
        semi(S1,choice(S2,S3)) = choice(semi(S1,S2), semi(S1,S3)),
        if(b,choice(S1,S2),S3) = choice(if(b,S1,S3), if(b,S2,S3)),
        if(b,S1,choice(S2,S3)) = choice(if(b,S1,S2), if(b,S1,S3)),
        choice(S1,S2) = choice(S2,S1),
```

We consider the following semantic equations involving the evaluation-operations eval, loops and elem (following /Broy, Wirsing 80b/ to specify the definedness of a term t by DEFINED(t)):

loops(nop) = ff, elem(nop,e,x) = (x ~ eval(e)),

DEFINED(abort), DEFINED(letrec(p,S)), DEFINED(if(B,S1,S2)),

eval(b) = tt ⇒ if(b,S1,S2) = S1,

eval(b) = ff ⇒ if(b,S1,S2) = S2,

eval(b) = error ⇒ if(b,S1,S2) = abort ,

loops(semi(S,assign(v,e1))) = loops(S),

elem(semi(S,assign(v,e1)),e2,x) = elem(S,e2[e1/v],x),

loops(semi(S,call(p))) = ff, elem(semi(S,call(p)),e,x) = (x ~ error),

DEFINED(semi(S1,S2)), DEFINED(choice(S1,S2)),

For our choice operation we require

(loops(S1) = ff ∧ loops(S2) = ff) ⇒ loops(choice(S1,S2)) = ff

elem(S1,e,x) = tt ⇒ elem(choice(S1,S2),e,x) = tt

Let us call this type CN. Every statement is defined in every model of CN whereas loops and elem may be partial functions. We indicate the undefinedness of the expression loops(S) by loops(S) = undefined (analogously for elem(S,e,x)). The theorems in /Broy, Wirsing 80a,b/ immediately give the following proposition.

Prop:

(1) The type CN is weakly sufficiently complete and every statement is defined

(2) The type CN has a reachably terminal model C with the following properties:

(a) C ⊨ loops(S) ∈ { ff,undefined }

(b) C is a minimally defined model:

- ∃ model M : M ⊨ loops(S) = undefined ⇒ C ⊨ loops(S) = undefined
- ∃ model M : M ⊨ elem(S,e,x) = undefined ⇒ C⊨ elem(S,e,x)= undefin

(c) C is a fully abstract model i.e.

C⊨ S1 = S2

iff for all b ∈ {tt,ff,undefined}, sta S, exp e, dom x:

⊢ loops(semi(S,S1)) = b ⇔ ⊢ loops(semi(S,S2)) = b

and

⊢ elem(semi(S,S1),e,x) = b ⇔ ⊢elem(semi(S,S2),e,x) = b

(3) The type CN has an initial model I_C which is minimally defined. The equality in I_C is determined by the equations STA:

I_C ⊨ S1 = S2 iff STA ⊢ S1 = S2

Therefore two statements are identical in the weakly terminal model C if they are visibly equivalent. From the "minimal definedness"-property we see that the weakly terminal models are equivalent to least fixed point semantics. The weak homomorphisms induce exactly the Egli-Milner-ordering (cf. e.g. /Nivat 80/) between semantic models:

<u>Prop.</u>

Let A, B be models of CN. If there exists a weak homomorphism from A to B then for every statement S

$$S^B \subseteq_{\text{Egli-Milner}} S^A$$

i.e. for all identifiers y and <u>dom</u> x :

$$B \models elem(S,y,x) = tt \Rightarrow A \models elem(S,y,x) = tt$$

and $$B \models loops(S) = ff \Rightarrow$$

$$(A \models loops(S)=ff \wedge (B \models elem(S,y,x) = tt \Leftrightarrow A \models elem(S,y,x) = tt)) .$$

The initial model I_C is minimally defined and I_C and C are extensionally equivalent , i.e. for all $b \in \{tt, ff, \text{undefined}\}$

$$C \models loops(S) = b \quad \text{iff} \quad I_C \models loops(S) = b$$

and $$C \models elem(S,e,x) = b \quad \text{iff} \quad I_C \models elem(S,e,x) = b$$

The equality between two statements in I_C is the strong equality: Two statements are identical in I_C if their equality is provable from the axioms STA .

The class of minimally defined models of CN coincides with the class of reachable models and forms a complete lattice w.r.t. to the usual homomorphisms as ordering relation (cf. /Wirsing, Broy 80/). The initial model I_C is initial in this class whereas the weakly terminal model is terminal. As in /Broy, Wirsing 80b/ one can define a partial order on the classes of extensionally equivalent models by

C1 $\leq$ C2 iff there exist models M1 $\in$ C1 and M2 $\in$ C2 such that
$loops^{M1}$ and $elem^{M1}$ are "less defined" than $loops^{M2}$ and $elem^{M2}$

where "less defined" reflects the usual ordering on flat domains (cf. e.g./Manna 74/). Then the minimally defined models are a minimum in this ordering. There does not exist a maximum, but every maximal class corresponds to maximal fixed point semantics (cf. figure 1).

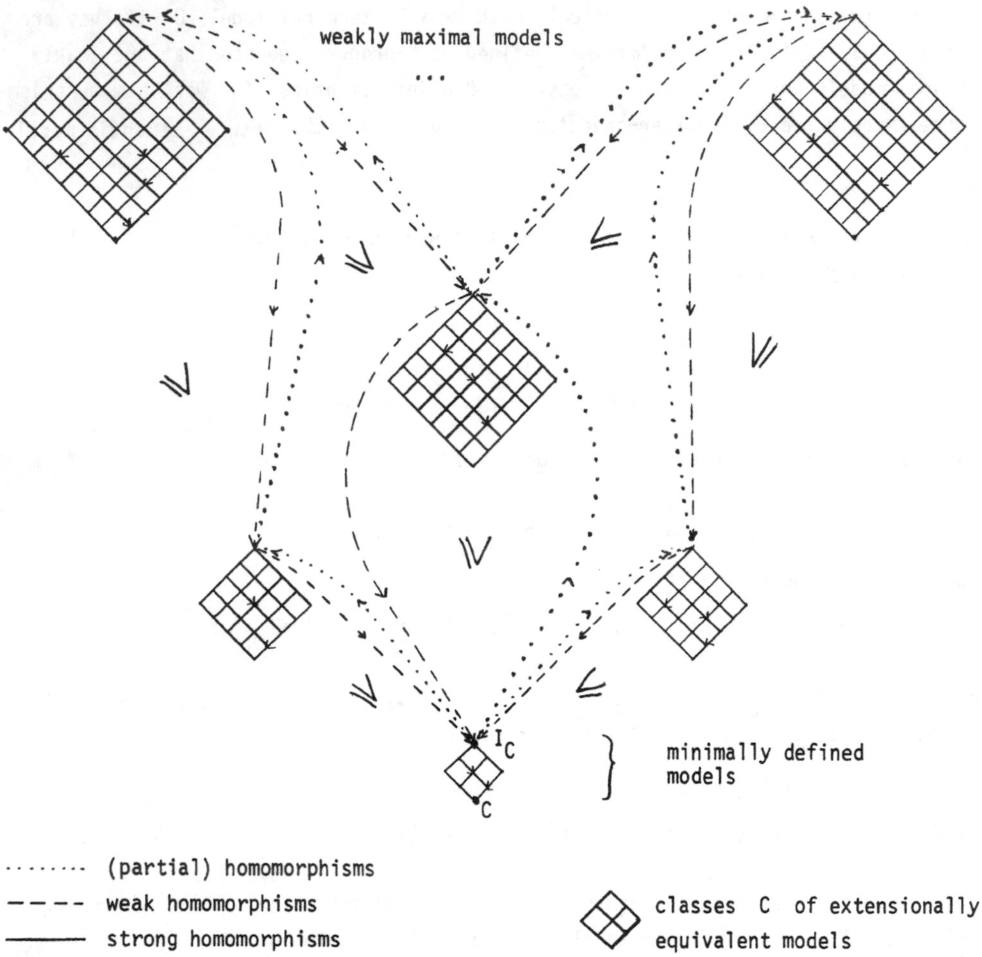

Figure 1: The structure of CN

In particular, for every minimally defined model M we have

$$M \models \text{loops}(S) \neq \text{tt},$$
$$M \models \text{loops}(S) = \text{ff} \Rightarrow \text{elem}(S,e,x) \in \{\text{tt},\text{ff}\} ,$$

and

$$M \models \text{loops}(S) = \text{undefined} \Rightarrow \text{elem}(S,e,x) \in \{\text{tt}, \text{undefined}\}.$$

According to the definition of the Egli-Milner ordering as defined for models we define for nondeterministic statements S1, S2:

$S1 \sqsubseteq_{\text{Egli-Milner}} S2$ iff

$\forall$ sta S : (loops(semi(S,S1)) = ff $\wedge$ $\forall$ exp e, dom x : elem(semi(S,S1),e,x) = elem(semi(S,S2,e,x))
 $\vee$ (loops(semi(S,S1)) $\neq$ ff $\wedge$ $\forall$ exp e, dom x : elem(semi(S,S1),e,x) = tt $\Rightarrow$
 elem(semi(S,S2),e,x) = tt)

This ordering is used to define a fixed point theory for nondeterministic programs. In minimally defined models of CN the (functionals associated with) recursive procedures are continuous wrt. to the Egli-Milner ordering (cf. /Nivat 80/). In particular this means that if elem(S,e,x) is tt for infinitely many x then loops(S) $\neq$ ff.

4. Backtrack Nondeterminism, Unbounded Nondeterminism and Loose Nondeterminism

Now we specify the further types AN, BN and LN based on the type CN.

```
type BN ≡ sort bsta ,
          bn : sta → bsta ,
          belem : bsta × exp × dom → {tt,ff} ,
          bloops: bsta → {tt,ff},

          bloops(bn(S)) = loops(S),
          belem(bn(S),e,x) = (not(loops(S)) and elem(S,e,x)),
          DEFINED(bn(S))
                                                  endoftype
```

```
type AN ≡ sort asta ,
          an : sta → asta ,
          aelem : asta × exp × dom → {tt,ff} ,
          aloops : asta → {tt,ff} ,
          loops(S) = ff ⇒ aloops(an(S)) = ff,
          aelem(an(S),e,x) = elem(S,e,x),
          loops (S1) = ff ⇒ aloops(an(choice(S1,S2))) = ff,
          DEFINED(an(S))
                                                  end of type
```

Following /Broy, Wirsing 80b/ we use a definedness predicate "DEFINED" to specify the definedness of all nondeterministic statements in the types AN, BN, and LN.

<u>type</u> LN ≡ <u>sort</u> <u>lsta</u> ,
ln : <u>sta</u> → <u>lsta</u> ,
lelem : <u>lsta</u> × <u>exp</u> × <u>dom</u> → {tt,ff} ,
lloops: <u>lsta</u> → {tt,ff},

loops(S) = ff ⇒ lloops(ln(S)) = ff,
lelem(ln(S),e,x) = tt ⇒ elem(S,e,x) = tt ,
(*) lloops(ls) = ff ⇒ ∃ <u>exp</u> e, <u>dom</u> x : lelem(ls,e,x) = tt ,
DEFINED(ln(S))

<u>endoftype</u>

Note, that we do not consider the type CN to be part of the types AN, BN and LN but as hidden. The same technique is applied e.g. in /Hennessy, Plotkin 80/. The axiom (*) must be required for LN but it holds in minimally defined models of AN, BN and CN.

The following propositions give some information about the types BN, AN and LN and their relationship to CN :

<u>Prop</u>: (1) The type BN is weakly sufficiently complete and every statement is defined.

(2) The type BN has a reachably terminal model B with the following properties

(a) B is a minimally defined, fully abstract model,
(b) for every two closed statements (i.e. statements without non-initialized variables) S1, S2: $B \models S1 = S2$ iff
$C \models$ loops(S1) = loops(S2) = undefined or $C \models S1 = S2$

(3) For every model N of CN there exists a model M of BN which is weaker than N .

(4) For every model M of BN there exists a model N of CN such that M is weaker than N.

(5) The type BN has an initial model I_B the restriction $I_B|_{STA}$ of which (to the constructor functions of statements) is isomorphic to the restriction $I|_{STA}$ of the initial model of CN

Therefore the equality between statements is the same in the initial models of BN and CN, whereas according to (2) the weakly terminal model B of BN is properly weaker than the weakly terminal model C of CN. The "natural" weak homomorphism $\varphi : N \to M$ (for models N of CN and M of BN) which is defined by

$$\varphi(S^N) =_{def} bn(S)^M, \quad \varphi(loops^N) =_{def} bloops^M \text{ and } \varphi(elem^N) =_{def} belem^M$$

is a surjective functor from CN onto BN.

Prop.: (1) The type AN is weakly sufficiently complete and every statement is defined.

(2) The type AN has a reachably terminal model A with the following properties

(a) A is minimally defined and fully abstract;

(b) for every two closed statements (i.e. statements without noninitialized variables) S1, S2 :

A ⊨ S1 = S2 iff C ⊨ choice(S1, letrec(p,call(p))) = choice(S2, letrec(p,call(p)))

(3) For every model N of CN there exists a model M of AN, such there is a partial homomorphism from N to M.

(4) For every model M of AN there exists a model N of CN such that there is a partial homomorphism from N to M.

(5) The type AN has an initial model I_A the restriction $I_A|_{STA}$ of which (to the constructor functions of statements) is isomorphic to the restriction $I|_{STA}$ of the initial model of CN.

Example: Let us consider the term S1 :
letrec(p, choice(nop, call(p)))

the term S2:
letrec(p,call(p))

and the term S3:
letrec(p, nop).

Then we have

C ⊨ loops(S1) = undefined,
C ⊨ loops(S2) = undefined,
C ⊨ loops(S3) = ff,
C ⊨ elem(S1,e,x) = (x ~ eval(e)),
C ⊨ elem(S2,e,x) = undefined,
C ⊨ elem(S3,e,x) = (x ~ eval(e)),

B ⊨ bloops(bn(S1)) = undefined,
B ⊨ bloops(bn(S2)) = undefined,
B ⊨ bloops(bn(S2)) = ff,
B ⊨ belem(bn(S1),e,x) = undefined,
B ⊨ belem(bn(S2),e,x) = undefined,
B ⊨ belem(bn(S3),e,x) = (x ~ eval(e)),

$A \models aloops(an(S1)) = ff,$

$A \models aloops(an(S2)) = undefined,$

$A \models aloops(an(S3)) = ff,$

$A \models elem(S1,e,x) = (x \sim eval(e)),$

$A \models elem(S2,e,x) = undefined,$

$A \models elem(S3,e,x) = (x \sim eval(e))$

According to this we have :

- S1, S2 and S3 are not visibly equivalent in C ,
- S1 and S2 are visibly equivalent in B ,
- S1 and S3 are visibly equivalent in A .

<u>end of example</u>

According to the axioms of BN we have for all nondeterministic statements S (we suppose that B and C have the same primitive models)

$$loops(S)^C = bloops(bn(S))^B ,$$

$$loops(S)^C = ff \Rightarrow elem(S,e,x) = belem(bn(S),e,c)$$

$$loops(S)^C = undefined \Rightarrow belem(S,e,x) = undefined$$

In particular wie have

$$belem(bn(S),e,x) \sqsubseteq elem(S,x,x)$$

where "$\sqsubseteq$" denotes Manna's "is less defined"-partial order (cf. /Manna 74/).

The reachably terminal models A and B of AN and BN resp. are incomparable.

The type LN , however, does not have initial nor weakly terminal models (cf. Fig.2):

Prop: (1) The type LN is not weakly sufficiently complete

(2) The type LN does not have any weakly terminal model nor any initial model

(3) For every model N of CN as well as of BN and AN there is a model M of LN which is isomorphic to A

For studying the relations between the models of LN we introduce the *implementation ordering* $\subseteq_I$ (cf. /Broy, Gnatz, Wirsing 78/):

$$L1 \subseteq_I L2 \quad \Leftrightarrow_{def}$$

for all <u>lsta</u> ls, <u>exp</u> e, <u>dom</u> x :

$$L2 \models \text{lloops(ls)} = \text{ff} \Rightarrow L1 \models \text{lloops(ls)} = \text{ff}$$
$$L1 \models \text{lelem(ls,e,x)} = \text{tt} \Rightarrow L2 \models \text{lelem(ls,e,x)} = \text{ff}$$

Then there does not exist unique minimum for $\subseteq_I$ in LN. But every $\subseteq_I$-minimal model L is a possible *deterministic mathematical semantics* for LN, i.e.

$$L \models (\text{lelem}(S,e,x) = \text{tt} \wedge x \neq y) \Rightarrow (\text{lloop}(S) = \text{ff} \wedge \text{lelem}(S,e,y) \neq \text{tt})$$

Furthermore the weakly terminal model C of CN is *optimal* in the following sense: C is the weakest model of LN which is $\subseteq_I$-greater than all minimal models of LN; or equivalently C is the weakest model of LN which is $\subseteq_I$ - greater than all $\subseteq_I$-minimal models of LN (cf. figure 2).

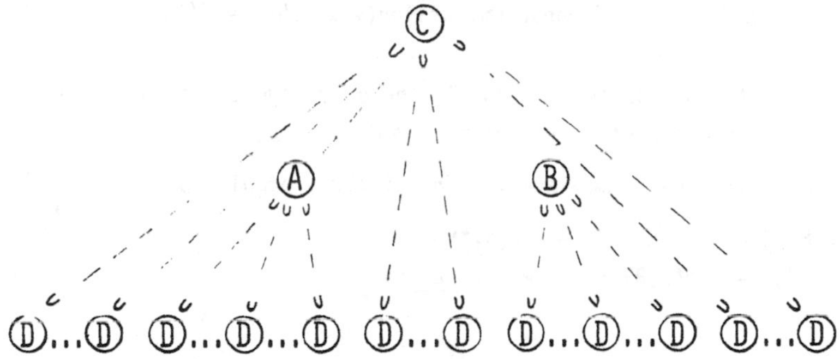

Figure 2: The implementation ordering $\subseteq_I$
A,B,C denote the classes of reachably terminal models of AN, BN and CN resp.
D denotes classes of extensionally equivalent, deterministic semantic models of LN.

Note: (1) In a deterministic mathematical semantics L for LN, i.e. in a $\subseteq_I$-minimal model L of type LN , we may introduce a partial function

$$\text{value} : \underline{\text{sta}} \times \underline{\text{exp}} \rightarrow \underline{\text{dom}}$$

such that

$$\text{value}(S,e) = x \quad \text{iff} \quad L \models \text{lelem}(\text{ln}(S),\ e,x) = \text{tt}$$

(2) The meaning of the nondeterministic language of guarded commands in /Dijkstra 76/, which is very similar to our language, is defined by the predicate transformers of the wp-calculus. If we define the predicate calculus as primitive subtype with the sort predicate, the semantic function (cf. also "dynamic logic" in /Harel, Pratt 78/):

$$\text{wp} : \underline{\text{sta}} \times \underline{\text{predicate}} \rightarrow \underline{\text{predicate}}$$

defines backtrack nondeterminism (as weakly terminal model). An appropriate definition of the wlp-predicate transformers (cf. /Broy et al. 80/), however, leads to angelic nondeterminism, while the considering wp/wlp together gives choice nondeterminism.

This remark becomes obvious, if we define for our language:

$$\text{wlp}(S,R) = \lambda\, x\ .\ \forall\ \underline{\text{dom}}\ y : (\text{elem}(\text{semi}(\text{assign}(v,x),S),v,y) = \text{tt} \Rightarrow R(y))$$
$$\text{wp}(S,R) = \text{wlp}(S,R) \wedge \lambda\, x.\ (\text{loops}(\text{semi}(\text{assign}(v,x),\ S)) = \text{ff})$$

where, for simplicity, we assume that v is the only ("generalized") program variable in the nondeterministic statement S and $x \in \underline{\text{dom}}$.

The axioms of BN and AN immediately give (cf./Broy et al. 80/):

$$\text{BN} \models \text{wlp}(S,R) = \text{wp}(S,R) \vee \neg\ \text{wp}(S,\underline{\text{true}})$$
$$\text{AN} \models \text{wp}(S,R) = \text{wlp}(S,R) \wedge \neg\ \text{wlp}(S,\underline{\text{false}})$$

end of note

We like to consider the minimally defined models of the types AN, BN, and CN resp. , i.e. models which are extensionally equivalent to the reachably terminal models, as *tight* semantic models, whereas the models of LN (especially the deterministic ones) which are less than these models in the implementation ordering may be considered as *loose* semantic models for these types.

5. Concluding Remarks

The four different types properly reflect the four different notions of nondeterminism:

Backtrack nondeterminism assumes the computation of the "whole set of possible values". If there is a possibility of nontermination then this nontermination must happen. Thus backtrack nondeterminism is nothing but an implicit notation for programs working with sets. The choice is made *after* the computation of the set between the possible *semantic* values.

Choice nondeterminism represents a particular abstraction of a couple of decisions deliberately left open to the executing instance. Thus it corresponds to choices *during* the course of execution between alternative *statements* (i.e. the executing instance has the option of choice which statement to execute).

Unbounded nondeterminism corresponds to a "prophetic" choice during evaluation, avoiding nonterminating branches. Obviously we cannot give an operational semantics such that all possible values can be results, but nonterminating branches are excluded. This is reflected by the fact that unbounded nondeterminism is not continuous in the Egli-Milner ordering (cf. /Apt, Plotkin 81/). Nevertheless we may give approximations for operational semantics, i.e. models of type LN which are weaker than the partial initial model of AN.

Loose nondeterminism represents a convienient notation for treating a couple of possible semantic models in one specification. Thus it corresponds to choices *before* the execution of the program (or more understandable to choices of particular implementations, i.e. between *semantic models* , of a language). This comprises the choice of particular scheduling strategies in a compiler or operating system.

All four notions of nondeterminism have their justification in different areas of applications. *Angelic nondeterminism* is the notion used in automata theory. *Backtrack nondeterminism* can be used as a convenient notation for certain search problems (cf. /Floyd 67/). *Choice nondeterminism* serves as a formal basis for modelling concurrent processes (cf. /Broy 80/). Furthermore it can be used as a design tool for representing "program families", for expressing "delayed design decisions" (cf. /Bauer, Wössner 81/) or for explicit formulation of backtrack algorithms (cf. /Broy, Wirsing 80c/).

Of course, there are still other notions of nondeterminism. If we want to accept only specific objects (or situations) as possible results of computations this leads to a mixture of choice and backtrack nondeterminism assuming backtracking only in the specific situations ("exceptions").

References

/Apt, Plotkin 81/
K.R. Apt, G.D. Plotkin: A Cook's Tour of Countable Nondeterminism. Submitted for publication

/Bauer, Wössner 81/
F.L. Bauer, H. Wössner: Algorithmische Sprache und Programmentwicklung. Berlin-Heidelberg-New York: Springer 1981, to appear

/Broy 80/
M. Broy: Transformational Semantics for Concurrent Programs. IPL 11:2, October 1980, 87-91

/Broy, Gnatz, Wirsing 78/
M. Broy, R. Gnatz, M. Wirsing: Semantics of Nondeterministic and Noncontinuous Constructs. In: F.L. Bauer, M. Broy (eds.): Program Construction, Marktoberdorf 78. LNCS 69

/Broy, Wirsing 80a/
M. Broy, M. Wirsing: Programming Languages as Abstract Data Types. In: M. Dauchet (ed.): Lille Colloque 80

/Broy, Wirsing 80b/
M. Broy, M. Wirsing: Initial Versus Terminal Algebra Semantics for Partially Defined Abstract Types. Techn. Universität München, Institut für Informatik, TUM-I 8018, Dezember 1980

/Broy, Wirsing 80c/
M. Broy, M. Wirsing: From Enumeration to Backtracking. IPL 10:4, July 1980, 193-197

/Broy et al. 80/
M. Broy, H. Partsch, P. Pepper, M. Wirsing: Semantic Relations in Programming Languages, IFIP Congress 80

/Dijkstra 76/
E.W. Dijkstra: A Discipline of Programming. Prentice Hall, Englewood Cliffs 1976

/Floyd 67/
R.M. Floyd: Nondeterministic Algorithms. J. ACM 14, 1967, 636-644

/Grätzer 68/
G. Grätzer: Universal Algebra. Princeton: Van Nostrand 1968

/Harel, Pratt 78/
D. Harel, V.R. Pratt: Nondeterminism in Logics of Programs. Proc. 5th ACM Symp. on Principles of Programming Languages. Jan. 1978, 203-213

/Hennessy, Plotkin 80/
M.C.B. Hennessy, G.D. Plotkin: A Term Model of CCS. In: P. Dembinski(ed.): MFCS 80. LNCS 88, 262-274

/Kennaway, Hoare 80/
J.R. K. Kennaway, C.A.R. Hoare: A Theory of Nondeterminism. In: J. de Bakker, J.v.d. Leuwen (eds.): ICALP 80, LNCS 85

/Manna 74/
Z. Manna: Mathematical Theory of Computation. New York: McGraw Hill 1974

/McCarthy 63/
J. McCarthy: A Basis for a Theory of Computation. In: B. Bradfort, D. Hirschberg (eds.): Computer Programming and Formal Systems. Amsterdam: North-Holland 1963, 33-70

/Milner 77/
R. Milner: Fully Abstract Models of Typed λ-calculi. TCS 4, 1977, 1-22

/Nivat 80/
M. Nivat: Nondeterministic Programs: An Algebraic Overview. Invited paper, IFIP Congress 80

/Park 80/
D. Park: On the Semantics of Fair Parallelism. In: D. Björner (ed.): Abstract Software Specification. LNCS 86, 504-526

/Wirsing, Broy 80/
M. Wirsing, M. Broy: Abstract Data Types as Lattices of Finitely Generated Models. In: P. Dembinski (ed.): MFCS 80. LNCS 88

APPLIED TREE ENUMERATIONS

Nachum Dershowitz*
Department of Computer Science
University of Illinois
Urbana, Illinois 61801
U.S.A.

Shmuel Zaks
Department of Computer Science
Technion
Haifa 32000
Israel

I. INTRODUCTION

In this paper we consider the class T_n of unlabelled ordered (plane-planted) trees with n edges and give combinatorial proofs to several enumeration formulae concerning T_n. In particular, closed-form expressions are given for (1) the number of trees in T_n with n_0 leaves, n_1 unary nodes, ..., n_d nodes with d children, and no restrictions on nodes with more than d children, and for (2) the number of nodes in T_n on level ℓ with d children. Several statistical results are derived from these.

The combinatorial tools we use to prove our results include one-to-one correspondences between ordered trees and other combinatorial objects, the Cycle Lemma, and lattice-path techniques. Many of these results could, alternatively, have been obtained using generating functions and the Lagrange inversion formula.

We demonstrate the use of these enumerations in analyzing the following applications: (1) a sorting problem, (2) the average height of a stack during tree-traversal, (3) algorithms for threaded binary trees, and (4) a pattern-matching problem.

II. CORRESPONDENCES

We consider ordered trees (see Knuth [1968] for definitions). Each node has a degree (the number of its children). A node of degree 0 is termed a leaf; otherwise it is called an internal node. The level of a node is its distance from the root (the root is on level 0). The number of trees in the set T_n of ordered trees with n edges is the well-known Catalan number

----

*Research supported in part by the National Science Foundation under Grant MCS 79-04897.

$$|T_n| = C_n = \frac{1}{n+1}\binom{2n}{n}$$

(see, for example, Gardner [1976]).

There are numerous one-to-one correspondences between elements of these sets of ordered trees and other combinatorial objects (see, for example, Kuchinski [1977]). Among them, the correspondences between the following sets help in our enumerations:

T_n: the set of ordered trees with n edges.

B_n: the set of binary trees with n internal nodes, each having exactly two children.

P_n: the set of sequences of n open parentheses and n close parentheses, where each open parenthesis has a matching close parenthesis.

I_n: the set of sequences $a_0a_1\cdots a_n$ of n+1 nonnegative integers summing to n, such that $\sum_{j=0}^{i} a_j > i$ for $i=0,1,\ldots,n-1$.

L_n: the set of shortest lattice-paths from (0,0) to (n,n) that do not go below the diagonal y=x (all steps are either up or to the right).

The correspondences between these five sets are illustrated in Figure 1. In general, from a tree t in T_n one gets a sequence p(t) in P_n by traversing t in preorder, writing "(" for each edge passed on the way down and ")" for each edge passed on the way up. (See Figure 1.2.) From p(t) one gets a lattice path $\ell(t)$ in L_n by starting at (0,0) and going up one coordinate for each open parenthesis and going right one coordinate for each close parenthesis. (See Figure 1.3.) From t one gets a sequence i(t) in I_n by reading the degrees of all the nodes of t in preorder. (See Figure 1.4.) From $\ell(t)$ a binary tree b(t) in B_n is built in preorder, each step up on the path corresponding to an internal node and each step to the right corresponding to a leaf (a final leaf is also added). (See Figure 1.5.)

A sequence of open and close parentheses is called <u>legal</u> if in each prefix the number of open parentheses is greater than the number of close parentheses. We use the following lemma:

<u>Cycle Lemma</u> (Dvoretzky and Motzkin [1947]): For any sequence $p_1p_2\cdots p_{m+n}$ of m open and n close parentheses, m>n, there are exactly m-n cyclic permutations

$$p_jp_{j+1}\cdots p_{m+n}p_1\cdots p_{j-1}$$

that are legal.

t = 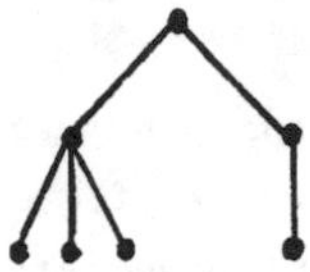

1.1 An ordered tree $t \in T_6$

$p(t) = (()()())(())$

1.2 A sequence $p(t) \in P_6$

$i(t) = 2300010$

1.4 A sequence $i(t) \in I_6$

$\ell(t) =$ 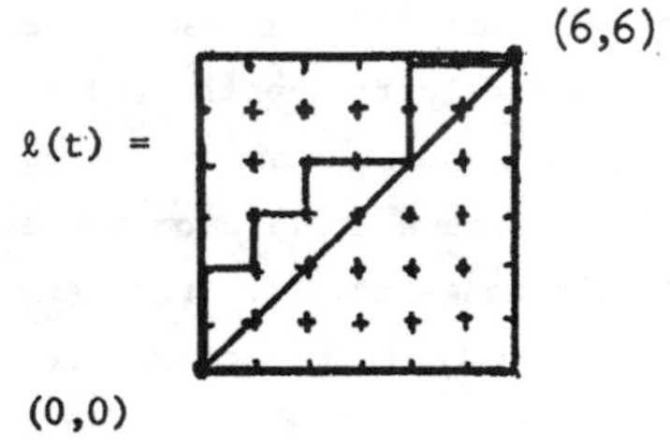

1.3 A lattice path $\ell(t) \in L_6$

$b(t) =$ 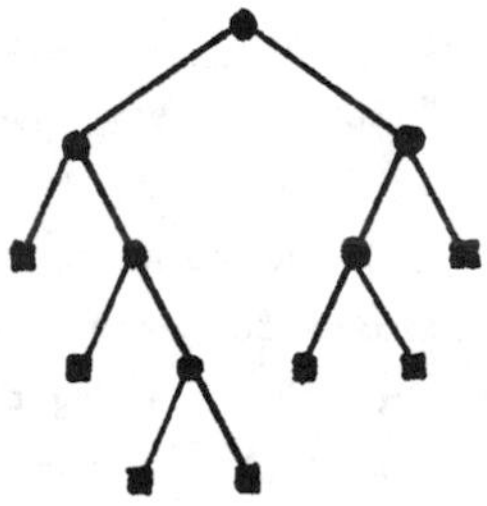

1.5 A binary tree $b(t) \in B_6$

Figure 1. Correspondences between T_n, B_n, P_n, I_n, and L_n.

From this lemma it follows that there is a one-to-one correspondence between ordered trees in T_n and cycles with n+1 open parentheses and n close parentheses, there being only one legal permutation of p(t) with an extra open parenthesis prepended. Furthermore, each such cycle of parentheses corresponds to a cycle of n+1 nonnegative integers summing to n (representing the number of close parentheses between pairs of open parentheses). These correspondences are the basis for our use of the Cycle Lemma in the enumerations.

III. ENUMERATIONS

In this section we present the main enumeration results and discuss some of their consequences. All trees are in T_n.

Theorem 1: The number $L_n(n_0,n_1,\ldots,n_d)$ of trees in T_n with n_j nodes of degree j, $0\leq j\leq d$, and no restrictions on nodes of degree greater than d is

$$\frac{1}{n+1}\binom{n-e-d(n-m+1)-1}{n-m}\binom{n+1}{n_0,n_1,\ldots,n_d,n-m+1},$$

where $m=\Sigma n_j$ (the total number of restricted nodes) and $e=\Sigma jn_j$ (the total number of edges accounted for).

This generalizes a result of Narayana [1959] for d=0 (proved in Dershowitz and Zaks [1980] using the Cycle Lemma) and the multinomial formula for the case d=n (Erdelyi and Etherington [1940], proved in Raney [1960] using the Cycle Lemma).

Proof: By the Cycle Lemma we have to count the cycles with n+1 nonnegative integers summing to n, of which n_i are i, $0\leq i\leq d$. To count the number of cycles with these restrictions, note that there are

$$\frac{1}{n+1}\binom{n+1}{n_0,\ldots,n_d,n+1-m}$$

ways of placing the degrees of the restricted nodes on a cycle with n+1 positions. The remaining n+1-m unrestricted nodes must have degrees ranging from d+1 to n and summing to n-e. The number of ways to place these degrees is the same as the number of ways of decomposing the integer n-e-(d+1)(n+1-m) into n+1-m integers ranging from 0 to n-(d+1), which is

$$\binom{n-e-d(n+1-m)-1}{n-m}. \qquad []$$

From this theorem, we can derive the following

Consequences:

1.1) The number $L_n(k)$ of trees with k leaves is

$$L_n(k) = \frac{1}{n+1}\binom{n-1}{n-k}\binom{n+1}{k} = \frac{1}{k}\binom{n-1}{k-1}\binom{n}{k-1}.$$

This is Narayana's [1959] result (see also Mohanty [1979]).

1.2) The number $L_n(k)$ of trees with k leaves is equal to the number $L_n(n+1-k)$ of trees with n+1-k leaves.

1.3) The expected number of leaves (or internal nodes) is $\frac{n+1}{2}$. (Here, and in the sequel, all trees in T_n are assumed equiprobable). This result has also been given by Dasarathy and Yang [1980].

1.4) The expected degree of an internal node is

$$\frac{n\,C_n}{\frac{n+1}{2}C_n} = \frac{2n}{n+1} \approx 2.$$

Theorem 2: The total number $N_n(\ell,d)$ of nodes in T_n of degree d on level ℓ is

$$N_n(\ell,d) = \binom{2n-d-1}{n+\ell-1} - \binom{2n-d-1}{n+\ell} = \frac{2\ell+d}{2n-d}\binom{2n-d}{n+\ell}.$$

This result was first proved in Dershowitz and Zaks [1980]. We give here a lattice-path proof.

Proof: The number of lattice paths from (0,0) to $(n-d-\ell, n+\ell-1)$ that do not go below the diagonal y=x is

$$\frac{2\ell+d}{2n-d}\binom{2n-d}{n+\ell}$$

(see, for example, Mohanty [1979]). We give a correspondence between each such path and each node in T_n of degree d on level ℓ. (The correspondence applies to the case $\ell>1$; for $\ell=0$ a simpler correspondence works.)

Consider a tree t in T_n and the corresponding lattice path $\ell(t)$ from (0,0) to (n,n). A node x of degree d on level ℓ of t corresponds to a path segment

$$(i_0,i_0+\ell-1)\to(i_0,i_0+\ell)\to\cdots\to(i_1-1,i_1+\ell)\to(i_1,i_1+\ell)\to\cdots$$
$$\to(i_j-1,i_j+\ell)\to(i_j,i_j+\ell)\to\cdots\to(i_d,i_d+\ell)\to(i_d+1,i_d+\ell) \tag{a}$$

that does not go below the diagonal $y=x+\ell$ (except at the two ends). This segment is preceded by another segment

$$(0,0)\to\cdots\to(i_0,i_0+\ell-1) \tag{b}$$

and is followed by a segment

$$(i_d+1,i_d+\ell)\to\cdots\to(n,n). \tag{c}$$

To get the desired lattice path from $(0,0)$ to $(n-d-\ell,n+\ell-1)$, we do the following (see Figure 2):

(a) The lattice path steps $(i_j-1,i_j+\ell)\to(i_j,i_j+\ell)$, $1\leq j\leq d$, as well as $(i_0,i_0+\ell-1)\to(i_0,i_0+\ell)$ and $(i_d,i_d+\ell)\to(i_d+1,i_d+\ell)$, are removed from segment (a), yielding a path segment from $(i_0,i_0+\ell-1)$ to $(i_d-d,i_d+\ell-1)$.

(b) Segment (b) is left intact.

(c) Segment (c) is inverted by reversing both the order and the direction of its steps, i.e. every step right becomes a step down and every step up becomes a step left, yielding a path segment that ends at $(n-d-\ell,n+\ell-1)$ and begins at $(i_d-d,i_d+\ell)$. The missing step $(i_d-d,i_d+\ell-1)\to(i_d-d,i_d+\ell)$ is added.

Since this correspondence between nodes and paths is one-to-one, the desired result is proved. (To get the corresponding node, given a path, note that ℓ and d are determined by the endpoint $(n-d-\ell,n+\ell-1)$ and that for all j, $0\leq j\leq d$, the path segment from $(i_j-j,i_j+\ell)$ to the endpoint does not return below the diagonal $y=x+\ell+j$. Both the deleted steps and the inverted path segment can be easily restored.) []

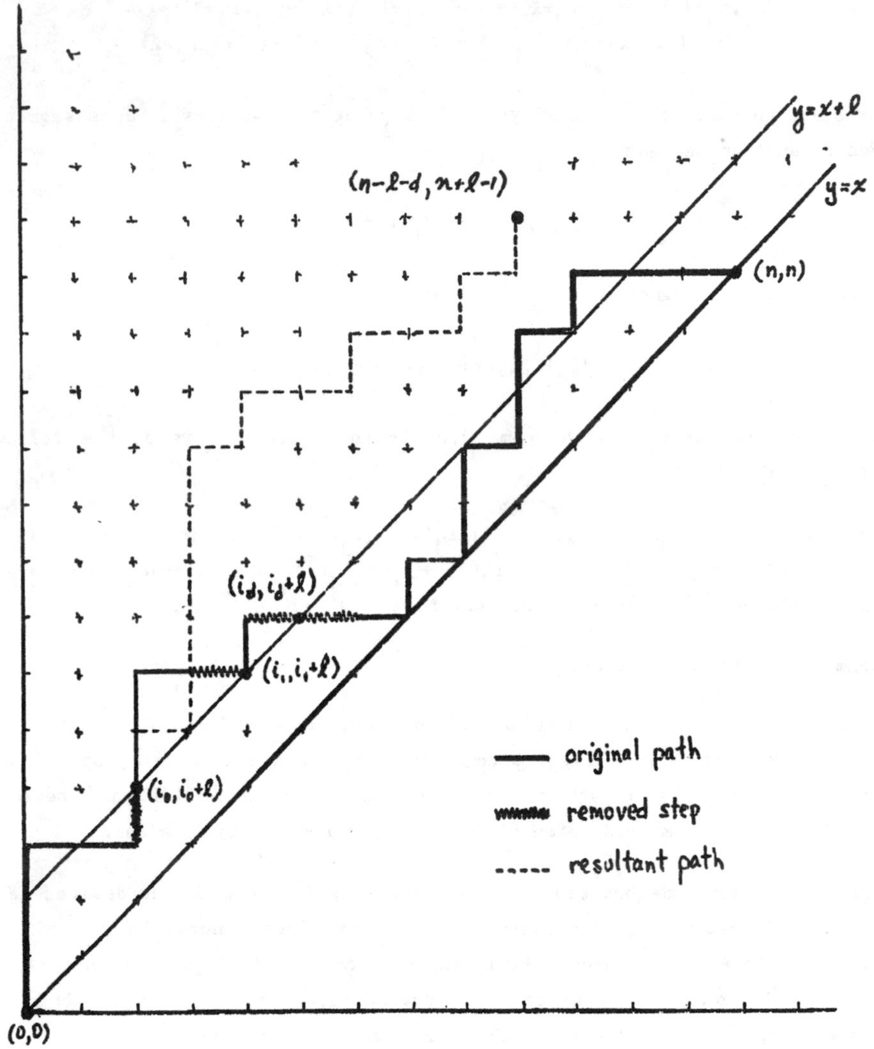

Figure 2. Proof of Theorem 2.

From this theorem, we can derive the following

Consequences:

2.1) The total number of nodes of degrees i through j on level ℓ is

$$\sum_{d=i}^{j} N_n(\ell,d) = \sum_{d=i}^{j} \frac{2\ell+d}{2n-d}\binom{2n-d}{n+\ell} = \frac{2\ell+i+1}{2n-i+1}\binom{2n-i+1}{n+\ell+1} - \frac{2\ell+j+2}{2n-j}\binom{2n-j}{n+\ell+1}.$$

In particular, the total number of nodes on level ℓ is

$$\frac{2\ell+1}{2n+1}\binom{2n+1}{n-\ell}.$$

2.2) The expected number of leaves on level ℓ of a tree in T_n is

$$\frac{N_n(\ell,0)}{C_n} = \frac{\frac{\ell}{n}\binom{2n}{n+\ell}}{\frac{1}{n+1}\binom{2n}{n}} = \frac{\ell\binom{2n}{n-\ell}}{\binom{2n}{n-1}} < \ell.$$

For small ℓ,

$$\frac{N_n(\ell,0)}{C_n} \approx \ell.$$

2.3) The expected level of a leaf is

$$\frac{\sum_{\ell} \ell N_n(\ell,0)}{\frac{n+1}{2} C_n} = \frac{2^{2n-1}}{\binom{2n}{n}} \approx \frac{\sqrt{\pi n}}{2}.$$

(This sum and subsequent ones may be evaluated using the identities in, for example, Riordan [1968].) The <u>external path length</u> of a tree (as defined in Knuth [1968]) is the sum of the levels of its leaves. Thus, the expected external path length is

$$\frac{2^{2n-1}}{\binom{2n}{n}}\frac{n+1}{2} \approx \frac{\sqrt{\pi n}(n+1)}{4}.$$

2.4) The expected number of internal nodes on level ℓ is

$$\frac{\sum_{d=1}^{n} N_n(\ell,d)}{C_n} = \frac{(\ell+1)\binom{2n}{n-\ell-1}}{\binom{2n}{n-1}} < \ell+1.$$

For small ℓ,

$$\frac{\sum_{d=1}^{n} N_n(\ell,d)}{C_n} \approx \ell+1.$$

2.5) The expected level of an internal node is

$$\frac{\sum_{\ell} \ell \frac{\ell+1}{n}\binom{2n}{n-\ell-1}}{\frac{n+1}{2}C_n} = \frac{2^{2n-1}}{\binom{2n}{n}} - 1 \approx \frac{\sqrt{\pi n}}{2} - 1.$$

The <u>internal path length</u> of a tree is the sum of the levels of its internal nodes. Thus, the expected internal path length of a tree is

$$\left[\frac{2^{2n-1}}{\binom{2n}{n}} - 1\right]\frac{n+1}{2} \approx \frac{\sqrt{\pi n}(n+1)}{4} - \frac{n+1}{2}.$$

2.6) The expected level of a node is

$$\frac{\sum_{\ell} \ell \frac{2\ell+1}{2n+1}\binom{2n+1}{n-\ell}}{\binom{2n}{n}} = \frac{2^{2n-1}}{\binom{2n}{n}} - \frac{1}{2} \approx \frac{\sqrt{\pi n}}{2} - \frac{1}{2}.$$

This result has been given by Volosin [1974], Meir and Moon [1978], and Dasarathy and Yang [1980]. Higher moments of the node level can be calculated in the same way.

2.7) The total number of nodes of degree d on levels i through j is

$$\sum_{\ell=i}^{j} N_n(\ell,d) = \sum_{\ell=i}^{j}\left[\binom{2n-d-1}{n+\ell-1} - \binom{2n-d-1}{n+\ell}\right] = \binom{2n-d-1}{n+i-1} - \binom{2n-d-1}{n+j}.$$

In particular, the total number $D_n(d)$ of nodes of degree d is

$$D_n(d) = \binom{2n-d-1}{n-1}.$$

It follows that the total number of nodes on levels i through j is

$$\binom{2n}{n-i} - \binom{2n}{n-j-1}$$

and the total number of nodes of degrees i through j is

$$\binom{2n-i}{n} - \binom{2n-j-1}{n}.$$

2.8) The expected number of nodes of degree d in a tree in T_n is

$$\frac{D_n(d)}{C_n} = \frac{\binom{2n-d-1}{n-1}}{\frac{1}{n+1}\binom{2n}{n}} < \frac{n+1}{2^d}.$$

For small d,

$$\frac{D_n(d)}{C_n} \approx \frac{n+1}{2^{d+1}}.$$

2.9) The number $R_n(r)$ of trees with root degree r is

$$R_n(r) = N_n(0,r) = \frac{r}{n}\binom{2n-r-1}{n-1}.$$

The expected root degree is

$$\frac{\sum_r r\,R_n(r)}{C_n} = \frac{\sum_r \frac{r^2}{n}\binom{2n-r-1}{n-1}}{\frac{1}{n+1}\binom{2n}{n}} = \frac{\frac{3}{n+2}\binom{2n}{n+1}}{\frac{1}{n}\binom{2n}{n-1}} = \frac{3n}{n+2} \approx 3.$$

These results are also given in Ruskey and Hu [1977]. In a similar manner higher moments can be calculated. For example, the variance of the root degree is

$$\frac{\sum_r r^2 R_n(r)}{C_n} - \left[\frac{3n}{n+2}\right]^2 = \frac{2\binom{2n}{n+3} + 3\binom{2n+1}{n+3} + \binom{2n+2}{n+3}}{\frac{n}{n+1}\binom{2n}{n}} - \left[\frac{3n}{n+2}\right]^2 \approx 1\tfrac{1}{4}.$$

IV. APPLICATIONS

In this section we illustrate how the formulae of the previous section can help in the analysis of various algorithms.

1. A sorting problem: Given an oriented tree with n edges, with a number stored at each node, we want to order the edges so that the children of each node will have increasing numbers. (The need to sort trees in this manner arose in an algorithm for computing the recursive path ordering on terms, as defined in Dershowitz [1981]). A node with d children needs no more than $d^2/3$ comparisons to sort them, hence the total number of comparisons is bounded by $\Sigma i^2 n_i/3$ for a tree with n_i nodes of degree i. To find the average $\bar{s}$ of this function over all trees in T_n, we make use of the closed form for the number $D_n(d)$ of nodes of degree d in T_n (Consequence 2.7):

$$\bar{s} = \frac{\frac{1}{3}\sum_d d^2 D_n(d)}{C_n} = \frac{\frac{1}{3}\sum_d d^2 \binom{2n-d-1}{n-1}}{\frac{1}{n+1}\binom{2n}{n}} = \frac{n^2}{n+2}.$$

Thus, on the average, less than one comparison per node is required. Considering that sorting d children actually requires only on the order of $d\cdot\log_2 d$ comparisons, a slightly better result should be possible.

2. Average height of a stack: The height of an ordered tree is the worst stack size that is formed while traversing the tree in, say, preorder. deBruijn, Knuth, and Rice [1972] and others have shown that the expected value of this height, over all trees in T_n, is about $\sqrt{\pi n}$. If, instead of studying the worst stack size, we look at the average while traversing the tree, we need the expected level of a node, which we have already seen (Consequence 2.6) is

$$\frac{2^{2n-1}}{\binom{2n}{n}} - \frac{1}{2} \approx \frac{\sqrt{\pi n}}{2} - \frac{1}{2}.$$

If the average is weighted by the number of times during the traversal that the stack is of that height, i.e. by the degree of the node plus one, then we get

$$\frac{\sum_{\ell,d} (d+1)\,\ell\, N_n(\ell,d)}{(2n+1)\,C_n} = \frac{2^{2n} - \binom{2n+1}{n}}{\binom{2n+1}{n}} \approx \frac{\sqrt{\pi n}}{2} - 1.$$

3. Threaded binary trees: In a recent paper by Brinck and Foo [1981], algorithms on threaded binary trees are investigated. The analyses are based on certain enumeration lemmas, proved using recurrence relations. Our enumeration techniques enable one to prove those lemmas using a direct combinatorial approach, which also sheds light on the structure of these trees.

For example, the expected number of left leaves (i.e. backward threads) in a binary tree in B_n, and the expected number of right leaves (i.e. forward threads), are equal to the expected number of leaves in an ordered tree (by the correspondence between T_n and B_n), which we have seen (Consequence 1.3) is $\frac{n+1}{2}$. Also, the expected distance from the root of a binary tree to the leftmost (or rightmost) leaf is equal to the expected root degree of an ordered tree (letting each digit in a sequence in I_n correspond to the number of internal leaves that precede a leaf in the preorder construction of a tree in B_n), which we have seen (Consequence 2.9) is $\frac{3n}{n+2}$.

4. Pattern matching: Flajolet and Steyaert [1980] have recently investigated tree-matching algorithms. Using our techniques one can deal with similar problems. For example, a pattern is an ordered tree some of whose leaves are designated open. A pattern p is said to occur in a tree x if x contains a subtree of the same form as p with arbitrary trees substituted for the open nodes. In order to know how many times a certain pattern p -- with e edges and d open leaves -- occurs as a subtree in some tree in T_n, we simply replace p by a node of degree d, and ask how many nodes of degree d are in T_{n-e+d}; this number is given (Consequence 2.7) by

$$D_{n-e+d}(d) = \binom{2n-2e+d-1}{n-e}.$$

Similarly, the number of occurrences of p on level ℓ is given by

$$N_{n-e+d}(\ell,d) = \frac{2\ell+d}{2n-2e+d}\binom{2n-2e+d}{n-e-\ell}.$$

More generally, the number of occurrences of the combination of m distinct nonoverlapping patterns $p_1,p_2,\cdots,p_m$ within T_n is

$$(n-e+d)^{\underline{m-1}} D_{n-e+d}(d),$$

where e is the total number of edges in the patterns, d is the total number of open leaves, and $x^{\underline{m}}=x(x-1)\cdots(x-m+1)$. To prove this, note that the number of occurrences of the patterns p_i in trees with n nodes is equal to the number of trees with m labelled nodes having degrees $d_1,\ldots,d_m$ (where d_i is the number of open leaves in p_i) and n-e+d edges. Each such tree corresponds (by the Cycle Lemma) to a cycle of n-e+d+1 degrees -- m of them specified and labelled -- summing to n-e+d. There are

$$(n-e+d)^{\underline{m-1}}$$

ways of placing the labelled nodes on the cycle and there are

$$D_{n-e+d}(d)$$

ways of specifying the degrees of the remaining nodes.

In a similar manner it can be shown that

$$(n-e+d)^{\underline{m-1}}\binom{2n-2e+d}{n-e}$$

is the number of occurrences of m patterns in the set B_n of binary trees, in which case e denotes the total number of internal nodes in the patterns. For example, the number of left nodes in B_n that have forward threads (i.e. the number of internal nodes that are left children and have right leaves) is

$$(n-2+2)^{\underline{1-1}}\binom{2n-4+2}{n-2} = \binom{2n-2}{n}$$

(cf. Brinck and Foo [1981]).

REFERENCES

[1] K. Brinck and N. Y. Foo [1981], Analysis of algorithms on threaded trees, to appear in Comp. J.

[2] N. deBruijn, D. E. Knuth, and O. Rice [1972], The average height of planted plane trees, in Graph Theory and Computing (R. C. Read, ed.), Academic Press, New York, 15-22.

[3] B. Dasarathy and C. Yang [1980], A transformation on ordered trees, Comp. J. 23 (2), 161-164.

[4] N. Dershowitz [1981], Orderings for term-rewriting systems, to appear in J. Theoretical Computer Science.

[5] N. Dershowitz and S. Zaks [1980], Enumerations of ordered trees, Discrete Math. 31 (1), 9-28.

[6] A. Dvoretzky and Th. Motzkin [1947], A problem of arrangements, Duke Math. J. 14, 305-313.

[7] A. Erdelyi and I. M. H. Etherington [1940], Some problems of non-associative combinations (2), Edin. Math. Notes 32, 7-12.

[8] Ph. Flajolet and J. M. Steyaert [1980], On the analysis of tree-matching algorithms, Proc. 7th Intl. Conf. Automata, Languages and Programming, Amsterdam, 208-220.

[9] M. Gardner [1976], Mathematical games: Catalan numbers, Scientific American 234 (6), 120-125.

[10] D. E. Knuth [1968], The Art of Computer Programming, Vol. 1: Fundamental algorithms, Addison-Wesley, Reading, MA.

[11] M. J. Kuchinski [1977], Catalan structures and correspondences, M.S. Thesis, Dept. of Mathematics, West Virginia Univ., Morgantown, WV.

[12] A. Meir and J. Moon [1978], On the altitude of nodes in random trees, Can. J. Math. 30 (5), 997-1015.

[13] S. G. Mohanty [1979], Lattice Path Counting and Applications, Academic Press, New York.

[14] T. V. Narayana [1959], A partial order and its applications to probability, Sankhya 21, 91-98.

[15] G. H. Raney [1960], Functional composition patterns and power series reversion, Trans. AMS 94, 441-451.

[16] J. Riordan [1968], Combinatorial Identities, Wiley, New York.

[17] F. Ruskey and T. C. Hu [1977], Generating binary trees lexicographically, SIAM J. Computing 6 (4), 745-758.

[18] Ju. M. Volosin [1974], Enumeration of the terms of object domains according to the depth of embedding, Sov. Math. Dokl. 15, 1777-1782.

EVALUATION D'ARBRE

POUR UN CALCUL FORMEL

(APPLICATION A L'ENUMERATION DE PROTEINES)

Marie-Pierre FRANCHI-ZANNETTACCI
Département d'Informatique
Université de Bordeaux II

Le travail présenté ici se situe dans le cadre d'une contribution à l'étude statistique de la configuration des protéines, objectif de nombreuses recherches biologiques (2), (4), (12). Les protéines étant codées par un mot écrit sur un alphabet à 20 lettres (les acides aminés), nous étudions le nombre d'apparitions de facteurs de longueur k .

Nous nous heurtons alors aux difficultés décrites par GUIBAS et ODLYZKO (6) qui donnent une évaluation du nombre de mots de longueur N , ne possédant pas un facteur donné.

Nous proposons pour notre part un algorithme général permettant d'obtenir ce nombre de mots chaque fois que la configuration recherchée est descriptible, en un sens précisé dans la suite, par un langage rationnel (et même linéaire).

A partir d'une grammaire **G(rationnelle ou linéaire)** non ambigüe, l'algorithme génère la série énumératrice (3) du langage engendré par G et fournit donc une formule (ou un programme) calculant le nombre de mots recherchés.

Pour automatiser les processus sur ordinateur nous utilisons une metagrammaire M décrivant les grammaires linéaires.

A partir d'une syntaxe abstraite (11) issue de l'analyse syntaxique d'un mot G engendré par M , nous calculons l'expression régulière représentant le langage $L(G)$. Nous effectuons un calcul par la méthode des attributs sur la syntaxe abstraite de l'expression régulière afin de donner les coefficients de la série énumératrice de $L(G)$. Ce dernier point est un processus de "compilation" simplifié par l'utilisation d'attributs hérités et synthétisés.

Dans la première partie, nous explicitons le passage de l'étude statistique des protéines à des problèmes combinatoires.

Après avoir donné quelques définitions et notations dans la seconde partie, nous donnons en troisième partie les algorithmes permettant d'obtenir à partir d'une grammaire non ambigüe, une expression régulière représentant le langage en image commutative.

Dans la quatrième partie, nous définissons une fonction de traduction des expressions régulières permettant d'obtenir une série énumératrice associée. Nous décrivons la mise en oeuvre par la méthode des attributs.

I - FORMULATION DU PROBLEME BIOLOGIQUE

Les protéines sont des macromolécules, constituées de molécules accolées les unes aux autres, les acides aminés.

Elles naissent sous une forme linéaire ; sous l'influence de forces attractives ou répulsives, elles prennent une forme tridimensionnelle qui définit leur fonction.

Ainsi, par exemple, une hémoglobine se replie sur elle-même de telle sorte que les acides aminés fixant l'oxygène se trouvent sur sa partie extèrieure. Il y a 20 acides aminés, chacun est codé par une lettre, la forme linéaire d'une protéine est ainsi codée par un mot sur un alphabet à 20 lettres.

Les forces qui s'exercent le long de la chaîne linéaire sont dues principalement aux acides aminés eux-mêmes. Par suite, un enchaînement de k acides aminés particuliers peut donner une certaine forme à la protéine qui les contient.

Ainsi, en admettant que les proximités entre acides aminés peuvent caractériser une protéine, une étude statistique permet de trouver les enchaînements d'acides aminés prépondérants.

Pour réaliser cette étude, nous formulons l'hypothèse suivante :

- si P est un ensemble de lettres (représentant les acides aminés) donné par la suite des lettres intervenant dans un mot.
- si E_P est l'ensemble de tous les mots que l'on peut former à partir de P, tous les éléments de E_P sont considérés comme équiprobables.

Ainsi, nous recherchons des enchaînements de lettres qui infirment cette hypothèse.

Variable aléatoire

La remarque faite précédemment nous conduit à étudier la statistique du nombre de facteurs f donné de longueur k. Nous étudions la variable aléatoire de Y_f de E_P dans $\mathbb{N}$ définie par $\forall g\ E_P \in Y_f(g) = i$ si et seulement si g contient i fois le facteur f.

Détection d'un enchaînement prépondérant

Soit i_o le nombre observé de facteurs f dans une protéine donnée P

Soit p_r la probabilité sur E_P

Soit $i_1 \in \mathbb{N}$ tel que $p_r(Y_f \geq i_1) = \alpha$ où α est un seuil d'erreur (par exemple 5%)

On dira que f est un enchaînement prépondérant de la protéine P si

$$i_o \in [i_1, +\infty[$$

Remarque 1 :

Cette affirmation équivaut à rejeter l'hypothèse d'équiprobabilité pour la protéine avec un risque d'erreur α .

Remarque 2 :

Si la longueur du facteur f est 2, on peut donner une formule énumérative simple (5). Des études ont par ailleurs été faites pour un ensemble de tels facteurs (8).

Cependant pour $k>2$, comme nous l'avons dit en introduction, une formule énumérative générale ne peut être obtenue. Il faut donc traiter cas par cas, ce qui justifie l'automatisation du processus.

Nous devons donc calculer $\mathrm{Card}\{W \in X^* / Y_f(W) = i\}$ pour chaque facteur f . On est ainsi amené à calculer les séries énumératrices de langages rationnels donnés par une grammaire non ambigüe (5).

Exemple 1 : Supposons que l'on veuille compter le nombre de mots écrits sur $X = \{a,b,x\}$ et possédant Na lettres a
Nb lettres b
Nx lettres x
i fois le facteur ab

Considérons la grammaire non ambigüe :

$$G = \langle \{L_0, L_1\}, \{a,b,x,y\}, R, L_0 \rangle$$

avec

$$R \left\{ \begin{array}{ll} \langle L_0\rangle ::= a\langle L_1\rangle & \langle L_1\rangle ::= a\langle L_1\rangle \\ \langle L_0\rangle ::= b\langle L_0\rangle & \langle L_1\rangle ::= y\langle L_0\rangle \\ \langle L_0\rangle ::= x\langle L_0\rangle & \langle L_1\rangle ::= x\langle L_0\rangle \\ \langle L_0\rangle ::= 1\!\!1 & \langle L_1\rangle ::= 1\!\!1 \end{array} \right.$$

Les mot engendrés par cette grammaire ne contiennent de lettres y que précédées de a ; les lettres a ne sont jamais suivies d'un b .
Le coefficient de la série énumératrice L_0 du langage engendré $(\underline{L}_0, a^{Na} b^{Nb-i} x^{Nx} y^{i})$ est le nombre recherché.

II - DEFINITIONS ET NOTATIONS

II-1 Notations

- On note $1\!\!1$ le mot vide
- Soit $W \in X^*$ où X est un alphabet, soit $x \in X$, nous notons $|W|_x$ le nombre d'occurrences de la lettre x dans W

- Nous notons $L(A)$ le langage engendré par le non terminal A d'une grammaire
- Nous noterons $\binom{n}{k_1,k_2,\ldots,k_l}$ le multinomial $\frac{n!}{k_1!\ldots k_l!}$

II-2 Définitions

Définition 1 (3) : *Soit* L *un langage sur un alphabet* $X = \{x_1,\ldots,x_r\}$, *la série énumératrice* $\underline{L}$ *du langage* L *de* $\mathbb{N}[\![X]\!]$ *donnée par :*

$$\underline{L} = \sum_{(i_1,i_2,\ldots,i_r)} N(i_1,i_2,\ldots,i_r)\, x_1^{i_1}\, x_2^{i_2} \ldots x_r^{i_r}$$

où $N(i_1,i_2,\ldots,i_r)$ *est le nombre de mots* W *de* L *tels que* $\forall\, j \in [1,r]\quad |W|_{x_j} = i_j$

Définition 2 (1) : *Une expression régulière sur un alphabet* X *est définie à partir des expressions suivantes :*
(i) $\emptyset, 1\!\!1, a\,(a \in X)$ *sont des expressions régulières*
(ii) si p *et* q *sont des expressions régulières (représentant des langages rationnels* p *et* q *) alors* $(p+q)$, $(p.q)$, p^* *sont des expressions régulières représentant respectivement* $P \cup Q$, $P.Q$ *et* P^*. *Nous associerons à ces expressions régulières une représentation arborescente*

$p_1+p_2+\ldots+p_n$ *sera représenté*

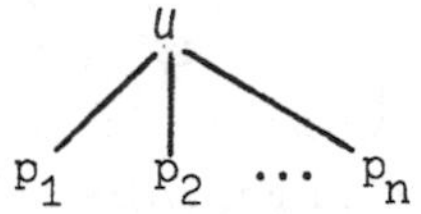

$p_1.p_2 \ldots .p_n$ *sera représenté*

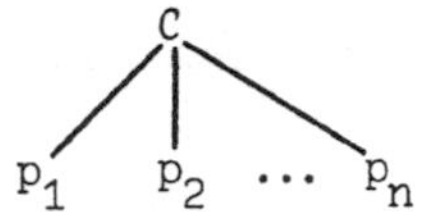

p^* *sera représenté*

U *représente l'union,* C *la concaténation,* E *l'étoile.*

Propriété 1 : Si A et B sont des langages représentés par les expressions régulières a et b et si A ne contient pas le mot vide, l'équation $X = aX+b$ a une solution unique $X = a^*b$.

II-3 Principes de l'algorithme

Notre cadre de travail étant les langages rationnels, le calcul d'une série énumératrice peut s'effectuer à partir d'une expression régulière solution d'une grammaire non ambigüe.

Exemple II.1 : reprenons l'exemple précédent.

D'après la propriété 1, on a

$$L(L_1) = a^*(X+Y)L(L_0) \pm a^*$$

donc

$$L(L_0) = ((b+X)+aa^*(X+Y))^*a^*$$

$$L(L_0) = (a^*X+b+aa^*Y)^*a^*$$

d'où, après passage à l'image commutative le coefficient de la série énumératrice de (L_0) est :

$$(\underline{L}(L_0),\ a^{N_a}\ b^{N_b-i}\ x^{N_x}\ y^{i}) = \binom{N_x+N_b}{N_x, N_b-i, i}\binom{N_x+N_a}{N_a-i}$$

Remarque : Du fait que le calcul de la série s'effectue en image commutative, il est possible d'accepter des grammaires linéaires, leur image étant rationnelle. Dans la suite nous autoriserons des grammaires linéaires étendues, c'est-à-dire des grammaires dont les règles sont de la forme :

$$\langle V\rangle ::= e_1\langle V'\rangle e_2 \quad \text{ou} \quad \langle V\rangle ::= e_1$$

avec e_1 et e_2 expressions régulières sur l'alphabet terminal.

L'analyse syntaxique d'une grammaire G dans une meta-grammaire M des grammaires linéaires étendues, produit une syntaxe abstraite S(G) représentant les règles de G .

A partir de S(G) nous calculons l'expression régulière associée à $L(G)$ en image commutative. Par une opération de traduction, nous associons à l'expression régulière ainsi obtenue une série énumératrice de $L(G)$.

III - OBTENTION D'UNE EXPRESSION REGULIERE REPRESENTANT UN LANGAGE LINEAIRE ETENDU

Ce travail s'effectue en deux temps :

(1) On fait l'analyse syntaxique (dans la meta-grammaire M) des règles d'une grammaire G afin de produire une syntaxe abstraite la représentant.

(2) On calcule l'expression régulière associée à la syntaxe ainsi obtenue.

III-1 Syntaxe abstraite d'une grammaire. Meta-grammaire des règles des grammaires linéaires étendues.

Nous définissons la meta-grammaire M suivante qui engendre les règles des grammaires linéaires étendues (ayant V_T et V_N comme alphabets terminaux et non terminaux).

$\langle \text{Règle} \rangle ::= \langle \text{Non terminal} \rangle \underline{\to} \langle \text{Expression} \rangle \langle \text{non terminal} \rangle \langle \text{Expression} \rangle$
$/ \langle \text{Expression} \rangle$

$\langle \text{Non terminal} \rangle ::= \underline{x} \in V_N$

$\langle \text{Expression} \rangle ::= \langle U \rangle / \langle U_1 \rangle$

$\langle U \rangle ::= \langle U_1 \rangle \pm \langle U_1 \rangle$

$\langle U_2 \rangle ::= \langle U_2 \rangle \pm \langle U_1 \rangle / \langle U_1 \rangle$

$\langle U_1 \rangle ::= \langle C_1 \rangle / \langle C \rangle$

$\langle C_1 \rangle ::= \langle C_2 \rangle \dot{-} \langle C_1 \rangle$

$\langle C_2 \rangle ::= \langle C_2 \rangle \dot{-} \langle C_1 \rangle / C_1$

$\langle C_1 \rangle ::= \langle E \rangle / \langle P \rangle$

$\langle E \rangle ::= \langle P \rangle \underline{\pm}$

$\langle P \rangle ::= \underline{(} \langle \text{Expression} \rangle \underline{)} / \underline{\mathbb{I}} / \underline{a} \in V_T / \varepsilon$ (mot vide)

Représentation arborescente d'une règle de grammaire G

On supprime dans l'arbre de dérivation dans M tous les sommets non étiquettés ⟨REGLE⟩, $\langle U \rangle$, $\langle C \rangle$, $\langle E \rangle$ ainsi que toutes les feuilles autres que $a \in V_T$, $x \in V_N$ et $\mathbb{I}$.

On obtient ainsi pour toute règle $\langle B \rangle :: e_1 \langle C \rangle e_2$ la représentation :

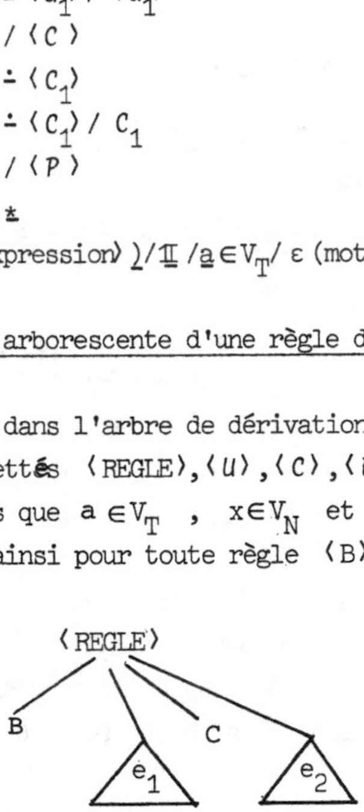

pour toute règle $\langle B \rangle :: = e_1$

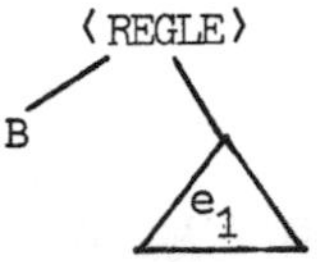

où les sous-arbres associés à e_1 et e_2 utilisent la représentation arborescente des expressions régulières donnée au paragraphe 2 .

Représentation d'une grammaire

Soit G la grammaire linéaire étendue $G = \langle V_T, V_N, R, A \rangle$.
Nous utilisons l'algorithme suivant :

Début pour tout $l \in V_N$ faire
Début soient $A_1, A_2, \ldots, A_{i_l}$ les arbres obtenus précédemment
pour $j = 1$ jusqu'à i_l faire
début remplacer ⟨REGLE⟩ par ⟨C⟩
supprimer le fils ainé de la racine
on obtient un nouvel arbre A'_j
fin
construire $A(x)$ comme suit :

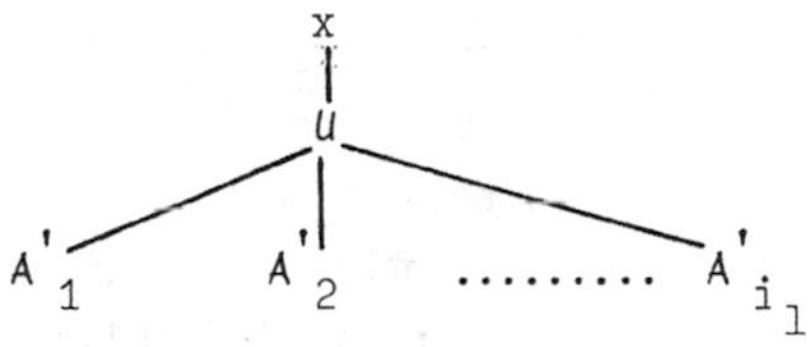

fin
fin

Exemple : Si nous reprenons l'exemple précédent on a :

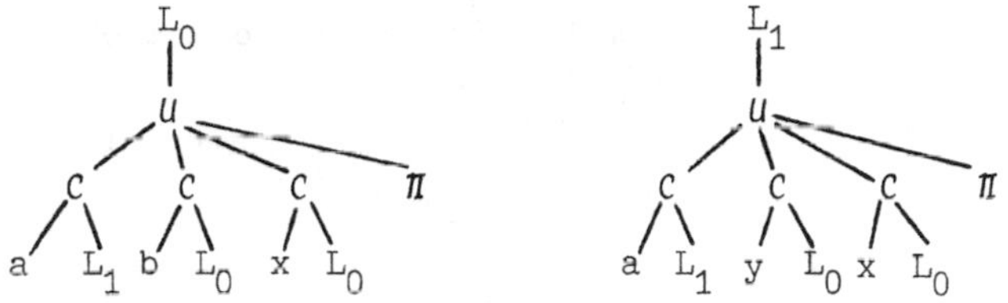

III-2 Calcul de l'expression régulière

Soit H_G la "forêt" représentant une grammaire linéaire étendue

$$G = \langle V_T, V_N, R, A \rangle$$

le premier arbre de la forêt est celui associé à l'axiome A

$$H_G = (A_0, \ldots, A_k)$$

l'algorithme que nous utilisons découle de la propriété annoncée au paragraphe 2.
L'opération consiste à remplacer

$$\langle A \rangle ::= e_1 \langle A \rangle e'_1 + \ldots + e_n \langle A \rangle e'_n + E_2$$

par

$$\langle A \rangle ::= (e_1 . e'_1 + \ldots . + e_n . e'_n)^* E_2$$

(ceci est cohérent car seule l'image commutative nous intéresse)

L'algorithme est le suivant :

<u>début</u> Pour i:= k pas 1 jusqu'à 0 faire

<u>début</u> j:= 0

<u>tant que</u> A_i contient une feuille f étiquetée comme sa racine r_i <u>faire</u> :

<u>début</u> . j:= j + 1

. soit $C = (r_i, c_1, \ldots, c_n, f)$ le chemin dans A_i de r_i vers f

. soit A'_j la copie de A_i obtenue en supprimant la racine r_i

. soit $C' = (c'_1, \ldots, c'_n, f')$ le chemin dans A'_i identique à C (à r près)

. dans A'_j $\forall l \in [1,n]$ si c'_l est étiqueté U détruire tous les sous-arbres issus de ' autres que c'_{l+1}

. dans A_i, $\exists l$ tel que c_l est étiqueté U et $\forall l' > l$ $c_{l'}$ n'est pas étiqueté U Détruire le sous-arbre de sommet c_{l+1}

. dans A'_j supprimer f

. dans A_i et A'_j contracter les noeuds non étiquetés E de degré 1

<u>fin</u>

Construire A_{RED} comme suit :

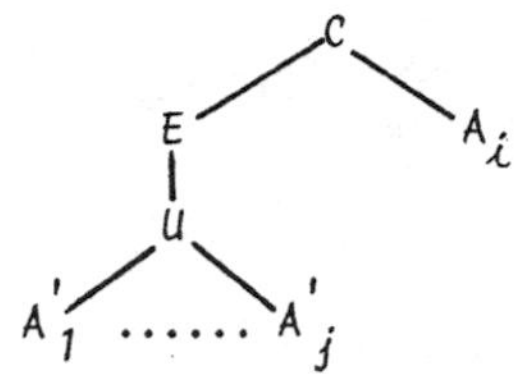

e:= étiquette r_i ; contracter r_i

<u>si</u> i ≠ 0 <u>alors pour</u> j = i-1 <u>pas</u> 1 <u>jusqu'à</u> 0 <u>faire</u>

remplacer dans A_j toutes les feuilles étiquetées e par une copie de A_{RED} ;

<u>fin</u>

<u>fin</u>

<u>Remarque</u>: Un algorithme supplémentaire éffectue la simplification aa*+1 en a* , lorsqu'elle ne nécessite pas de factorisation.

Exemple

* Dans l'exemple on obtient:

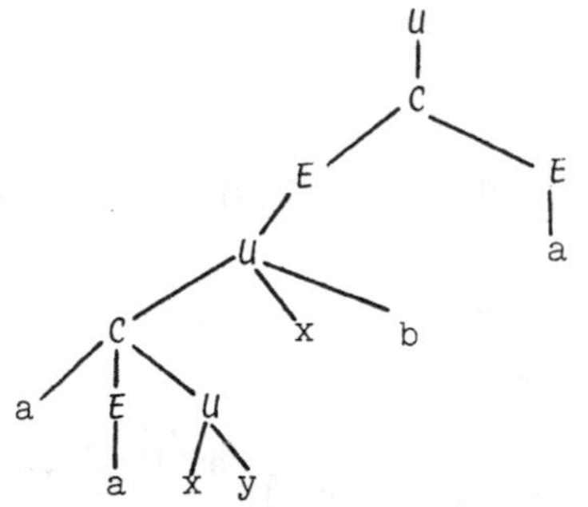

ce qui correspond à l'expression régulière solution :

$$(aa^*(x+y)+x+b)^*a^*$$

Remarquons que la simplification effectuée au paragraphe 1

$$aa^*x + x = a^*x$$

n'est pas exécuté par le système donné

Dans la suite, nous nous intéresserons d'une part, à la traduction d'une expression régulière, et d'autre part à la mise en oeuvre de cette traduction à partir de la représentation arborescente obtenue.

IV-CALCUL D'UNE SERIE ENUMERATRICE ASSOCIEE A UNE EXPRESSION REGULIERE

Par une opération de traduction, nous associons à une expression régulière représentant un langage linéaire étendu L , une série énumératrice de L .

Une expression régulière sur un alphabet X est un mot d'un langage R écrit sur l'alphabet

$$X \cup \{(\ , \ ),*,+,.\}$$

Nous lui faisons correspondre un mot écrit sur

$$X \cup \{K_i\}_{i\geq 0} \cup \{\Sigma,(\ , \ ),\geq,=,1,-\}$$

La mise en oeuvre de cette traduction relève des techniques de compilation. Nous y reviendrons au paragraphe IV.2

IV-1 Définition d'une fonction de traduction

Soit $Trad^n$ l'application définie récursivement sur les expressions régulières, sur un alphabet X , à image dans le monoïde d'alphabet:

$X \cup \{K_i\}_{i\geq 0} \quad \cup\{\Sigma, +, (\ ,\ ), \geq, =\ , 1\ , -\}$

telle que $n \in \{K_i\}_{i\geq 0} \quad \cup\{1\}$

$Trad^n(A)$ est la traduction de A^n et vérifie :

(1) $Trad^n(\mathbb{1}) = 1$

(2) $Trad^n(a) =$ si $n = 1$ alors a sinon a^n

(3) $Trad^n(e^*) =$ si $n = 1$ alors $\sum_{K_1\geq 0} Trad^{K_1}(e)$

sinon $\sum_{K_1\geq 0} \binom{K_1+n-1}{K_1} Trad^{K_1}(e)$

(4) $Trad^n(e_1+e_2+\ldots+e_p) =$ si $n = 1$ alors

$Trad^n(e_1)+ \ldots + Trad^n(e_p)$

sinon $\sum_{K_1+K_2+\ldots+K_{i_p} = n} \binom{n}{K_1,K_2,\ldots,K_{i_p}} Trad^{K_1}(e_1)\ldots Trad^{K_1}(e_p)$

(5) $Trad^n(e_1.e_2\ldots e_p) = Trad^n(e_1)\ldots Trad^n(e_p)$

avec la convention suivante :

Tous les K_i intervenant dans $Trad^n(A)$ sont distincts.

La traduction d'une expression régulière e sera le résultat de $trad^1(e)$.

Remarque 1 : pour la mise en oeuvre de l'algorithme, on utilisera la méthode des attributs, et"l'exposant"de Trad sera communiqué par un attribut hérité.

Remarque 2 : La forme proposée pour la traduction n'est pas tout à fait satisfaisante. En effet $\forall\ x\in X$ le monome apparaît sous la forme $x^{k_{i_1}} x^{k_{i_2}} \ldots x^{k_{i_p}}$. Ceci ne correspondant pas à la forme donnée pour une série énumératrice (paragraphe 2).

Il serait préférable que la lettre x intervienne sous la forme x^{N_x} . Pour cette raison, la traduction se déroulera en 3 étapes :

(1) Calcul des exposants des lettres

(2) Réduction des exposants

(3) production de la traduction

IV-2 <u>Réduction de l'exposant d'une variable</u>

Soit e une expression régulière, x une variable intervenant dans un monome de $Trad^1(e)$, $A(x)$ l'expression arithmétique de l'exposant de x dans le monome.

On peut montrer par induction que $A(x)$ est une fonction linéaire des K_i

$$A(x) = \alpha_0 + \alpha_1 K_{N_1} + \alpha_2 K_{N_2} + \ldots + \alpha_1 K_{N_1}$$

Ainsi cette fonction permet de remplacer K_{N_1} par :

(T) $1/\alpha_1(N_x-(\alpha_0 + \alpha_2 K_{N_2} + \ldots + \alpha_l K_{N_l}))$

L'algorithme que nous utilisons pour effectuer cette opération manipule des expressions arithmétiques ayant reçu une représentation arborescente.

Cette représentation (notée $A(x)$) peut être construite par la fonction Trad^n en remplaçant la règle (2) par

$\text{Trad}^n(x)$ = si $A(x)$ = vide alors $A(x) := \bullet^{\,n}$

sinon $A(x) :=$ 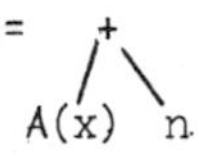

Il est alors aisé d'appliquer la transformation (T) sur les arbres ainsi obtenus. Le lecteur intéressé trouvera l'algorithme détaillé dans (5).

IV-3 Mise en oeuvre par la méthode des attributs

Comme nous l'avons dit en introduction nous sommes amenés à un problème de compilation. Plusieurs techniques sont envisageables, en particulier la traduction dirigée par la syntaxe (1), la méthode des attributs (9),(10). La première méthode n'est pas utilisable ; en effet la traduction d'une expression régulière ne peut s'effectuer indépendamment de son contexte comme le montre l'exemple suivant :

$b + c$ se traduit $b + c$

$(b + c)^*$ se traduit $\sum_{n\geq 0} \sum_{m\geq 0} \binom{m+n}{n} b^n c^m$

En d'autres termes, si A_1 et A_2 sont les représentations de deux expressions régulières et si T est un sous-arbre commun à A_1 et A_2 :

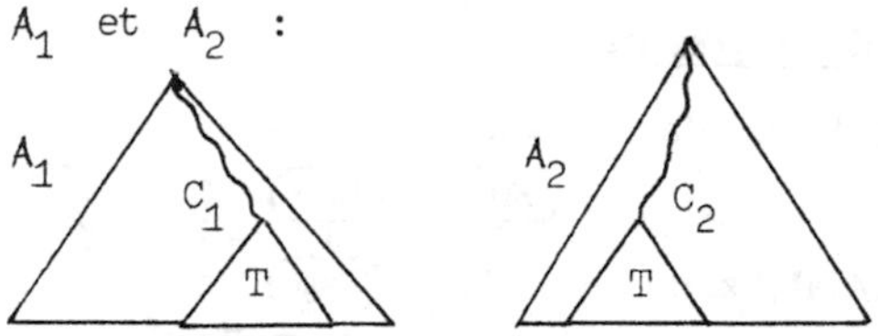

la traduction de T dans A_i dépend du chemin c_i (elle dépend en particulier de l'existence d'un opérateur E dans c_i)

Pour ces raisons il nous a paru préférable d'effectuer la traduction en utilisant la méthode des attributs. Ceci permet de prendre en compte facilement tout le contexte d'un sous-arbre quelconque.

Nous ne donnons ici qu'une partie du système d'attribut utilisé (5). Nous le définissons sur les arbres représentant les expressions régulières

Nous utilisons 2 attributs issus de la fonction Trad:

- Trad :fournit la traduction (<u>synthétisé</u>)
- Exp : fournit l'exposant (<u>hérité</u>); d'après le paragraphe IV.2, c'est un arbre.

Deux attributs permettent d'assurer que les K_j sont différents et donnent donc la valeur j maximum des K_i utilisés (ainsi K_{j+1} sera disponible).

- KD :<u>attribut hérité</u> donne cette information en descendant dans l'arbre.
- KM : <u>attribut synthétisé</u> donne cette même information en montant.

Nous utilisons les <u>fonctions</u> suivantes :

- CREE(K) crée un sommet étiqueté (K)
- PLUS(A,A') ajoute A' à l'arbre A pour obtenir

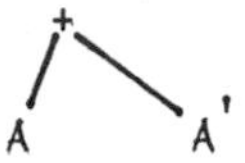

- MOINS(A,A') produit

- REP(A) (où A est l'arbre d'une expression arithmétique) calcule la chaîne alphanumérique associée à A .
- ARBRE (i) renvoie l'arbre correspondant à l'indice K_i .

L'application de l'algorithme cité au paragraphe IV-2 permet de réduire les exposants et d'obtenir ainsi les exposants des lettres d'une part, les expressions des indices d'autre part (ceci s'effectue par un calcul par attribut).

<u>1ère étape</u> : <u>production des exposants</u>

- Le calcul pour une feuille de l'arbre est :

KM(s) : = KD(s)

Exp(a) := PLUS (A(a),Exp(s))

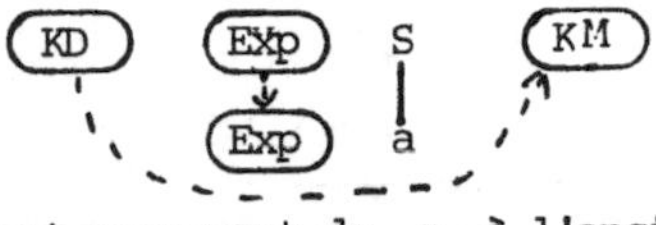

ceci signifie que l'on ajoute l'exposant provenant de s à l'ancien exposant de a

- pour un sous arbre de sommet E on utilise un nouvel indice donc

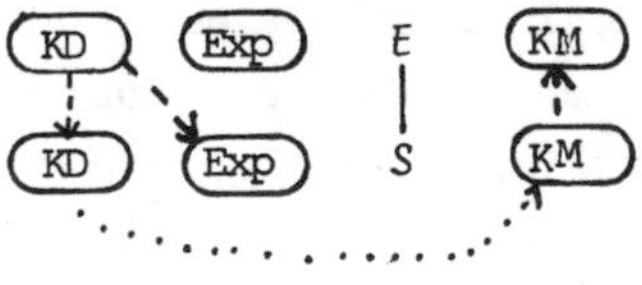

KD(s) = KD (E) + 1

l'exposant de s est le dernier indice libre donc

$$Exp(s) = CREE (KD (E))$$

On a de plus

$$KM(E) = KM(s)$$

<u>Exemple</u> L'attribut Exp est représenté sur l'arbre donne au §III.

$$A(a) = \begin{array}{c} + \\ K_2 \; K_5 \; K_8 \end{array}$$

$$A(b) = K_1 \begin{array}{c} - \\ \; \; + \\ K_2 \; K_3 \end{array}$$

$$A(y) = \begin{array}{c} - \\ K_2 \; K_6 \end{array}$$

$$A(x) = \begin{array}{c} + \\ K_3 \; K_6 \end{array}$$

$$A(K_4) = K_1 \begin{array}{c} - \\ \; \; + \\ K_2 \; K_3 \end{array}$$

$$A(K_7) = \begin{array}{c} - \\ K_2 \; K_6 \end{array}$$

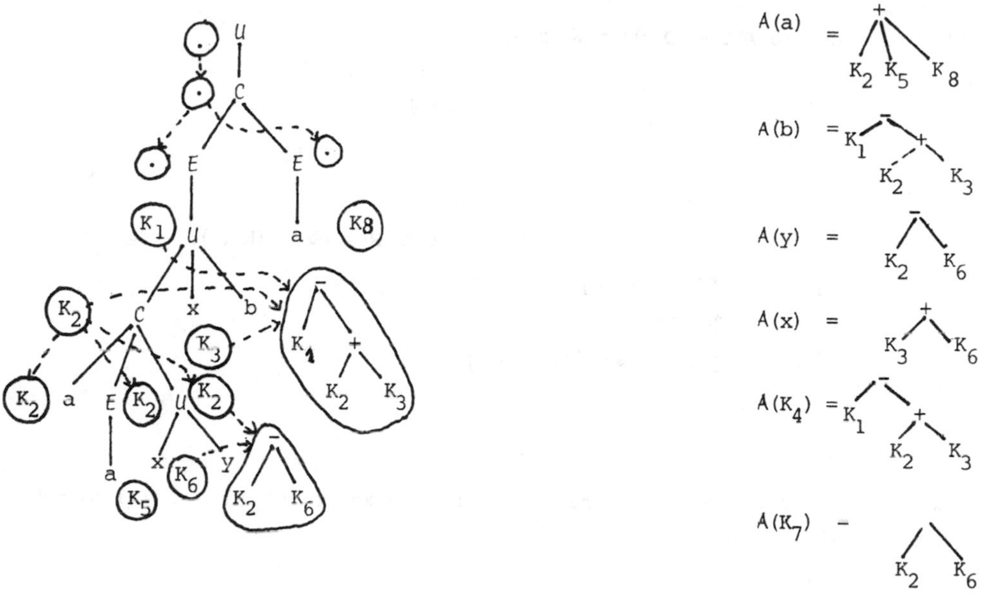

L'application de l'algorithme cité au paragraphe 4.2 permet de réduire les exposants et d'obtenir ainsi les exposants des lettres d'une part, les expressions des indices d'autre part (ceci s'effectue par un calcul par attribut).

<u>Exemple</u>

On obtient dans l'exemple précédent:

$$A(x) = .^{N}a \qquad A(b) = .^{N}b \qquad A(x) = .^{N}x \qquad A(y) = .^{N}y$$

$$A(K_1) = \begin{array}{c} + \\ N_b \; N_x \; N_y \end{array} \qquad A(K_2) = \begin{array}{c} + \\ N_y \; K_6 \end{array} \qquad A(K_3) = \begin{array}{c} - \\ N_x \; K_6 \end{array}$$

$$A(K_4) = .^{N}b \qquad A(K_5) = \begin{array}{c} - \\ N_x \; K_6 \end{array} \qquad A(K_7) = .^{N}y$$

$2^{\text{ème}}$ étape : <u>production de la traduction</u>

Pour un sous arbre de racine E on doit (d'après la définition de la fonction de traduction paragraphe 4.1) écrire un signe somme, un binomial et modifier

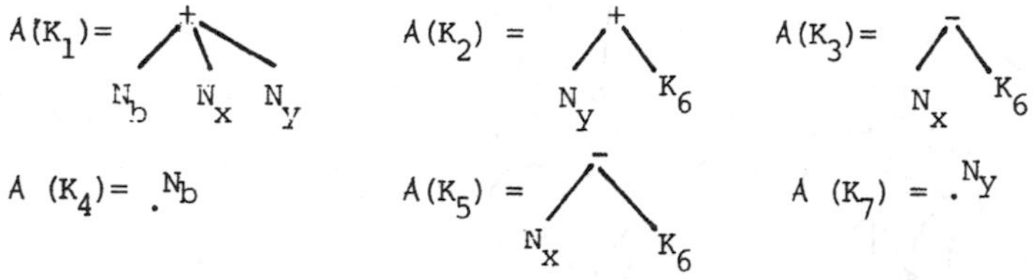

l'exposant

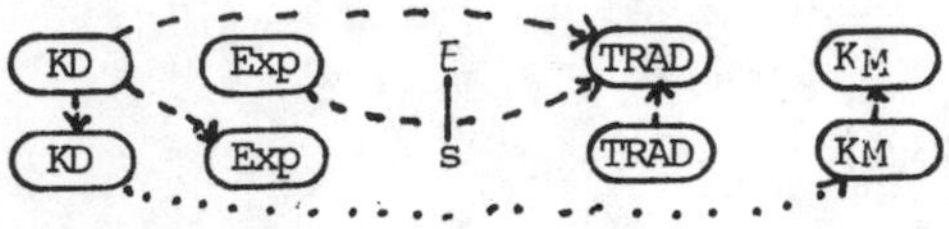

* Exp(s) = si ARBRE (KD(E)) = Ø alors CREE (KD(E)) sinon ARBRE(KD(E))

* TRAD(E)= début si ARBRE(KD(E)) = Ø alors début

R := CREE(KD(E))

écrire

$$\sum_{REP(R) \geq 0}$$

fin

sinon R:= ARBRE (KD(E))

si Exp(E) ≠ 1 alors

écrire $\begin{pmatrix} REP(R) \\ REP(R)+REP(\ Exp(E))-1 \end{pmatrix}$

TRAD (s)

Fin

On peut aussi pour chaque sous arbre donner une traduction associée en utilisant la fonction donnée au paragraphe 4.1.

<u>Exemple</u>

Pour l'exemple de départ on a la circulation de Exp suivante et la traduction associée

A =

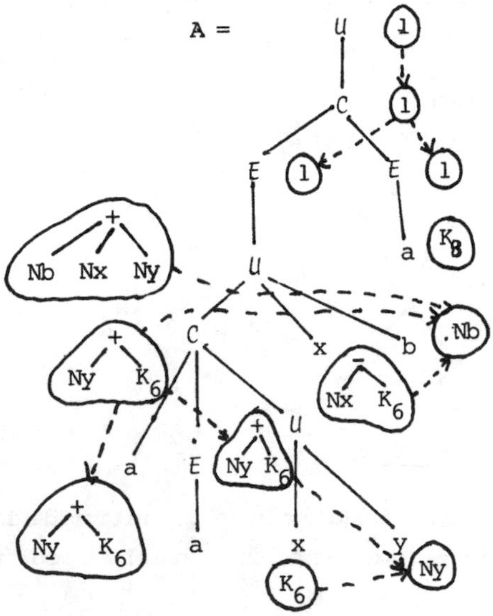

$$\mathrm{Trad}(A)=\sum_{k_6\geq 0}\ \sum_{k_8\geq 0}\binom{Nb+Nx+Ny}{Ny+K_6,Nx-K_6,Nb}\binom{Na-K_8-1}{Na-Ny-K_6-k_8}\binom{Ny+k_6}{k_6,N_y}$$

qui est le coefficient recherché c'est-à-dire celui de

$$a^{Na}\quad b^{Nb}\quad x^{Nx}\quad y^{Ny}$$

Remarque.-

Nous obtenons d'une manière générale des formules assez complexes. Cependant dans les applications qui nous occupent la formule n'a que peu d'importance en regard de son calcul effectif. Actuellement cette méthode est utilisée pour faire une investigation systématique (dans la mesure du possible) d'un échantillon de 118 protéines et par ailleurs pour tester des modèles d'évolution concernant d'autres cellules biologiques les t-RNA.

Tout le système est programmé en Pascal 80 et utilise le système d'édition automatique de J. HARDOUIN DUPARC [7].

B I B L I O G R A P H I E

[1] AHO et ULLMAN : The theory of parsing, translation and compiling, Ed. Prentice Hall, (1972)

[2] J.A. BLACK, R.N. HARKINS et P. STENZEL : Non-random relationships among amino acids in protein sequences. International Journal in Peptide and Protein Research, 8, 125-130, (1976).

[3] N.CHOMSKY et M.P. CHUTZENBERGER : The algebric theory of context free languages. Ed. North Holland, Amsterdam, (1963).

[4] D. DEBOUVERIE et J. SHLUSSELBERG : Research of specific constraints in pair association of amino acids. Biochimie, 56, 1045-1051, (1974).

[5] M.P. FRANCHI-ZANNETTACCI : Etude statistique des enchainements d'acides aminés par des méthodes combinatoires et l'utilisation du calcul formel. Thèse 3° cycle, Bordeaux , (1980).

[6] L.J. GUIBAS et A.M. ODLYZKO : Strings overlaps, pattern matching and nontransitive games. Manuscrit non publié.

[7] J. HARDOUIN DUPARC : Une méthode pour la résolution de certains problèmes combinatoires, Thèse d'état Bordeaux, 1980.

[8] J.P. HUTCHINSON et H.S. WIRLF : On Eulerian circuits and words with described adjency patterns ,Journal of combinatorial theory (A) 18,80-87 (1975).

[9] D.E. KNUTH : Examples of formal semantics, Lecture Notes in Mathematics, n° 188, Springer Verlag, (1971).

[10] L I V E R C Y : Théorie des programmes, shémas, preuves, sémantique, Edition Dunod, Paris, (1978).

[11] P. LUCAS et K. WALK : On the formal definition of PL/I. Annual Review in automatic Programming, 105-182 ,(1969).

[12] S. TANAKA et H.A. SHERAGA : Statistical treatment of protein conformation. Macromollecules, Vol. 9, n°5, 812-833, (1976).

ON PUSHDOWN TREE AUTOMATA

Irène GUESSARIAN

LITP CNRS - UER de Mathématiques
Université Paris 7 - T 55 56 1er Et.
2, Pl. Jussieu - 75251 PARIS Cédex05

INTRODUCTION

Many structures in computer science can be represented by trees: derivation trees, syntax directed translations, search in files, etc... . There was thus developed, at first, a theory of recognizable tree languages, tree grammars and tree transducers /AD,E1,RA,T/. Tree language theory has since been helpful in a broad range of domains, e.g. decision problems in logics /RA/, formal language theory /A,T/ and program scheme theory /B, C, G, GU, N/. In order to be applicable to such a wide range of problems, the tree language theory had to be extended to allow context free tree languages; it thus became possible to model program scheme theory (which is our motivation) and also more general language theory (e.g. Aho's indexed languages /A, DO, E2, F, R/ - since an indexed language is the yield of a context free tree language). Now, any language theory usually presents three complementary aspects: grammatical, set-theoretical or algebraic and automaton theoretical. For recognizable tree languages, all three aspects have been well studied, even for infinitary languages /P, RA/, and comprehensive accounts can be found in the above given references /AD, E1, T,.../. For context free languages, the grammatical point of view was first studied in /DO, F, R/, the algebraic aspects were considered in /B, ES, GU, N/, and comprehensive and exhaustive surveys can be found in /B, E2, S/. But the automaton theoretical aspect has not been studied at all; it is merely hinted at in /B, R/, but we can say that the prevailing point of view is grammatical and no attempt has been made to find a more operational way of looking at context free tree languages. This paper is a first step in that direction. It is organized as follows: we first introduce pushdown tree automata (in short PDTA's), i.e. top down tree automata with a pushdown store which is also a tree. We prove that the class of languages recognized by such automata (either by empty store or by final state) is exactly the class of context free tree languages, and that each such language can be recognized by a restricted automaton (whose pushdown store consists only of strings). We also prove the equivalence of various kinds of PDTA's and give numerous examples and counter-examples.

NOTATIONS

Let F (resp. Φ) be a finite ranked alphabet of base function symbols (resp. of variable function symbols). The rank of a symbol s in $F \cup \Phi$ is denoted by $r(s)$. Symbols in F are denoted f,g,h,... if they have rank ≥ 1, and a,b,... if they have rank 0; F_i denotes the symbols of rank i in F. Symbols in Φ are denoted G,H,... . Let V be a set of variables; the variables have rank 0 and are denoted by u,v,w,... possibly with indices.

The notions of tree, node in a tree, occurrence of a variable or a subtree in a tree, substitution, etc... are supposed to be known (see /GU/). The set of finite trees (resp. finite trees with variables in V) on F is denoted by A(F) (resp. A(F,V)); it is called the *free* F-*magma*, or F-*algebra* (resp. the free F-magma over V), and is denoted, in the ADJ terminology FT_F (resp. $FT_F(V)$).

$A_i(F,V)$ denotes the trees having i variables , and $A^i(F)$ (resp. $A^i(F,V)$) denotes the trees of depth i; clearly, $\forall i$: $A^i(F) \subset A^i(F,V)$ and $A^i(F) \subset A(F) \subset A(F,V)$, and similar relations hold for the A_i's.

PUSHDOWN TREE AUTOMATA

Definition 1: A *context free tree grammar* G, also called a rewriting system, is a system G of n equations:

$$G: \quad G_i(v_1,\ldots,v_{r(G_i)}) = T_i$$

where for $i=1,2,\ldots,n$, $G_i \in \Phi$ and $T_i = \{t_i^j , j=1,\ldots,n_i\}$ is a finite subset of $A(F \cup \Phi, \{v_1,\ldots,v_{r(G_i)}\})$. □

Each pair $(G_i(\vec{v}),t_i^j)$ is called a production of G (notice the vector shorthand notation $G_i(\vec{v})$ for $G_i(v_1,\ldots,v_{r(G_i)})$).

Rewritings and derivations according to G are defined as usual. We denote by $L(G,t)$ the language generated by G with axiom t (or $L(G)$ when the axiom t is explicit by the context).

Example 1: Let G be defined by:

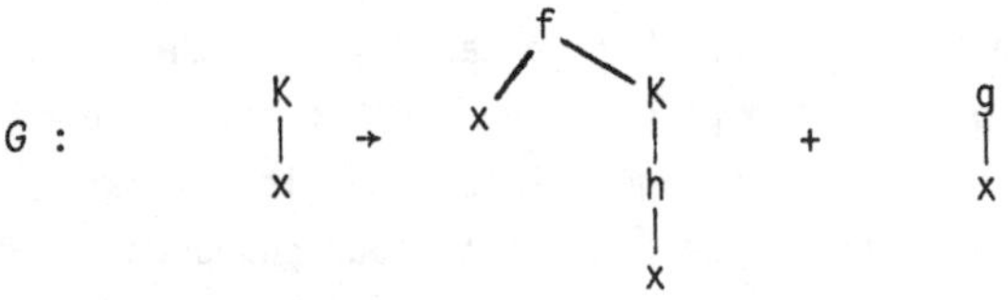

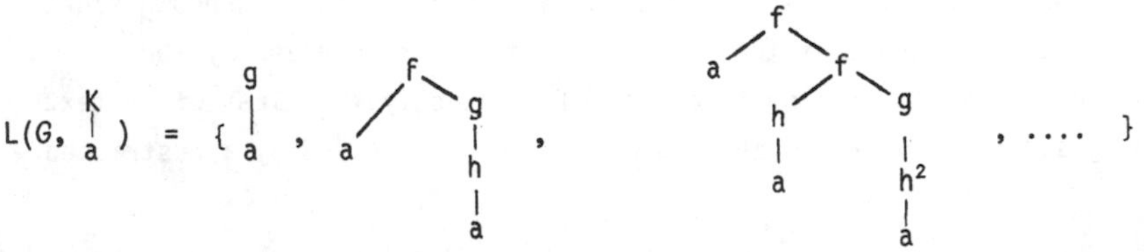
□

A pushdown tree automaton is a tree automaton /AD,E1/, together with a pushdown store consisting of a tree. We shall consider *top down* automata and see that, like top down tree recognizers, they have the essential ability of duplicating the pushdown store. The case of bottom up automata would be similar but is less suited to our purposes.

Definition 2: A *pushdown tree automaton (PDTA)* M is a seven-tuple (Q,F,X,q_0,Z_0,Q_f,R), where:

- Q is a finite set of states
- F is a finite ranked alphabet, called the input alphabet
- X is a finite ranked alphabet, called the pushdown alphabet. We may suppose, w.l.g. that X and F are disjoint alphabets.
- q_0 is the initial state
- Z_0 is the starting tree, appearing initially on the pushdown store, and of the form:

$$Z_0 = E(B_1, \ldots, B_{r(E)}) \quad , \quad E \in X \text{ , and } B_1, \ldots, B_{r(E)} \in X_0$$

- Q_f is the set of final states $(Q_f \subset Q)$
- R is a finite set of rules (or moves), such that each rule is a pair of one of the following forms:

  (i) read move

$$\left(f{:}q(v_1, \ldots, v_{r(f)}) \, , \, E(x_1, \ldots, x_{r(E)}) \right) \vdash f(v_1{:}q_1,t_1, \ldots, v_{r(f)}{:}q_{r(f)},t_{r(f)})$$

  (ii) ε-move

$$\left(\varepsilon{:}q \, , \, E(x_1, \ldots, x_{r(E)}) \right) \vdash (\varepsilon{:}q' \, , \, t')$$

  where $t', t_1, \ldots, t_{r(f)}$ are elements of $A(X,\{x_1,\ldots,x_{r(E)}\})$, and E is in X.

The computation relations $\vdash$ and $\overset{*}{\vdash}$ are defined as usual, as well as the notion of deterministic automaton (DPDTA): M is said to be *deterministic* iff for any pair (q,E), either there exists at most one ε-move in state q with pushdown root E and no read move, or there exists no ε-move and at most one read move for each f in F (in state q with pushdown root E).

The elements of X are denoted by $B,C,D,\ldots$ (symbols of rank 0), $E,G,H,K,\ldots$ (symbols of rank ≥ 1), or $x_1,x_2,\ldots$ (variables). The elements of F are denoted by the corresponding lower case letters and the variables representing elements of F are denoted by $v_1,v_2,\ldots$.

The tree language accepted by final state by M is the set of trees $T(M)$ defined as follows:

$$T(M) = \{ t \in A(F,V) \; / \; (t{:}q_0,Z_0) \overset{*}{\vdash} t' \in t(P(\vec{S}_0)/\vec{S}_0) \text{ , where } \vec{S}_0 = F_0 \cup V \text{ , and }$$
$$P(\vec{S}_0) = \{ \{ s(\varepsilon) \} \times Q_f \times A(X) \; / \; s \in \vec{S}_0 \} \;\}$$

where:

- $(t:q_0,Z_0)$ denotes the initial configuration where the root of t is scanned by M in state q_0 with pushdown store Z_0
- $t(P(\vec{S}_0)/\vec{S}_0)$ denotes the set of final configurations corresponding to the input tree t : in any such configuration, M should have processed the whole of t and arrived, after reading each leaf of t, in a state belonging to Q_f, with some pushdown store belonging to $A(X)$. So, $t(P(\vec{S}_0)/\vec{S}_0)$ is the set of trees deduced from t by substituting to every leaf of t labeled by an s in $\vec{S}_0$ an element of $\{\overset{s}{\underset{\varepsilon}{|}}\} \times Q_f \times A(X)$.

The tree language accepted by empty store by M is the set of trees $N(M)$ defined as follows:

$$N(M) = \{t \in A(F,V) \ / \ (t:q_0,Z_0) \overset{*}{\vdash} t' \in t(P'(\vec{S}_0)/\vec{S}_0) \text{ where } \vec{S}_0 = F_0 \cup V \text{ and } P'(\vec{S}_0) = \{ \{\overset{s}{\underset{\varepsilon}{|}}\} \times Q \times \varepsilon \ / \ s \in \vec{S}_0\} \ \}$$

A tree t is in $N(M)$ if M processes the whole of t and arrives in a configuration where the pushdown store is empty after reading each leaf of t.

Variants of PDTA's which do not extend the class of defined languages (e.g. several initial states, a start symbol instead of a starting tree, a starting tree of depth greater than 2, etc...) may be introduced; these variants do not give additional power to the corresponding PDTA's. Let us check, for instance, that the ability to read in a single move input trees of depth greater than two does not extend the class of languages recognized by PDTA's.

Formally, let $F'=F-F_0$, $t(v_1,\dots,v_n) \in A^i_n(F',V)$ where $i \geq 3$, $v_1,\dots,v_n$ label the leaves of t, no two leaves have the same label and no leaf is labeled by a constant symbol; let us say that a PDTA is *extended* if it is allowed to make extended read moves of the following more general form:

$$t(v_1,\dots,v_n):q,E(x_1,\dots,x_{r(E)}) \quad \vdash \quad t(\ (v_1:q_1,t_1)/v_1\ ,\dots,(v_n:q_n,t_n)/v_n\ )$$

where E, the x_i 's and t_j 's are as in definition 2 and $t(w_1/v_1,\dots,w_n/v_n)$ denotes the tree obtained by substituting the expression w_i to the leaf labeled by v_i, for $i=1,\dots,n$.

Proposition 1: L is accepted by empty store (resp. final state) by some extended PDTA M iff it is accepted by empty store (resp. final state) by some PDTA M'.

Sketch of proof: The "if" part is trivial; for the "only if" part, we have to simulate each extended read move of M by a sequence of moves of M'. Formally, let:

$$m_i: \quad t_i(v_1,\dots,v_{n_i}):q^i,E_i(x_1,\dots,x_{r(E_i)}) \quad \vdash \quad t_i(\ (v_1:q^i_1,t^i_1)/v_1\ ,\dots,\ (v_{n_i}:q^i_{n_i},t^i_{n_i})/v_{n_i}\ )$$

for $i=1,\dots,p$ and t_i of depth greater than 2 be the extended read moves of M. Then M' is defined as follows:

$Q' = Q \cup \{(m_i,o) \,/\, o$ is an internal occurrence in $t_i(\vec{v})$, for some $i=1,\dots,p$ $\}$
(Recall that an occurrence o is said to be internal if it is not a leaf, namely if its label $t_i(\vec{v})(o)$ has rank >0)
$F' = F$, $X' = X$, $q_o' = q_o$, $Z_o' = Z_o$, $Q_f' = Q_f$ and R' is defined as follows:

- ε-moves and non-extended read moves of M are in R'
- each extended read move m_i of M is replaced by the following set of moves:
  - let $f_\varepsilon^i = t_i(\vec{v})(\varepsilon) \in F'$ be the label of the root of $t_i(\vec{v})$; then R' contains the move :

    $$f_\varepsilon^i(v_1,\dots,v_{r(f_\varepsilon^i)}):q^i,E_i(\vec{x}) \quad \vdash \quad f_\varepsilon^i(w_1,\dots,w_{r(f_\varepsilon^i)})$$

    where $w_j = (v_j:(m_i,j),E_i(\vec{x}))$ if j is internal in $t_i(\vec{v})$, and otherwise: $t_i(\vec{v})(j) = v_k \in V$, for some k , and $w_j = (v_j:q_k^i,t_k^i)$
  - for each internal occurrence $o \neq \varepsilon$ in $t_i(\vec{v})$, let $f_o^i = t_i(\vec{v})(o) \in F'$ be the label of o ; then R' contains the move:

    $$f_o^i(v_1,\dots,v_{r(f_o^i)}):(m_i,o),E_i(\vec{x}) \quad \vdash \quad f_o^i(w_1,\dots,w_{r(f_o^i)})$$

    where $w_j = (v_j:(m_i,oj),E_i(\vec{x}))$ if oj is internal in $t_i(\vec{v})$, and otherwise: $t_i(\vec{v})(oj) = v_k \in V$, for some k, and $w_j = (v_j:q_k^i,t_k^i)$.

For lack of space, we leave it to the reader to check that $N(M') = N(M)$ and $T(M') = T(M)$. □

Example 2: Consider the following automaton:

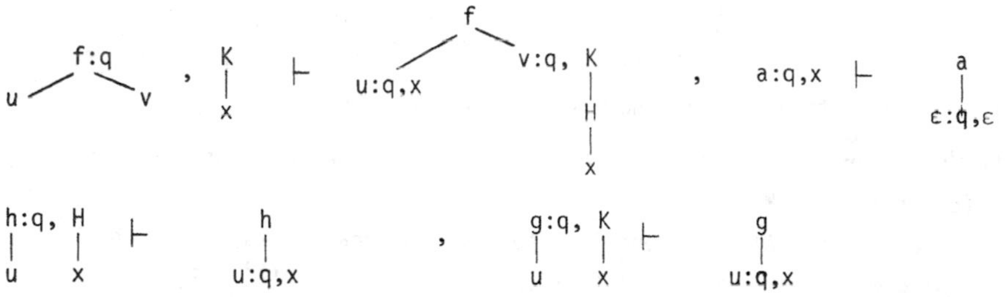

with $q_0=q$, $Z_0=$ K(x) .

This automaton recognizes the language $L(G,K(a))$ of example 1. Moreover, this is an example of a real-time (without ε-moves), deterministic , and restricted automaton (i.e. its pushdown alphabet consists only of words, see definition 3 below). □

Theorem 1: L is $T(M)$ for some PDTA M iff L is $N(M')$ for some PDTA M'.

Sketch of proof: Show for instance that every $T(M)$ is an $N(M')$: M' simulates M until it reaches a final state, when it has the choice, either of going on simulating M , or of erasing the pushdown store with ε-moves. Formally, it suffices to add a state q_{er} to M together with the rules:

$$\forall q_f \in Q_f \ , \ \forall E \in X \ : \quad \varepsilon{:}q_f \, , \ E(x_1 \cdots x_{r(E)}) \vdash \varepsilon{:}q_{er} \, , \ x_1$$

$$\forall E \in X \ : \quad \varepsilon{:}q_{er} \, , \ E(x_1 \cdots x_{r(E)}) \vdash \varepsilon{:}q_{er} \, , \ x_1 \qquad \square$$

Theorem 2: Every context free tree language is an $N(M)$ for some PDTA M, and conversely, every $N(M)$ is context free.

Proof: The construction of M is quite straightforward for the first part of the theorem. Let G be a grammar with terminal alphabet F, non terminal alphabet Φ, and an axiom which we may suppose w.l.g. of the form $t_0 = G(a_1,\ldots,a_{r(G)})$.

Let $\underline{F}$ be a copy of F, with say underlined letters having the same ranks as the corresponding letters of F. Then M has a single state q, input alphabet F, pushdown alphabet $X = \Phi \cup \underline{F}$, initial pushdown store $Z_0 = \underline{t}_0 = G(\underline{a}_1,\ldots,\underline{a}_{r(G)})$. For each t in $A(F \cup \Phi)$, let $\underline{t}$ denote the element of $A(\underline{F} \cup \Phi) = A(X)$ obtained by replacing each f in F by the corresponding $\underline{f}$ in $\underline{F}$. Then the productions of M are defined as follows:

- to each production $K(x_1 \cdots x_{r(K)}) \to t$ of G is associated an ε-move of M having the form: $\varepsilon{:}q \, , \ K(x_1 \cdots x_{r(K)}) \vdash \varepsilon{:}q \, , \ \underline{t}$

- for each f in F, we have a read move of M, having the form:

$$f{:}q(u_1 \cdots u_{r(f)}) \, , \ \underline{f}(x_1 \cdots x_{r(f)}) \vdash f\big((u_1{:}q,x_1) \cdots (u_{r(f)}{:}q,x_{r(f)})\big)$$

Then the following facts are clear and give the desired result:

Fact 1: For each derivation $t \overset{*}{\to} t'$ of G, there is a sequence of ε-moves: $(\varepsilon{:}q,\underline{t}) \overset{*}{\vdash} (\varepsilon{:}q,\underline{t}')$ of M.

Fact 2: For each t in $A(F)$, there is a sequence of read moves of M which accepts t when beginning with pushdown store $\underline{t}$ (i.e. $(t{:}q,\underline{t}) \overset{*}{\vdash} t(P'(\vec{S}_0)/\vec{S}_0)$).

The construction, for the reverse part of the theorem, of some grammar G generating $N(M)$ is somewhat more complicated. We first give the idea of it: essentially, the changes of the pushdown store will be modeled by the productions of the grammar, but we will have also to model the additional ability of changing the state; hence, the non terminals of G will encode pairs (state, top pushdown symbol). Moreover, the arguments of each non terminal (q,K) should render all possible "next states" after K has been popped. First, we have to reduce M to a simpler form.

Notation: In the sequel, $\phi(t_1,\ldots,t_{r(\phi)})$ will be abbreviated in $\phi(\vec{t})$ for ϕ in $F \cup X$.

Lemma 1: If a tree language L is accepted by empty store by a PDTA M, then L is also accepted by empty store by a simplified PDTA M' whose only read moves on input symbols of rank 0 are of the following form: $(a:Q,B) \vdash \begin{array}{c} a \\ | \\ (\varepsilon:q,\varepsilon) \end{array}$.

Sketch of proof: Intuitively, a leaf is accepted whenever it is read, and this can only be done while popping a leaf of the store. M' has the same alphabets as M, and a set of states $Q_{M'} = Q_M \cup \bigcup_{a \in F_0} \{ Q_M^a \cup \bar{Q}_M^a \}$, where for each symbol a of rank 0 in the input alphabet F, Q_M^a and $\bar{Q}_M^a$ are two duplicate copies of the set Q_M of states of M. Then the moves of M' are obtained by adding to the moves of M the following set of moves:

- for each move m: $(f(\vec{u}):q,G(\vec{x})) \vdash f((u_1:q_1,t_1),\ldots,(u_{r(f)}:q_{r(f)},t_{r(f)}))$, add the set of moves m^a, for any a in F_0, m^a: $(f(\vec{u}):q,G(\vec{x})) \vdash f((u_1:q_1^a,t_1),\ldots,(u_{r(f)}:q_{r(f)}^a,t_{r(f)}))$
- for each ε-move: $(\varepsilon:q,t_0) \vdash (\varepsilon:q',t)$, with $t_0=G(\vec{x})$ or $t_0=B$, add the corresponding moves, for every a in F_0: $(\varepsilon:q,t_0) \vdash (\varepsilon:q'^a,t)$ and $(\varepsilon:\bar{q}^a,t_0) \vdash (\varepsilon:\bar{q}'^a,t)$
- for each read move on a leaf, $(a:q,G(\vec{x})) \vdash \begin{array}{c} a \\ | \\ (\varepsilon:q',t) \end{array}$ add the move: $(\varepsilon:q_a,G(\vec{x})) \vdash (\varepsilon:\bar{q}'^a,t)$
- finally, for each ε-move emptying the store : $(\varepsilon:q,B) \vdash (\varepsilon:q,\varepsilon)$, add the read moves: $(a:\bar{q}^a,B) \vdash \begin{array}{c} a \\ | \\ (\varepsilon:q,\varepsilon) \end{array}$, for every a in F_0.

M' simulates M, but delays reading the leaves by making a "guess" stored in the state which is checked when the store is emptied. □

Suppose now M is of the simplified form given in the previous lemma. Let $k = \mathrm{card}(Q)$, and $\Phi = \{G^q \ / \ q \in Q, \ G \in X\}$ where each G^q in Φ has rank $k \times r(G)$; similarly, if Y is the set of pushdown variable symbols, let $Y^Q = \{ x^q \ / \ x \in Y, \ q \in Q \}$ be a set of variables (x^q is intuitively intended to correspond to the selection of variable x and next state q). Define σ by: $\sigma: Q \times A(X,Y) \to A(\Phi,Y^Q)$ satisfies $\sigma(q,s) = s^q$ if s has rank 0 or is a variable, and, using a vector notation $\vec{\sigma}(t)$ instead of $(\sigma(q_1,t),\sigma(q_2,t),\ldots,\sigma(q_k,t))$, define by induction: $\sigma(q,G(t_1,\ldots,t_{r(G)})) = G^q(\vec{\sigma}(t_1),\ldots,\vec{\sigma}(t_{r(G)}))$ (in short $\sigma(q,G(\vec{t}))=G^q(\vec{\sigma}(\vec{t}$

The grammar G generating N(M) is now defined as follows: G has terminal alphabet F and non terminal alphabet Φ, and:

- to each ε-move: $(\varepsilon:q,t_0) \vdash (\varepsilon:q',t)$ with $t_0=G(\vec{x})$ or $t_0=B$, is associated a production: $\sigma(q,t_0) \to \sigma(q',t)$ of G
- to each read move: $(f(\vec{u}):q,G(\vec{x})) \vdash f((u_1:q_1,t_1),\ldots,(u_{r(f)}:q_{r(f)},t_{r(f)}))$, is associated a production: $\sigma(q,G(\vec{x}))=G^q(\vec{\sigma}(\vec{x})) \to f(\sigma(q_1,t_1),\ldots,\sigma(q_{r(f)},t_{r(f)}))$ of G

- to each read move: $(a{:}q,B) \vdash \begin{matrix} a \\ | \\ (\varepsilon{:}q',\varepsilon) \end{matrix}$, is associated a terminal production $B^q=\sigma(q,B) \to a$ of G .

We then have:

Lemma 2: For every T in $A(X)$, t in $A(F)$, and q in Q : $\sigma(q,T) \overset{*}{\to} t$ iff there exists an accepting computation sequence $(t{:}q,T) \overset{*}{\vdash} t(P'(\vec{S}_0)/\vec{S}_0)$.

Proof: "only if" part: by induction on the length n of the derivation $\sigma(q,T) \overset{*}{\to} t$.
- If $n=1$ then: either $T=G(\vec{x})$ and $G^q(\vec{x}) \to \varepsilon = t$, which corresponds to the accepting move: $(\varepsilon{:}q,G(\vec{x})) \to (\varepsilon{:}q,\varepsilon)$.

  or $T=B$ and $B^q \to a$, which corresponds to the accepting move:

  $(a{:}q,B) \to \begin{matrix} a \\ | \\ (\varepsilon{:}q,\varepsilon) \end{matrix}$

- Inductive step: if $\sigma(q,T) \xrightarrow{n+1} t$, then:
  either $T=B$ and $\sigma(q,B)=B^q \to \sigma(q',T') \overset{n}{\to} t$, and by the induction $(t{:}q',T) \overset{*}{\vdash} t(P'(\vec{S}_0)/\vec{S}_0)$ is an accepting computation sequence , and by the construction of G there is a move $(\varepsilon{:}q,B) \vdash (\varepsilon{:}q',T')$, whence an accepting computation sequence starting from $(t{:}q,B)$
  or $T=G(\vec{T}')$ and:
  - either $\sigma(q,G(\vec{T}')) \to \sigma(q',T_1(\vec{T}')) \overset{n}{\to} t$, where we conclude as above since $(\varepsilon{:}q,G(\vec{x})) \vdash (\varepsilon{:}q',T_1)$ and $(t{:}q',T_1(\vec{T}')) \overset{*}{\vdash} t(P'(\vec{S}_0)/\vec{S}_0)$
  - or $\sigma(q,G(\vec{T}')) \to f(\sigma(q_1,t_1(\vec{T}')),\dots,\sigma(q_{r(f)},t_{r(f)}(\vec{T}'))) \xrightarrow{n} t$, whence we conclude that:
    1) $(f(\vec{u}){:}q,G(\vec{x})) \vdash f((u_1{:}q_1,t_1),\dots,(u_{r(f)}{:}q_{r(f)},t_{r(f)}))$ is a move of M
    2) $t = f(t'_1, \dots ,t'_{r(f)})$
    3) $\forall\, i=1,\dots,r(f)$ $\sigma(q_i,t_i(\vec{T}')) \xrightarrow{p} t'_i$ for some $p \leq n$

    whence by induction an accepting computation sequence of $(t{:}q,T)$ starting with the move

    $(f(\vec{t}'){:}q,G(\vec{T}')) \vdash f((t'_1{:}q_1,t_1(\vec{T}')), \dots ,(t'_{r(f)}{:}q_{r(f)},t_{r(f)}(\vec{T}')))$

"if" part: similar proof using an induction on the maximum number of moves applied during the computation on each branch of t . □

Lemma 2 implies that G with axiom $\sigma(q_0,Z_0)$ generates $N(M)$. □

Example 3: Let M be defined by: $Q = \{q_0,q_1,q_2\}$, $F = \{b,c_1,c_2\}$, $X = \{G,C\}$ with $r(b)=2$, $r(G)=1$, $r(C)=r(c_1)=r(c_2)=0$. The moves are the following:

(i) $(\varepsilon{:}q_0,G(x)) \vdash (\varepsilon{:}q_0,G^2(x))$

(ii) $(b(u,v){:}q_0,G(x)) \vdash b((u{:}q_1,x),(v{:}q_2,x))$

(iii) for $i=1,2$ $(b(u,v){:}q_i,G(x)) \vdash b((u{:}q_i,x),(v{:}q_i,x))$

(iv) $(c_i{:}q_i,C) \vdash \begin{matrix} c_i \\ | \\ (\varepsilon{:}q_i,\varepsilon) \end{matrix}$ for $i=1,2$

and $Z_0=G(C)$.

It is associated with G having non terminals G^0, G^1, G^2, of rank 3, and C^0, C^1, C^2, of rank 0, axiom $G^0(C^0,C^1,C^2)$, and productions:

(i) $G^0(x^0,x^1,x^2) \to G^0(G^0(x^0,x^1,x^2),G^1(x^0,x^1,x^2),G^2(x^0,x^1,x^2))$

(ii) $G^0(x^0,x^1,x^2) \to b(x^1,x^2)$

(iii) for $i=1,2$ $\quad G^i(x^0,x^1,x^2) \to b(x^i,x^i)$

(iv) for $i=1,2$ $\quad C^i \to c_i$

and $L(G) = N(M)$ is the set of binary trees of the form:

$t_n =$

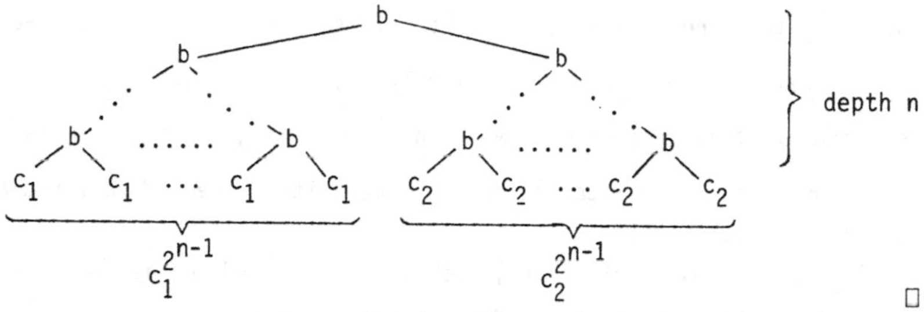

□

The characterization given in theorem 2 is quite similar to the ones given in /B/ or /R/. The spirit of the proof is quite similar to the one in Rounds' paper. However, these papers stress the grammatical aspects of the problem, whereas we want to consider its operational and automaton theoretical aspects. This is why we introduce linear stacks and prove the subsequent simplification theorem, different from the ones in /B,R/, who reduce the number of states whereas we simplify the stack.

Definition 3: A restricted PDTA (in short RPDTA) is defined as a PDTA, except for:
- X which is an ordinary (non ranked) alphabet, the pushdown store consisting of words in the free monoid X^*.
- the set of moves which are now of the simpler forms:

(i) read move $\quad \left(f{:}q(v_1,\ldots,v_{r(f)}),\ Ew \right) \vdash f(v_1{:}q_1,w_1w,\ \ldots,\ v_{r(f)}{:}q_{r(f)},w_{r(f)}w)$

(ii) ε-move $\quad (\varepsilon{:}q,Ew) \vdash (\varepsilon{:}q',w'w)$

where $E \in X$, $\quad w,w',w_1,\ldots,w_{r(f)} \in X^*$. □

Theorem 3: For every PDTA M, there exists a restricted PDTA M' such that $N(M)=N(M')$.

To prove this theorem, one would expect to simply use the fact that, if L is context-free, then the set of branches of L is a context-free language, hence is recognized by a pushdown automaton. Then one could construct the corresponding PDTA (by "sticking" together productions of the pushdown automaton). Unfortunately this PDTA will recognize a context-free tree language L' which is usually strictly greater than L (except when L is "branch closed", see A. Saoudi, mémoire de DEA Paris 7). We thus have to find a direct construction. It will be inspired from Gallier's paper /G/; for lack of space, we sketch this construction without actually proving it.

We first give the idea of the construction. We have to show that every context-free tree language is an $N(M)$ for some restricted PDTA M. We can no more have a single state PDTA where the whole tree which remains to be derived is stored in the pushdown store. Hence we shall use both the state and the pushdown store to code (or "remember") the derivations which remain to be done: i.e., the state remembers at which occurrence we are currently located in the right-hand side trees at the present moment of the derivation, and the pushdown store remembers which occurrences of variable function symbols still have to be derived. The state and the pushdown store are then used interactively to reconstruct the derivation of a tree.

Let G: $G_i(v_1,\dots,v_{r(G_i)}) = T_i = \{ t_i^j \,,\ j=1,\dots,n_i \}$ $\quad i=1,\dots,n$

be a context-free tree grammar with an axiom having the form $G_1(a_1,\dots,a_{r(G_1)})$ (this may be supposed without loss of generality). Then M defined as follows is such that $L(G)=N(M)$:

$Q = \{q_0\} \cup \{(i,j,o) \,/\, i=1,\dots,n \,,\ j=1,\dots,n_i$ and o is an occurrence in $t_i^j\}$

The input alphabet F is the terminal alphabet of G

$X = \{(i,j,o) \,/\, i=1,\dots,n \,,\ j=1,\dots,n_i$ and o is an occurrence of a $G_k \in \Phi$ in $t_i^j\} \cup \{ \$ \}$

$q_0=q_0$, $Z_0=\$$ and the moves are defined by:

- initializations: $(\varepsilon{:}q_0,\$) \vdash (\varepsilon{:}(1,j,\varepsilon),\$)$ $\quad \forall j=1,\dots,n_1$
- to each occurrence o of a base function symbol f in a t_i^j corresponds a read move: $\forall E \in X$ $\quad \left(f{:}(i,j,o)(v_1 \dots v_{r(f)}) , E \right) \vdash f\left((v_1{:}(i,j,o1),E) \dots (v_{r(f)}{:}(i,j,or(f)),E)\right)$

  Intuitively, reading f in a state corresponding to the position o in the tree t_i^j, we have to go down in the input tree without changing the store.
- to each occurrence o of a variable function symbol G_k in a t_i^j (i.e. $t_i^j(o)=G_k$), correspond push moves where we store G_k in the pushdown, thus remembering the recursive call which shall be done later, and reposition ourselves in a state which corresponds to beginning the derivation of G_k (i.e. at the roots of the right-hand sides $t_k^{j'}$ corresponding to G_k):

  $\forall E \in X$ $\quad (\varepsilon{:}(i,j,o),E) \vdash (\varepsilon{:}(k,j',\varepsilon),(i,j,o)E)$ $\quad j'=1,\dots,n_k$
- to each occurrence of a variable v_m indicating that the current recursive call has been completed, corresponds a pop move, i.e. going to the next recursive call and repositioning (by means of the state) to the v_m argument of each occurrence of the popped symbol in the t_i^j 's :

  $\forall w \in X^*$ $\quad (\varepsilon{:}(i,j,o),(i',j',o')w) \vdash (\varepsilon{:}(i',j',o'm),w)$ $\quad$ for any (i,j,o) , (i',j',o') such that $t_i^j(o)=v_m \in V$ and $t_{i'}^{j'}(o')= G_i \in \Phi$
- for each leaf of the axiom, a popping of the bottom marker $\$$ together with a checking of the leaf labels:

  $(a_m{:}(i,j,o),\$) \vdash (\varepsilon{:}(i,j,o),\varepsilon)$ $\quad$ for any (i,j,o) such that $t_i^j(o)=v_m$.

□

Let us apply this construction to the following simple example :

Example 4: Let G be the following grammar:

G : 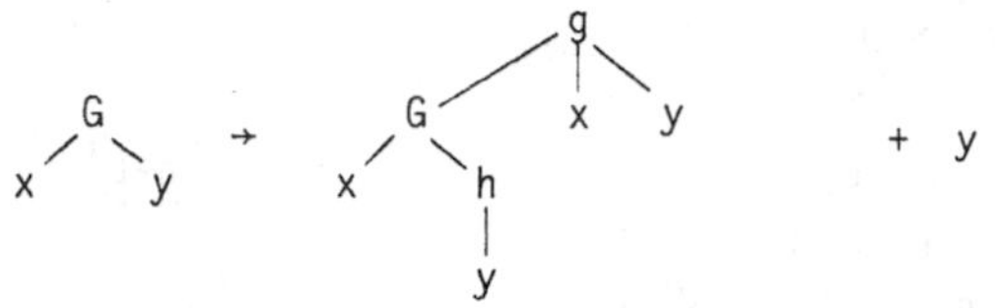

Since $n=1$ and $n_1=2$, we shall omit the components i,j in the states ($(1,1,\varepsilon)$ is denoted by ε and $(1,2,\varepsilon)$ is denoted by ε'); since there is but one occurrence of G in the t_i^j 's we simplify the pushdown alphabet X into $\{G,\$\}$.

The moves of M are then defined as follows:

- $(\varepsilon:q_0,\$) \vdash (\varepsilon:\varepsilon,\$) + (\varepsilon:\varepsilon',\$)$
- 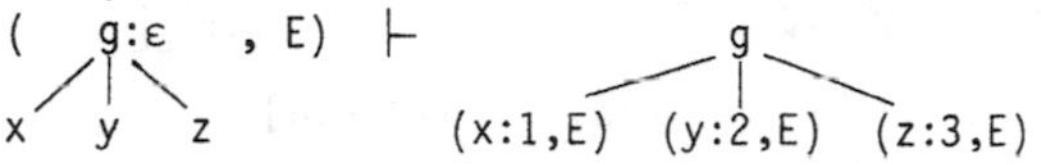 $\forall E \in X$

  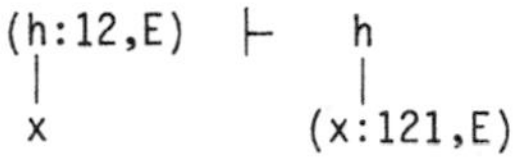 $\forall E \in X$

- $(\varepsilon:1,E) \vdash (\varepsilon:\varepsilon,GE)$ $\forall E \in X$

  $(\varepsilon:1,E) \vdash (\varepsilon:\varepsilon',GE)$
- $(\varepsilon:11,G) \vdash (\varepsilon:11,\varepsilon)$ and $(\varepsilon:2,G) \vdash (\varepsilon:11,\varepsilon)$

  $(\varepsilon:121,G) \vdash (\varepsilon:12,\varepsilon)$, $(\varepsilon:3,G) \vdash (\varepsilon:12,\varepsilon)$ and $(\varepsilon:\varepsilon',G) \vdash (\varepsilon:12,\varepsilon)$
- $(a:11,\$) \vdash (\varepsilon:11,\varepsilon)$ and $(a:2,\$) \vdash (\varepsilon:2,\varepsilon)$

  $(b:121,\$) \vdash (\varepsilon:121,\varepsilon)$, $(b:3,\$) \vdash (\varepsilon:3,\varepsilon)$ and $(b:\varepsilon',\$) \vdash (\varepsilon:\varepsilon',\varepsilon)$.

□

The previous results can be extended to higher-type tree languages /D/ , and, to some extent, to pushdown tree-transducers /BD/; this will be the subject of a forthcoming paper /DG/. However, this last extension is by no means trivial, as can be guessed from the following proposition and example.

Let us say that a tree language is

- *real-time* iff it is an N(M) for some M without ε-moves
- *deterministic* iff it is an N(M) for some deterministic M.

Proposition 2: The intersection of deterministic real-time languages may be non context-free.

This is shown by the following example:

Let M_1 be defined by: $Q=\{q,\bar{q}\}$, $F=\{\bot,f,g,h\}$, $X=\{E,\$\}$, $Z_0=\$$, and:

$\forall E' \in X$: 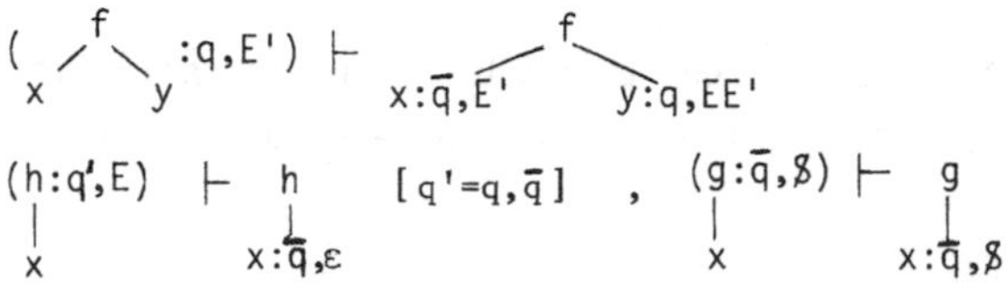

$$\bot:\bar{q},\$\ \vdash\ \varepsilon:\bar{q},\varepsilon$$

Then, $N(M_1) = \{ \ \dots \ / \ n,q_0,\dots,q_{n+1} \in \mathbb{N} \ \}$

Let M_2 be defined by: Q , F , X , Z_0 are as in M_1 and the moves are given by:

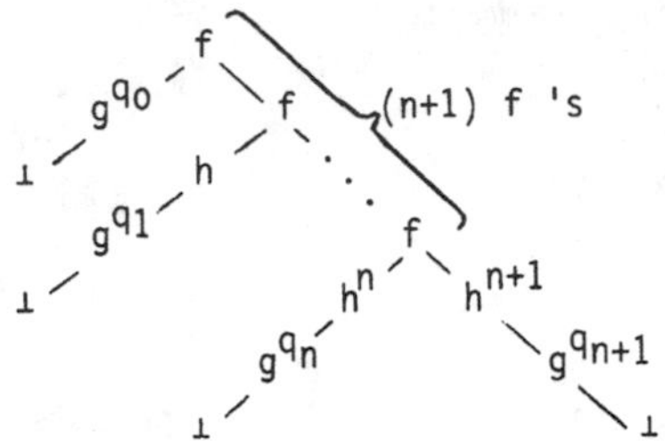

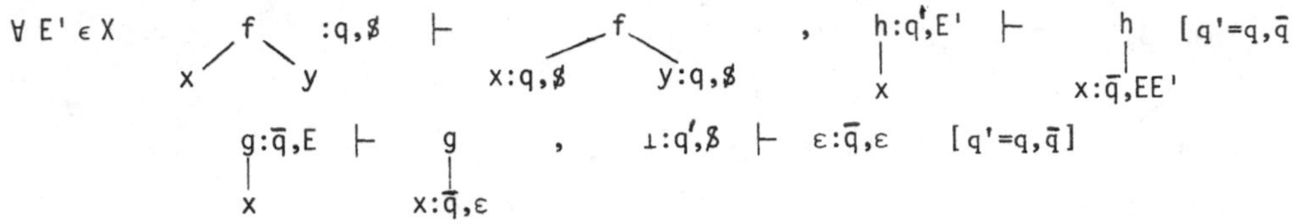

$$\forall E' \in X$$

$$\frac{f}{x \quad y}:q,\$\ \vdash\ \frac{f}{x:q,\$ \quad y:q,\$} \quad , \quad h:q',E'\ \vdash\ \frac{h}{x:\bar{q},EE'} \quad [q'=q,\bar{q}]$$

$$\frac{g}{x}:\bar{q},E\ \vdash\ \frac{g}{x:\bar{q},\varepsilon} \quad , \quad \bot:q',\$\ \vdash\ \varepsilon:\bar{q},\varepsilon \quad [q'=q,\bar{q}]$$

Then, $N(M_2) = \{ \ \dots \ / \ n_1,\dots,n_p \in \mathbb{N} \ \}$

and, clearly:

$$N(M_1) \cap N(M_2) = \{ \ \dots \ / \ n \in \mathbb{N} \ \}$$

which is non context-free (consider its branch language). $\qquad \square$

Corollary: The intersection of deterministic, realtime, prefix-free context-free word languages may be non context-free. $\qquad \square$

Acknowledgments: It is a pleasure to thank W. Damm for stimulating discussions, the (anonymous) referees, J. Engelfriet and H. Straubing for helpful comments.

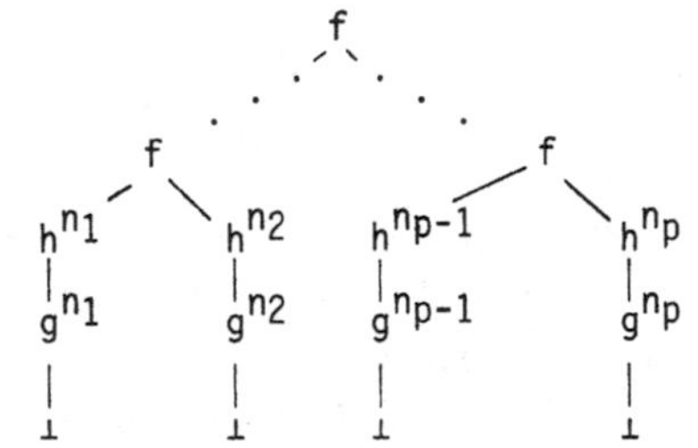

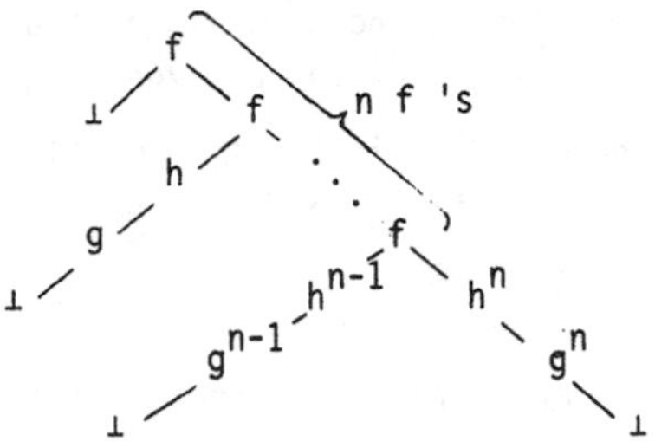

REFERENCES

/A/ A.V. Aho, *Indexed grammars: an extension of the context free case*, JACM 15 (1968), 647-671.

/AD/ A. Arnold, M. Dauchet, *Transductions de forêts reconnaissables monadiques. Forêts corégulières*, RAIRO Inf. Théor. 10 (1976), 5-28.

/BD/ J. Bilstein, W. Damm, *Top-down tree transducers for infinite trees*, this Conference.

/B/ G. Boudol, *Langages algébriques d'arbres*, LITP Report, to appear.

/C/ B. Courcelle, *On jump deterministic pushdown automata*, MST 11 (1977), 87-109.

/D/ W. Damm, *The IO and OI hierarchies*, Thesis, Report n°41 (1980), RWTH Aachen.

/DG/ W. Damm, I. Guessarian, *Combining T and n*, submitted for publication.

/DO/ P. Downey, *Formal languages and recursion schemes*, Ph. D. Harvard (1974).

/E1/ J. Engelfriet, *Bottom-up and top-down tree transformations. A comparison.* MST 9 (1975), 198-231.

/E2/ J. Engelfriet, *Some open questions and recent results on tree transducers and tree languages*, Proc. Int. Symp. on For. Lang. Theor., Santa Barbara , Academic Press, to appear.

/ES/ J. Engelfriet, E.M. Schmidt, *IO and OI*, JCSS 15 (1977), 328-353 , and JCSS 16 (1978), 67-99.

/F/ M.J. Fischer, *Grammars with macro-like productions*, 9th SWAT (1968), 131-142.

/G/ J.H. Gallier, *Alternate proofs and new results about recursion schemes and DPDA's*, Report MS-CIS-80-7, Dpt of Comp. Sc., Univ. of Pennsylvania (1980), to appear.

/GU/ I. Guessarian, *Algebraic semantics*, Lect. Notes in Comp. Sc. n°99, Springer-Verlag, Berlin (1981).

/N/ M. Nivat, *On the interpretation of recursive polyadic program schemes*, Symp. Mat. 15, Rome (1975), 255-281.

/P/ N. Polian, *Langages infinis engendrés par les grammaires algébriques d'arbres*, Th. 3ème Cycle, in preparation, Poitiers (1981).

/RA/ M. Rabin, *Automata on infinite objects and Church's problem*, CBSM regional Conf. series in Math. n°13, AMS (1969).

/R/ W.C. Rounds, *Mappings and grammars on trees*, MST 4 (1970), 257-287.

/S/ J.M. Steyaert, *Lemmes d'itération pour les familles d'arbres*, Actes du Séminaire d'Inf. Théor. 1977-1978, LITP Report, Paris (1979).

/T/ J.W. Thatcher, *Tree automata: an informal survey*, in Currents in Theory of Comp., Aho (ed.), Prentice-Hall, London (1973).

****

INITIAL AND TERMINAL ALGEBRA SEMANTICS OF PARAMETERIZED ABSTRACT DATA TYPE SPECIFICATIONS WITH INEQUALITIES

Günter Hornung and Peter Raulefs

Institut für Informatik III

Postfach 2220

D-5300 Bonn 1

Abstract. A generalization of our terminal algebra semantics approach is presented. We then give a uniform initial/terminal semantics of parameterized data type specifications.

1. Introduction

Terminal algebra semantics of algebraic data type specifications has turned out to provide an adequate basis for describing the semantics of module constructs in programming languages ([Hor 79],[HR 79, 80],[Gun 80]), which is difficult to obtain with initial algebra semantics ([ADJ 73, 75]), see also [GHM 76],[Kam 80], and [Wan 78]. Bergstra and Tucker [BT 80] showed a certain duality of initial and terminal algebra semantics with respect to computability. Final algebras were also considered in [WB 80] for model theoretic reasons. In [HR 79,80], foundations of a theory for terminal algebra semantics have been presented. In this paper, we extend this work towards parameterized specifications of abstract data types.

For initial algebra semantics, a first approach towards parameterized data types has been developed in [EKT 80,80a]. They are, however, only interested in the transformation of *initial* algebras given by some algebraic specification. Thus, passing compatibility is required only for the initial algebra of the actual parameter specification.

This paper develops an alternative mechanism for applying parameterized specifications, which is uniform for both initial and terminal algebra semantics: We define a *parameterized specification* of an abstract data type to consist of a formal parameter specification SPEC and a target specification SPEC1 s.t. SPEC1 is an initial resp. terminal extension of SPEC (see [HR 79, 80]). Then, the specified construction is an *initial* resp. *terminal persistent functor* mapping the category of SPEC-models to the category of SPEC1-models. An actual parameter specification consists of a specification SPEC' together with a parameter morphism (renaming) $h: SPEC \to SPEC'$ s.t. all SPEC'-algebras satisfy the axioms of SPEC.

Our initial/terminal algebra semantics of data type specifications is introduced in Section 2. A notion of parameterized abstract data types, which is dual for initial and terminal algebra semantics, and somewhat more general than the notion presented in [EKT 80, 80a] for the initial case, is given in Section 3. Also in Section 3, we introduce our approach to parameterized specifications. The parameter passing mechnism is then presented in Section 4.

Notation:

$\omega := \{0,1,2\ldots\}$ $\qquad \omega_+ := \{1,2\ldots\}$

For any $n\in\omega$, (n) denotes *both* the set $\{1,2,\ldots,n\}$ *and* the sequence $<1,\ldots,n>$,
and $[n]$ denotes *both* the set $\{0,1,\ldots,n\}$ *and* the sequence $<0,1,\ldots,n>$.

Analogously, for any $n\in\omega$, $t_{(n)}$ denotes *both* $\{t_1,\ldots,t_n\}$ *and* $<t_1,\ldots,t_n>$, and
$t_{[n]}$ denotes *both* $\{t_0,t_1,\ldots,t_n\}$ *and* $<t_0,t_1,\ldots,t_n>$.

2. Specifications with Inequalities under Initial and Terminal Algebra Semantics

2.1 Basic Notions

In algebraic specifications of abstract data types, we admit *positive conditional* and *unconditional negative equations*.

2.1.1. Definition [specification]

A *specification* (S,Σ,E) consists of a signature (S,Σ) and a set E of axioms s.t. E can be divided into two disjoint subsets EQ and NE so that

EQ is a set of positive conditional equations of the form

e_1 & e_1 &...& e_n => e_{n+1}

where $n\in\omega$ and $e_{(n+1)}$ is a set of S-sorted equations.

NE is a set of (unconditional) S-sorted inequalities.

For the remainder of this paper, we take SPEC=(S,Σ,E) to be a specification with $E = EQ \cup NE$. T_Σ $(T_{\Sigma(X)})$ denotes the algebra of Σ-terms (with variables from X), $Subst_{\Sigma(X)}$ the set of substitutions of Σ-terms for variables from X. The axioms in E induce the following relations ρ_{EQ} and ρ_{NE} on the term algebra T_Σ.

2.1.2. Definition [ρ_{EQ}, ρ_N, ρ_{NE}]

For any specification (S,Σ,E) with $E = EQ \cup NE$,

1. ρ_{EQ} is the least (under set inclusion) Σ-congruence on T_Σ s.t.
$\forall\; l_1=r_1$ & $l_2=r_2$ &...& $l_n=r_n$ => $l_{n+1}=r_{n+1} \in$ EQ.
let $l_{(n+1)}, r_{(n+1)} \in \underset{i\in(n+1)}{\times} T_{\Sigma(X),s_i}$ *with* $\forall i\in(n+1).\ s_i\in S$ *in*
$\forall\sigma\in Subst_{\Sigma(X)}$.
$\{\forall i\in(n).(\sigma l_i,\sigma r_i)\in\rho_{EQ}\}$ => $(\sigma l_{n+1},\sigma r_{n+1})\in\rho_{EQ}$
2. ρ_N is the least (under set inclusion) relation on T_Σ s.t.
$\forall s\in S.\ \forall l,r\in T_{\Sigma(X),s}.\ \forall\sigma\in Subst_{\Sigma(X)}.\ l=r\in NE$ => $(\sigma l,\sigma r)\in\rho_N$.

3. ρ_{NE} is the least (under set inclusion) relation on T_Σ s.t.
(1) $\rho_N \subseteq \rho_{NE}$.
(2) ρ_{NE} is symmetric.
(3) $\forall(p,q)\in\rho_{NE}.\forall(q,r)\in\rho_{EQ}.\ (p,r)\in\rho_{NE}$.

ρ_{EQ} identifies those terms the equality of which is derivable from the axioms EQ. Two terms belong to ρ_{NE} iff their inequality is directly derivable from NE *and* EQ. For induction proofs, the following equivalent definition of ρ_{NE} is more convenient.

2.1.3. Lemma [inductive definition of ρ_{NE}]

For any specification (S,Σ,E) with $E = EQ \cup NE$, let $(\rho_i \mid i\in\omega)$ be the following sequence of relations on T_Σ:

$$\rho_0 := \rho_N$$
$$\forall i\in\omega.\rho_{i+1} := \rho_i^{-1} \cup \rho_i * \rho_{EQ}.$$

Then, $$\rho_{NE} = \bigcup_{i\in\omega}\rho_i.$$

2.1.4. Definition [distinguishing sorts]

Let $SPEC=(S,\Sigma,E)$ be a specification. Then, $DIS := \{s\in S \mid \rho_{NE,s} \neq \emptyset\}$ is the set of all *distinguishing sorts* of SPEC.

The sorts in DIS are those sorts the terms of which are directly distinguished by NE-inequalities.

2.1.5. Definition [consistent, complete, simple]

A specification (S,Σ,E) with $E = EQ \cup NE$ is
1. *consistent* iff $\rho_{EQ} \cap \rho_N = \emptyset$
2. *complete* iff $\forall s\in DIS.\rho_{EQ,s} \cup \rho_{NE,s} = T^2_{\Sigma,s}$
3. *simple* iff $\forall l_1=r_1 \& \ldots \& l_n=r_n \Rightarrow l_{n+1}=r_{n+1} \in EQ.\forall i\in(n).l_i,r_i \in \bigcup_{d\in DIS} T_{\Sigma(X),d}$

A specification is *consistent* if the equations in EQ and the inequalities in NE do not contradict each other. Below it will be shown (Lemma 2.2.3.) that consistent specifications have a model. In *complete* specifications, the equality and inequality of terms of a distinguishing sort is fully defined, i.e. two terms are either derivably equal or derivably unequal. The conditions of equations in *simple* specifications are of distinguishing sorts only, so if a specification is simple and complete it can be decided for any substitution whether the condition of any conditional equation holds or does not hold.

2.2. Models of Specifications

The initial and the terminal view are to some extent dual approaches to abstract data types. First, we consider the initial approach.

2.2.1. Definition [SPEC-models]

For any specification SPEC=(S,Σ,E) the category Mod_{SPEC} of SPEC-*models* is defined by

$|Mod_{SPEC}| := \{A | A$ is a Σ-generated Σ-algebra s.t.

$\forall s \in S.\ \forall p,q \in T_{\Sigma,s}.$ (1) all positive equations of EQ are satisfied by A.

and (2) $(p,q) \in \rho_{NE} \Rightarrow p_A \neq q_A\}$

$/Mod_{SPEC}/$ is the set of all Σ-homomorphisms on $|Mod_{SPEC}|$.

Remark. For any Σ-algebra A and term $p \in T_\Sigma$, p_A denotes the interpretation of p in A i.e. the result of applying the initial homomorphism $\phi_A : T_\Sigma \to A$ to p.

SPEC-models are those Σ-generated Σ-algebras which satisfy the axioms of E.

2.2.2. Corollary

Let SPEC=(S,Σ,E) a specification, $A \in |Mod_{SPEC}|$. Then, $\rho_{EQ} \subseteq \equiv_A$, where $\equiv_A$ is the Σ-congruence on T_Σ induced by A.

The converse - that any Σ-generated Σ-algebra A for which $\rho_{EQ} \subseteq \equiv_A$ is a SPEC-model - is not true, however, as conditional equations and inequalities are involved. The following lemma shows that consistency amounts to the exsistence of models.

2.2.3. Lemma

For any specification (S,Σ,E), $|Mod_{SPEC}| \neq \emptyset \Leftrightarrow$ SPEC is consistent.

Proof:

"=>": Let $A \in |Mod_{SPEC}|$ and assume $(p,q) \in \rho_{EQ} \cap \rho_N \neq \emptyset$. Then, $p_A = q_A$ *and* $p_A \neq q_A$ is contradictory.

"<=": Assume $\rho_{EQ} \cap \rho_N = \emptyset$. Then, $\rho_{EQ} \cap \rho_{NE} = \emptyset$ as well implying $T_{\Sigma/\rho_{EQ}} \in |Mod_{SPEC}|$:

(1) all positive equations hold in $T_{\Sigma/\rho_{EQ}}$

(2) $\forall s \in S.\ \forall p,q \in T_{\Sigma,s}.\ (p,q) \in \rho_{NE} \Rightarrow [(p,q) \notin \rho_{EQ}$ *and* $p_{T_{\Sigma/\rho_{EQ}}} \neq q_{T_{\Sigma/\rho_{EQ}}}]$

Remark. If SPEC is consistent, then $T_{\Sigma/\rho_{EQ}}$ is initial in the category Mod_{SPEC} (cf. Corollary 2.2.2.). It is the algebra specified by SPEC in initial algebra semantics.

Next, we generalize the terminal algebra semantics presented in [HR 79,80]. To develop this approach, we start out with a more flexible view on "contexts of interest".

2.2.4. Definition [contexts]

Let SPEC=(S,Σ,E) be a specification. Then, for all s, s'$\in$S,

$C_\Sigma(s,s') := \{ct \in T_{\Sigma(X),s'} | ct$ contains exactly one variable x_s of sort s$\}$.

$\forall s,s' \in S. \forall ct \in C_\Sigma(s,s'). \forall p \in T_{\Sigma,s}. ct[p] := \{p/x_s\}ct.$

For any sorts s,s', the elements of $C_\Sigma(s,s')$ are called *contexts of sort s' for sort s*.

Contexts are used in order to distinguish terms by their *behaviour*. Two terms p and q have the same behaviour iff there is no context ct s.t. ct[p] and ct[q] are derivably unequal. Therefore, only contexts of distinguishing sorts are interesting in that respect.

2.2.5. Definition

Let (S,Σ,E) be a specification as above. The family $\sim_E = \{\sim_{E,s} \mid s \in S\}$ of S-sorted relations on T_Σ is defined by

$\forall s \in S. \forall d \in DIS. \forall p,q \in T_{\Sigma,s}. \; p \sim_{E,s} q :\Leftrightarrow \forall ct \in C_\Sigma(s,d). \; (ct[p],ct[q]) \notin \rho_{NE}$

$\sim_E$ identifies terms which are indistinguishable by E in their behaviour with respect to sorts DIS. Next, we show that the relation $\sim_E$ is indeed a Σ-congruence if (S,Σ,E) is consistent and complete.

2.2.6. Lemma [congruence property of $\sim_E$]

Let (S,Σ,E) be a consistent and complete specification, then $\sim_E$ is a Σ-congruence with $\rho_{EQ} \subseteq \sim_E$.

Proof:

(1) As ρ_{EQ} is a Σ-congruence, $\sim_E$ is reflexive by consistency.

(2) Symmetry of ρ_{NE} implies that $\sim_E$ is symmetric.

(3) Transitivity of ρ_{EQ}, consistency and completeness imply the transitivity of $\sim_E$.

(4) The congruence property is just. "$p \sim_{E,s} q \Rightarrow ct[p] \sim_{E,s} ct[q]$" for any context ct, sort s, and terms $p,q \in T_{\Sigma,s}$. This is immediate from Definition 2.2.4.

(5) $\rho_{EQ} \subseteq \sim_E$ by consistency.

Our next theorem justifies the phrase "terminal congruence" for $\sim_E$.

2.2.7. Theorem [terminality of $T_{\Sigma,\sim_E}$]

Let SPEC=(S,Σ,E) be a consistent, simple and complete specification, and $T_{SPEC,\sim} := T_{\Sigma/\sim_E}$. Then, $T_{SPEC,\sim}$ is a terminal object of Mod_{SPEC}.

Proof:

(1) Clearly, $T_{SPEC,\sim}$ is a Σ-generated Σ-algebra. We still have to show that the E-axioms hold in $T_{SPEC,\sim}$.

(1.1) Let $l_1=r_1 \;\&\ldots\&\; l_n=r_n \Rightarrow l_{n+1}=r_{n+1} \in EQ$, $\sigma \in Subst_{\Sigma(X)}$. Assume that $\forall i \in (n). \; \sigma(l_i) \sim_E \sigma(r_i)$. Then,

$\forall i \in (n). \; (\sigma(l_i),\sigma(r_i)) \in \rho_{EQ}$ implying $(\sigma(l_{n+1}),\sigma(r_{n+1})) \in \rho_{EQ}$ and

$\sigma(l_{n+1}) \sim_E \sigma(r_{n+1})$

(1.2.) For any sort $s \in S$, let p,q $T_{\Sigma,s}$ be terms with $(p,q) \in \rho_{NE}$. Hence for ct := x_s, $(ct[p],ct[q]) \in \rho_{NE}$ i.e. $p \not\sim_{E,s} q$.

By (1.1) and (1.2), $T_{SPEC,\sim} \in |Mod_{SPEC}|$.

(2) Take any algebra $A \in |Mod_{SPEC}|$. As both A and $T_{SPEC,\sim}$ are Σ-generated, there is at most one Σ-homomorphism $\Psi_A : A \to T_{SPEC,\sim}$. Let $\Phi_A : T_\Sigma \to A$ be the initial Σ-homomorphism and define

$$\forall t \in T_\Sigma . \Psi_A(\Phi_A(t)) := [t]_{\sim_E}.$$

Ψ_A is well-defined: Let $s \in S$, $p,q \in T_{\Sigma,s}$ with $\Phi_A(p) = \Phi_A(q)$. Then, $\forall s' \in DIS. \forall ct \in C_\Sigma(s,s').\ \Phi_A(ct[p]) = \Phi_A(ct[q])$, and $p \sim_E q$ as $\equiv_A \subseteq compl(\rho_{NE})$.

For complete, consistent and simple specifications, $T_{SPEC,\sim}$ is the algebra specified by SPEC in terminal algebra semantics.

2.2.8. Definition [initial/terminal abstract data type]

Let $SPEC=(S,\Sigma,E)$ be a consistent specification.

1. $T_{\Sigma,E} := T_{\Sigma/\rho_{EQ}}$ is called the *initial abstract data type* specified by SPEC.
2. If SPEC is also complete and simple, then $T_{SPEC,\sim}$ is called the *terminal abstract data type* specified by SPEC.

When extending a specification by adding new sorts, operation symbols and axioms, we expect the semantics of the "old" sorts and operations to be preserved in the extended specification.

2.2.9. Definition [i-/t-extension]

Let $SPEC=(S,\Sigma,E)$ and $SPEC'=(S',\Sigma',E')$ be consistent specifications with $S \subseteq S', \Sigma \subseteq \Sigma', E \subseteq E'$.

1. SPEC' is an *i-extension* of SPEC iff $T_{\Sigma',E'|\Sigma} \cong T_{\Sigma,E}$.
2. SPEC' is a *t-extension* of SPEC iff SPEC and SPEC' are complete and simple and $T_{SPEC',\sim|\Sigma} \cong T_{SPEC,\sim}$

3. Parameterized Data Types and Specifications

A parameterized data type is actually a *data type transformation* mapping actual parameter types to target types. In this section, we first develop the notions of initial and terminal parameterized data types to be functors mapping a category of of models to another model category. Next, this is generalized to parameterized specifications, and we show how to associate an initial and terminal algebra semantics with parameterized specifications.

3.1. Data Types

A parameterized data type maps algebras which are models of some specification (S,Σ,E) possibly extended by some other sorts and operation symbols to an extension of this algebra. For making this precise, we introduce two technical notions.

3.1.1. Definition/Lemma [combination, $*$-operation on model-categories]

1. Let SPEC=(S,Σ,E) be a specification, S1 a set of sorts, and Σ1 an (S+S1)-sorted family of operation symbols with $S\cap S1=\emptyset=\Sigma\cap\Sigma 1$, and let E1 be a set of axioms on (S+S1,Σ+Σ1). Then, SPEC + (S1,Σ1,E1) := (S+S1,Σ+Σ1,E∪E1) is a *combination* of SPEC and (S1,Σ1,E1).
2. For any family $\{C_i \mid i\in I\}$ of categories with $|C_i|\cap|C_j| = /C_i/\cap/C_j/ =\emptyset$, for $i\neq j (i,j\in I)$, we define $\bigoplus \{C_i \mid i\in I\}$ to be the category given by
   $|\bigoplus \{C_i \mid i\in I\}| := \cup\{|C_i| \mid i\in I\}$,
   $/\bigoplus/ \{C_i \mid i\in I\}/ := \cup\{/C_i/ \mid i\in I\}$.
3. If (S+S1,Σ+Σ1,E) is a combination of SPEC=(S,Σ,E) and (S1,Σ1,∅), we define the category
   $Mod^*_{SPEC,S1,\Sigma 1} := \bigoplus \{Mod_{(S+S',\Sigma+\Sigma',E)} \mid (S+S',\Sigma+\Sigma',E)$ is a combination of (SPEC) and $(S',\Sigma',\emptyset)$ with $S'\cap S1 = \Sigma'\cap\Sigma 1 = \emptyset\}$.

   Elements of this category are also called models of SPEC.

Parameterized data types are defined to be functors mapping the category of models of specification (S,Σ,E) to models of an extended specification (S+S1,Σ+Σ1,E) so that

(1) algebras a parameterized type is applied to may contain additional sorts and operations which are distinct from those in S1 and Σ1.

(2) the "structure " of an algebra a parameterized type is applied to is preserved in the resulting algebra (*persistency*, see [ADJ 78], [EKT 80,80a]).

3.1.2. Definition [parameterized data type]

A *parameterized data type* is a triple <(S,Σ, E), (S1,Σ1,E1),F> s.t.

(1) SPEC=(S,Σ,E) is a consistent specification and SPEC1=(S+S1,Σ+Σ1,E) is a combination of (S,Σ,E) and (S1,Σ1,E1).

(2) $F: Mod^*_{SPEC,S1,\Sigma 1} \to Mod^*_{SPEC1,S1,\Sigma 1}$ is a persistent functor so that $\forall(S',\Sigma').\ \forall A\in|Mod^*_{SPEC,S1,\Sigma 1}|\cap|Alg_{S+S',\Sigma+\Sigma'}|.\ FA \in |Alg_{S+S1+S',\Sigma+\Sigma 1+\Sigma'}|$.
    where $Alg_{S,\Sigma}$ denotes the category of all Σ-algebras.

$PDAT_{SPEC,SPEC1}$ is the category of parameterized data types <SPEC,SPEC1,F> with the natural transformations of the functors as morphisms.

3.2. Parameterized Specifications

Next, we consider parameterized specifications and develop a semantics which is essentially dual for the initial and terminal case: an initial/terminal persistent functor from the category of parameter models into the category of target models. First, we have to define two congruences.

3.2.1. Definition [$\equiv_{E1,A}$, $\approx_{E1,A}$]

Let SPEC=(S,Σ,E) and SPEC1=SPEC+(S1,Σ1,E1) be consistent specifications.

Let $A \in |Mod_{SPEC}|$.

1. $\equiv_{E1,A}$ is the smallest $(\Sigma+\Sigma1)$ congruence s.t.

(1.1) $\equiv_A \subseteq \equiv_{E1,A}$

(1.2) $\forall l_1=r_1 \,\&\ldots\&\, l_n=r_n => l_{n+1}=r_{n+1} \in E1.\ \forall \sigma \in Subst_{\Sigma(X)}.$
$(\forall i \in (n).\ \sigma(l_i) \equiv_{E1,A} \sigma(r_i) => \sigma(l_{n+1}) \equiv_{E1,A} \sigma(r_{n+1}))$

2. $\approx_{E1,A}$ on $T_{\Sigma+\Sigma1}$ is defined by
$\forall s \in S.\ \approx_{E1,A,s} = \equiv_{E1,A,s}$
$\forall s' \in S1.\ \forall p,q \in T_{\Sigma+\Sigma1,s'}.\ (p,q) \in \approx_{E1,A} <=> \forall s \in S. \forall ct \in C_{\Sigma+\Sigma1}(s',s).\ ct[p] \equiv_{E1,A} ct[q]$

3.2.2. Definition [parameterized specification]

Let SPEC=(S,Σ,E) and SPEC1=SPEC+(S1,Σ1,E1) be specifications s.t. E1 does not contain inequalities. <SPEC,SPEC1> is a *parameterized specification* iff
$\forall (S',\Sigma'). \forall A \in |Mod^*_{SPEC,S1,\Sigma1}| \cap |Alg_{(S+S',\Sigma+\Sigma')}|.\ T_{\Sigma+\Sigma'+\Sigma1/\equiv_{E1,A}}|_{\Sigma+\Sigma'} \cong A.$

In analogy to the unparameterized case, we now define the i-parameterized abstract data type.

3.2.3. Theorem/Definition [i-parameterized abstract data type]

Let SPEC=(S,Σ,E) and SPEC1=(S1,Σ1,E1) be specifications s.t. <SPEC,SPEC1> is a parameterized specification. Then there exists a functor I s.t. <SPEC,SPEC1,I> is initial in $PDAT_{SPEC,SPEC1}$. <SPEC,SPEC1,I> is the *i-parameterized abstract data type* specified by <SPEC,SPEC1>.

Proof:

Define $I: Mod^*_{SPEC,S1,\Sigma1} \to Mod^*_{SPEC1,S1,\Sigma1}$ as follows.
Let SPEC'=SPEC+(S',Σ',∅) a consistent specification, $A,B \in |Mod^*_{SPEC,S1,\Sigma1}| \cap |Alg_{SPEC'}|$, $H:A \to B \in /Mod^*_{SPEC,S1,\Sigma1}/$. Then $IA := T_{\Sigma+\Sigma1+\Sigma'/\equiv_{E1,A}}$ and IH is defined to be the unique $(\Sigma+\Sigma1+\Sigma')$-homomorphism $IH:IA \to IB$.

(1) $I\,A \in |Mod^*_{SPEC1,S1,\Sigma1}|$.
IA satisfies the equations E1 by Definition 3.2.1.1. As $IA|_{\Sigma+\Sigma'} \cong A$ and A satisfies the E-equations and inequalities,so does IA.

(2) I is persistent by Definition 3.2.2. So, <SPEC,SPEC1,I> is a parameterized data type.

(3) <SPEC, SPEC1,I> is initial in $PDAT_{SPEC,SPEC1}$.
Let $\langle SPEC,SPEC1,F\rangle \in |PDAT_{SPEC,SPEC1}|$. We define the natural transformation $\tau: I \dot{\to} F$ as follows. Let $A \in |Mod^*_{SPEC,S1,\Sigma1}| \cap |Alg_{S+S',\Sigma+\Sigma'}|$ for suitable S',Σ'.
$\forall t \in T_{\Sigma+\Sigma1+\Sigma'}.\ \tau A([t]_{\equiv_{E1,A}}) := \Phi_{FA}(t)$, where Φ_{FA} is the initial homomorphism $\Phi_{FA}: T_{\Sigma+\Sigma1+\Sigma'} \to FA$.

(3.1) τ is well defined.
FA satisfies all E1 equations as $FA \in |Mod_{SPEC1,S1,\Sigma1}|$. $\equiv_A \subseteq \equiv_{FA}$, as F is persistent. Thus, $\equiv_{E1,A} \subseteq \equiv_{FA}$, and τA is well-defined for any $A \in |Mod^*_{SPEC,S1,\Sigma1}|$.

(3.2) τ is a natural transformation.

$$\begin{array}{ccccc} A & & T_{\Sigma+\Sigma'+\Sigma1}/\equiv_{E1,A} & \xrightarrow{\tau A} & FA \\ H\downarrow & & IH\downarrow & \circlearrowleft & \downarrow FH \\ B & & T_{\Sigma+\Sigma1+\Sigma'}/\equiv_{E1,B} & \xrightarrow{\tau B} & FB \end{array}$$

The diagram commutes as all algebras are $(\Sigma'+\Sigma+\Sigma1)$-generated.

(3.3) τ is unique as both IA and FA are $(\Sigma+\Sigma'+\Sigma1)$-generated for any algebra $A \in |Mod^*_{SPEC,S1,\Sigma1}|$ and suitable Σ'.

The i-parameterized abstract data type identifies two terms of a newly introduced sort iff their equality can be derived using the equations E1 and the equalities given by the actual parameter algebra A. Thus, initial abstract data types are transformed into initial abstract data types, as the following corollary asserts.

3.2.4. Corollary

Let SPEC=(S,Σ,E) and SPEC1=SPEC+$(S1,\Sigma1,E1)$ s.t. <SPEC,SPEC1> is a parameterized specification. Let SPEC'=SPEC+$(S',\Sigma',\emptyset)$ a consistent specification with $T_{\Sigma+\Sigma',E} \in |Mod^*_{SPEC,S1,\Sigma1}|$. Then, $I(T_{\Sigma+\Sigma',E}) = T_{\Sigma+\Sigma'+\Sigma1,E+E1}$ and SPEC'+$(S1,\Sigma1,E1)$ is an i-extension of SPEC', where I is defined as in Definition 3.2.2.

The theorem and corollary above have their dual analogue in terminal algebra semantics.

3.2.5. Theorem/Definition [t-parameterized abstract data type]

Let SPEC=(S,Σ,E) and SPEC1=SPEC+$(S1,\Sigma1,E1)$ be simple specifications s.t. <SPEC,SPEC1> is a parameterized specification. Then there exists a functor T s.t. <SPEC,SPEC1,T> is terminal in $PDAT_{SPEC,SPEC1}$. <SPEC,SPEC1,T> is the *t-parameterized abstract data type* specified by <SPEC,SPEC1>.

Proof:

Define $T: Mod^*_{SPEC,S1,\Sigma1} \to Mod^*_{SPEC1,S1,\Sigma1}$ as follows.

Let SPEC'=SPEC+$(S',\Sigma'\emptyset)$ a consistent specification, $A,B \in |Mod^*_{SPEC,S1,\Sigma1}| \cap |Alg_{SPEC'}|$. $H: A \to B \in /Mod^*_{SPEC,S1,\Sigma1}/$. Then $TA := T_{\Sigma+\Sigma'+\Sigma1}/\approx_{E1,A}$, and TH is defined to be the unique $(\Sigma+\Sigma'+\Sigma1)$-homomorphism $TH: TA \to TB$.

(1) $TA \in |Mod^*_{SPEC1,S1,\Sigma1}|$.

TA satisfies the positive equations of E+E1 as $\equiv_{E1,A} \subseteq \approx_{E1,A}$ and SPEC1 is simple.
$\forall s \in DIS.\ \rho_{NE,s} \subsetneq \equiv^C_{E1,A,s} = \approx^C_{E1,A,s}$ as $DIS \subseteq S$ and according to the proof of Theorem 3.2.3.

(2) T is persistent by Definition 3.2.2. and therefore, <SPEC,SPEC1,T> is a parameterized data type.

(3) <SPEC,SPEC1,T> is terminal in $PDAT_{SPEC,SPEC1}$.

Let <SPEC,SPEC1,F> $\in |PDAT_{SPEC,SPEC1}|$. We define the natural transformation $\tau: F \dot\to T$ as follows. Let $A \in |Mod^*_{SPEC,S1,\Sigma1}| \cap |Alg_{S+S',\Sigma+\Sigma'}|$ for suitable S',Σ'.

$\forall t\in T_{\Sigma+\Sigma'+\Sigma 1}.\ \tau A(\phi_{FA}(t)):=[t]_{\approx_{E1,A}}$.

(3.1) τ is well-defined.

Let $s\in S+S'+S1$, $t,t'\in T_{\Sigma+\Sigma'+\Sigma 1,s}$ with $t\equiv_{FA}t'$. Let now $s'\in S+S'$ and $ct\in C_{\Sigma}(s,s')$. There exist $\bar{t},\bar{t}'\in T_{\Sigma+\Sigma',s'}$ s.t. $ct[t]\equiv_{E1,A}\bar{t}\equiv_A\bar{t}'\equiv_{E1,A}ct[t']$ implying $t\approx_{E1,A}t'$.

(3.2) τ is a natural transformation } analogous to the proof
τ is unique } of Theorem 3.2.3.

The t-parameterized data type identifies all terms of newly introduced sorts the behaviour of which with respect to the old sorts is equal. The following corollary states that for complete and simple specifications the terminal abstract data type of SPEC is transformed into the terminal abstract data type of SPEC1 by the t-parameterized abstract data type.

3.2.6. Corollary

Let SPEC=(S,Σ,E) and SPEC1=SPEC+(S1,Σ1,E1) complete and simple specifications s.t.<SPEC,SPEC1> is a parameterized specification. Let <SPEC,SPEC1,T> the t-parameterized abstract data type specified by <SPEC,SPEC1>. Then $T(T_{\Sigma,\sim_E})=T_{\Sigma+\Sigma 1,\sim_{E+E1}}$, and SPEC1 is a t-extension of SPEC as well as i-extension of SPEC.

Proof:

As $T_{\Sigma,\sim_E}\in|Mod^*_{SPEC,S1,\Sigma 1}|$, $T(T_{\Sigma,\sim_E})=T_{\Sigma+\Sigma 1/\approx_{E1,T_{\Sigma,\sim_E}}}$

(1) $\sim_{E+E1}\subseteq\approx_{E1,T_{\Sigma,\sim_E}}$

Let $s\in S+S1$, $t,t'\in T_{\Sigma+\Sigma 1,s}$ s.t. $t\sim_{E+E1}t'$. Let $s'\in S$, $ct\in C_{\Sigma+\Sigma 1}(s,s')$. Then there exist $\bar{t},\bar{t}'\in T_{\Sigma,s'}$ s.t. $ct[t]\equiv_{E1,T_{\Sigma,\sim_E}}\bar{t}\sim_E\bar{t}'\equiv_{E1,T_{\Sigma,\sim_E}}ct[t']$ implying $ct[t]\equiv_{E1,T_{\Sigma,\sim_E}}ct[t']$.

(2) $\approx_{E1,T_{\Sigma,\sim_E}}\subseteq\sim_{E+E1}$

Let $s\in S+S1$, $t,t'\in T_{\Sigma+\Sigma 1,s}$ s.t. $t\approx_{E1,T_{\Sigma,\sim_E}}t'$. Let $dis\in DIS$, $ct\in C_{\Sigma+\Sigma 1}(s,dis)$.
$ct[t]\approx_{E1,T_{\Sigma,\sim_E}}ct[t']$ implies $(ct[t],ct[t'])\notin\rho_{NE}$

as $T_{\Sigma+\Sigma 1/\approx_{E1,T_{\Sigma,\sim_E}}}\in|Mod_{SPEC1}|$. So $t\sim_{E+E1}t'$.

From (1) and (2) we get $T(T_{\Sigma,\sim_E})=T_{\Sigma+\Sigma 1/\sim_{E+E1}}$.

Persistency of T implies that SPEC1 is a t-extension of SPEC. In 3.2.4. we showed that SPEC1 is an i-extension of SPEC.

4. Parameter Passing

A parameterized data type <SPEC,(S1,Σ1,∅),F> maps an actual parameter algebra to a target algebra. As an actual parameter algebra, we admit any algebra in the category of SPEC-models. It may, in addition to sorts and operations in SPEC, incorporate further sorts and operations different from those in S1 and Σ1. Taking SPEC to satisfy, for example, the data type INTEGER, the type of rational numbers constitutes an acceptable actual parameter algebra which F e.g. maps to arrays of rational numbers.

A more general passing mechanism is achieved by admitting not just particular algebras, but specifications as actual parameters. However, this mechanism requires establishing a relationship between formal and actual parameter specification. As in [EKT 80,80a], we admit actual parameter specifications s.t.

(1) all sorts and operation symbols of the formal parameter SPEC are renamed sorts and operation symbols of the actual parameter SPEC' by some *parameter morphism* which must be given with the actual parameter,

(2) SPEC' may contain additional sorts and operation symbols which are distinct from those in S1 and Σ1.

It turns out that requiring that all algebras satisfying the actual parameter axioms satisfy the formal parameter axioms as well suffices to ensure correct parameter passing. By this parameter passing mechanism any specification SPEC' "satisfying" the formal parameter axioms is transformed into an extension SPEC1' of SPEC'.

4.1. Definition [parameter morphism, forgetful functor U_h]

Let SPEC=(S,Σ,E) and SPEC'=(S',Σ',E') be specifications. A parameter morphism h:SPEC→SPEC' is a pair $h=(h_s,h_\Sigma)$ of functions $h_s:S\to S'$ and $h_\Sigma:\Sigma\to\Sigma'$ with unique extension to $h:T_\Sigma\to T_{\Sigma'}$ s.t.

(1) $\forall f\in\Sigma[f:(s_1\dots s_n,s)].\ h_\Sigma(f):(h_s(s_1)\dots h_s(s_n),h_s(s))$ and

(2) all SPEC'-algebras satisfy the axioms h(E).

The parameter morphism h defines a forgetful functor

$U_h:Mod^*_{SPEC',S,\Sigma}\to Mod^*_{SPEC,S',\Sigma'}$ as follows:

Let $A\in|Mod^*_{SPEC',S,\Sigma}|\cap|Alg_{(S'+S\ ,\Sigma'+\Sigma\ )}$. Then $U_hA\in|Alg_{(S+S'',\Sigma+\Sigma'')}|$

$$\forall s\in S.\quad (U_nA)_s:=A_{h_s}(s)$$
$$\forall s\in S''.\quad (U_hA)_s:=A_s$$
$$\forall\sigma\in\Sigma.\quad \sigma_{U_hA}:=h(\sigma)_A$$
$$\forall\sigma\in\Sigma''.\quad \sigma_{U_hA}:=\sigma_A$$

In contrast to [EKT 80a], we need the strong condition (2) in order to guarantee that *all* SPEC'-models are also h(SPEC)-models (see Theorem 4.3 below).

We now give the syntactical basis for the insertion of specifications into parameterized specifications.

4.2. Definition [$PSPEC_h$ (SPEC')]

Let SPEC=(S,Σ,E) and SPEC1=SPEC+(S1,Σ1,E1) specifications s.t. PSPEC=<SPEC,SPEC1> is a parameterized specification. Let SPEC'=(S',Σ',E') a specification with S'∩S1=∅=Σ'∩Σ1 and h:SPEC→SPEC' a parameter morphism. Let h':SPEC1→SPEC'+(S1,Σ1,E1) the extension of h s.t.

$\forall s \in S1.\ h'_s(s)=s$

and $\forall \sigma \in \Sigma 1\ \ h'_\Sigma(\sigma)=\sigma$. Then $PSPEC_h(SPEC')=SPEC'+(S1,\Sigma 1,E1)$

is defined by (1) $S1'=S1$

(2) $\Sigma 1'=h'(\Sigma 1)$

(3) $E1'=h'(E1)$

We now show that the insertion of a specification into a parameterized specification via a parameter morphism renders a new parameterized specification.

4.3. Theorem

Let $SPEC=(S,\Sigma,E)$ and $SPEC1=SPEC+(S1,\Sigma 1,E1)$ specifications s.t. PSPEC=<SPEC,SPEC1> is a parameterized specification. Let $SPEC'=(S',\Sigma',E')$ a specification with $S'\cap S1=\emptyset=\Sigma'\cap\Sigma 1$ and $h:SPEC\to SPEC'$ a parameter morphism. Let $SPEC1':=PSPEC_h(SPEC')+SPEC'+(S1,\Sigma 1,E1)$ as above. Then <SPEC',SPEC1> is a parameterized specification and the extension of h $h':SPEC1\to SPEC1'$ is a parameter morphism. s.t. the following diagram commutes:

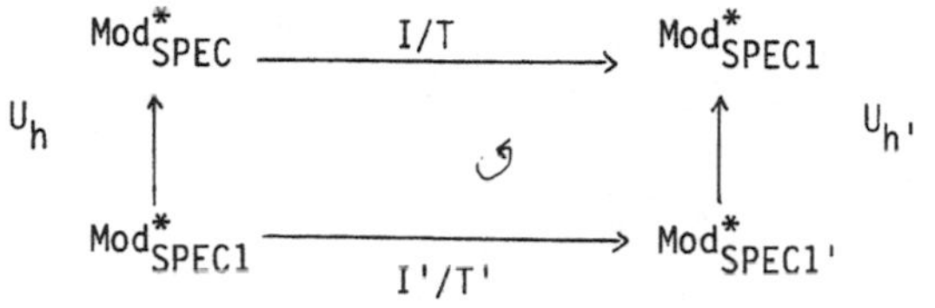

Proof:

Clear, as only renaming is involved. For the exact proof, see [HR 81].

The following theorem draws the connection with [EKT 80a]. It is shown that our approach guarantees the criteria of passing consistency, actual parameter protection and passing compatability.

4.4. Theorem

Let $SPEC=(S,\Sigma,E)$ and $SPEC1=SPEC+(S1,\Sigma 1,E1)$ specifications s.t. PSPEC=<SPEC,SPEC1> is a parameterized specification. Let $SPEC'=(S',\Sigma',E')$ a specification with $S'\cap S1=\emptyset=\Sigma'\cap\Sigma 1$ and $h:SPEC\to SPEC'$ a parameter morphism. Let $SPEC1':=PSPEC_h(SPEC')$ and $h':SPEC1\to SPEC1'$ the extension of h. Let <SPEC,SPEC1,I> and <SPEC',SPEC1',I'> the i-parameterized abstract data types specified by PSPEC and <SPEC',SPEC1'> resp. Then the semantical requirements for standard parameter passing as in [EKT 80] hold i.e.,

(1) $U_h(T_{SPEC'})\in|Mod_{SPEC,S1,\Sigma 1}|$ (passing consistency)

(2) $T_{SPEC1'|\Sigma'}=T_{SPEC'}$ (actual parameter protection)

(3) $I(U_h(T_{SPEC'}))=U_{h'}(T_{SPEC1'})$ (passing compatability)

Proof:

These properties follow immediately from the assertions in paragraph 3 together with Definitions 4.1. and 4.2.

5. Final Remarks

In our previous development of terminal algebra semantics [HR 79,80], the construction of a terminal congruence relation modelling equality of behaviour was based on a special sort 'dis' s.t. there are at least two distinct dis-constants. Although considering objects to be distinct only if their inequality is derivable from the axioms of a specification is basic to any terminal algebra semantics, our new construction based on the relation ρ_{NE} does away with the earlier syntactic requirements imposed on "terminal specifications".

In conclusion, we note that terminal algebra semantics has now reached a similar stage of development as the initial approach. As many, though not all, constructions and results are dual for both approaches which very often complement each other, we now have a strong basis for further developments and applications.

6. References

[ADJ 73] J.A. Goguen - J.W. Thatcher - E.G. Wagner - J.B. Wright,"A Junction Between Computer Science and Category Theory", IBM Research Report RC-4526, 1973

[ADJ 75] J.A. Goguen - J.W. Thatcher - E.G. Wagner - J.B. Wright, "Initial Algebra Semantics and Continuous Algebras", JACM 24(1977), 68-95

[ADJ 78] J.W. Thatcher - E.G. Wagner - J.B. Wright,"Data Type Specification: Parameterization and the Power of Specification Techniques", Proceedings of the 10th STOC, 1978

[BT 80] J.A. Bergstra - J.V. Tucker, "A Natural Data Type With a Finite Equational Final Semantics Specification but no Effective Equational Initial Semantics Specification", Report IW 133/80, Stichting Mathematisch Centrum, Afdeling Informatica, 1980

[EKT 80] H. Ehrig - H.J. Kreowski - J.W. Thatcher - E.G. Wagner - J.B. Wright, "Parameterized Data Types in Algebraic Specification Languages", Proceedings of the 7th ICALP, Noordwijkerhout 1980, 157-168

[EKT 80a] H. Ehrig - H.J. Kreowski - J.W.Thatcher - E.G. Wagner - J.B. Wright, " Parameter Passing in Algebraic Specification Languges", Draft Paper 1980

[GHM 76] J.V. Guttag - E. Horrowitz - D.R. Musser, "The Design of Data Type Specifications", Technical Report ISI/RR-76-49, Information Sciences Institute, University of Southern California, 1976

[Gun 80] U. Guntram, "Korrekte Implementierung abstrakter Datentypen durch Moduln in höheren Programmiersprachen", Memo SEKI-BN-80-09, Institut für Informatik III, Universität Bonn,1980

[Hor 79] G. Hornung, "Einige Probleme der Algebrasemantik abstrakter Datentypen", Memo SEKI-BN-79-07, Institut für Informatik III, Universität Bonn, 1979

[HR 79] G. Hornung - P. Raulefs, "Terminal Algebra Semantics and Retractions for Abstract Data Types", Memo SEKI-BN-79-06, Institut füt Informatik III, Universität Bonn, 1979

[HR 80] G. Hornung - P. Raulefs. "Terminal Algebra Semantics and Retractions for Abstract Data Types", Proceedings of the 7th ICALP, Nordwijkerhout 1980, 310-323 (Summary of [HR 79]).

[HR 81] G. Hornung - P. Raulefs, "A Uniform Algebra Semantics of Parameterized Data Type Specifications with Inequalities", Memo SEKI-BN-81-01, Institut für Informatik III, Universität Bonn, 1981

[Kam 80] S. Kamin, "Final Data Type Specifications: A New Data Type Specification Method", Proceedings of the 7th POPL Conference, Las Vegas 1980, 131-138

[Wan 78] M. Wand, "Final Algebra Semantics and Data Type Extensions", Technical Report No. 65, Computer Science Department, Indiana University, revised 1978

[WB 80] M. Wirsing - U. Broy, "Abstract Data Types as Lattices of Finitely Generated Models", Proceedings of the 9th International Symposium on MFCS, Springer 1980

CALCUL DU RANG
DES Σ-ARBRES INFINIS REGULIERS

G. JACOB

LITP et Université de Lille I,
59655 VILLENEUVE D'ASCQ CEDEX

RESUME :

Soit Σ un alphabet gradué. Les Σ-arbres infinis réguliers peuvent être codés sous forme de suites de données, grâce à une écriture sous forme d'expressions itératives scalaires. Cette écriture permettant un transfert aisé d'une représentation informatique dans une autre.

On peut donner plusieurs définitions de la complexité du décodage d'une telle expression itérative. La plus naturelle est donnée par le "rang" de l'expression, notion qui fut introduite par R. Kosaraju [9] sous une autre appellation (GRE_n-chartes) dans le cas particulier des arbres syntaxiques des schémas de Ianov. Il a posé à ce sujet plusieurs questions qui restent ouvertes.

C'est une réponse partielle à une de ces questions que nous apportons, en donnant un algorithme qui calcule le rang (ou le "rang strict") d'un Σ-arbre infini régulier, donné par une représentation de l'un des types (1) à (4) donnés ci-dessous dans l'introduction.

Il s'agit ici d'un exposé de nature informelle, l'exposé formel complet de l'algorithme et de sa preuve pouvant difficilement se réduire à la taille d'un exposé de colloque.

ABSTRACT :

Let Σ be a finite graded set. The regular Σ-trees can be encoded into data sequences, using the scalar iterative expressions (as in EXEL-language [1]). The complexity of scalar iterative expressions can be defined in various way and so it is for regular Σ-trees. Here, we present a method for calculating the "rank" of such a tree, with and without concatenation.

In the flow chart case, our algorithm allows to decide if a chart G is (syntactically) reducible to some GRE_n-chart. Recall that the request of Kosaraju [9] for a "structural characterization" of the GRE_n-charts is till now an open question.

INTRODUCTION

Représentations

Les Σ-arbres infinis réguliers (cf. G. Cousineau [4]), C. C. Elgot, S.L. Bloom and R. Tindell [6]...) sont ceux qui n'ont qu'un nombre fini de sous-arbres suffixes distincts.

Ils peuvent être décrits par différentes représentations. Notamment :

(1) Ce sont les Σ-arbres qui décrivent la structure des successions de certains Σ-graphes orientés finis, que nous appelerons Σ-graphes de contrôle, et qui sont en fait la représentation graphique des Σ-automates finis.

(2) On peut les calculer par certains ensembles de sous-arbres initiaux finis, ou d'approximants finis, ensembles que l'on peut définir comme forêts rationnelles. (G. Cousineau [4]).

(3) On peut les représenter par des expressions itératives "vectorielles" (cf. C.C. Elgot, S.L. Bloom and R. Tindell [6]) définies dans une théorie algébrique (S.L. Bloom and C.C. Elgot [5]).

(4) On peut enfin les représenter par des expressions itératives "scalaires" (S.L. Bloom and C.C. Elgot [5], G. Cousineau [4]).

Situons ces 4 représentations du point de vue de la programmation des Σ-arbres réguliers. La formulation (2) permet de manipuler l'arbre T par ses approximants finis, et donne ainsi une bonne description d'une "sémantique opérationnelle". Les formulations géométrique (1) et algébrique (3) fournissent un bon modèle pour la description des Σ-arbres infinis réguliers par des "machines abstraites". La description (4) peut se décrire linéairement sous forme d'une suite de données, et se prête donc très bien au transfert d'information d'une représentation à une autre, ou d'une machine abstraite à une autre.

La notion de rang

Le décodage d'une expression itérative scalaire nécessite une gestion des sorties d'itération. Celle-ci peut se faire par l'allocation dynamique d'un tableau d'indices, qui, à chaque entrée d'itération, répertorie les "niveaux de sortie" de cette itération.

Une expression itérative scalaire est *de rang au plus* n ∈ ℕ si et seulement si elle peut être décodée par allocation dynamique d'un tableau de gestion des sorties d'itération, dont la taille est n.

Un Σ-arbre régulier infini est *de rang au plus* n ∈ ℕ si et seulement si il peut être codé par une expression itérative scalaire de rang au plus n.

Kosaraju [9] appelle "GRE_n-chartes" les chartes qui sont de rang au plus n. Pour tout entier n, il a donné une GRE_{n+1}-charte qui n'est réductible (pour la "weak equivalence", ou égalité pour toute interprétation des fonctions partielles calculées) à aucune GRE_n-charte.

On en déduit aisément (pour tout alphabet Σ ayant au moins deux symboles distincts dont l'un de arité au moins 2) que pour tout entier n, il existe un Σ-arbre régulier de rang n.

Aucun résultat n'a été publié à ce jour à propos de la question posée par Kosaraju dans le texte cité [9] : existe-t-il une caractérisation structurelle (i.e. en termes de structures des boucles dans le graphe des chemins) des GRE_n-chartes. Pour rendre efficace l'étude que nous avons ébauchée dans G. Jacob [10]), il nous fallait en effet posséder une caractérisation algorithmique, ce qui nous a conduit au présent travail.

Notons enfin que la notion de complexité étudiée par G. Cousineau [3], est celle de la hauteur d'étoile. Il donne un algorithme pour la calculer. Par contre, la définition du rang qu'il introduit, qui diffère au plus de 1 de la nôtre, ne fait l'objet d'aucun algorithme.

I. EXPANSIONS DES Σ-GRAPHES DE CONTROLE

Nous introduisons les notions sur un exemple, avant de les présenter plus formellement.

Σ-graphes de contrôle

Soit T le Σ-arbre régulier infini défini par le système propre d'équations régulières

$$(S) \quad T = T_1 = a \overset{\displaystyle T_2}{\underset{\displaystyle T_2}{\overset{T_3}{\Longleftarrow}}} \qquad T_2 = b \overset{\displaystyle T_3}{\underset{\displaystyle T_2}{\overset{T_1}{\Longleftarrow}}} \qquad T_3 = c \overset{\displaystyle d - T_2}{\underset{\displaystyle T_3}{\diagdown}}$$

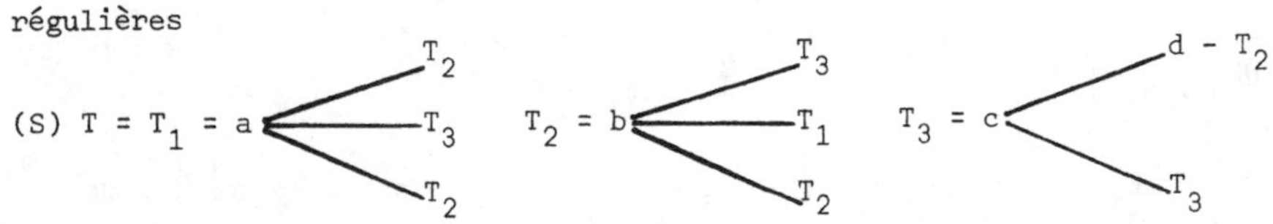

Les sous-arbres infinis distincts de T sont $T = T_1$, T_2, T_3 et $T_4 = d - T_2$.

En développant ces équations par substitution, nous obtenons des approximants finis de (S), que nous appelons *expansions* de (S). Par exemple :

(E_1)

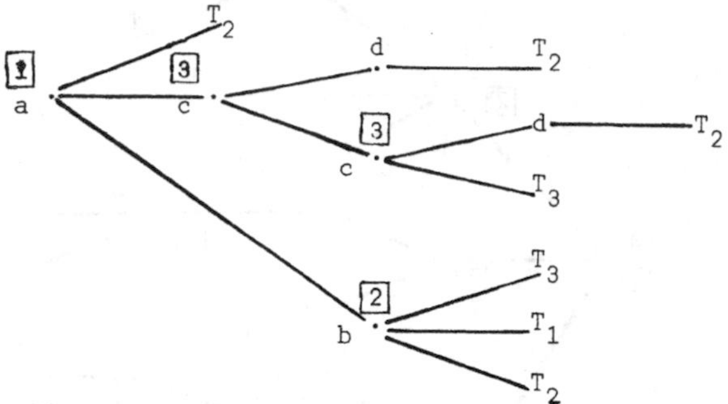

Les feuilles d'une expansion de S ont pour "valeur", ou étiquette les noms des arbres T_1, T_2 et T_3. Certains sommets, qui ne sont pas des feuilles, portent en *référence*, dans un □, l'indice de l'un des arbres T_1, T_2, T_3. Si T_i est la valeur d'une feuille, alors i est la référence d'un sommet au moins. On en déduit que toute expansion de (S) contient un codage de l'arbre infini $T = T_1$.

On peut choisir de réaliser l'expansion E_1 en remplaçant chaque feuille de valeur T_i par l'appel d'un sommet de référence i. En représentant cet appel par un arc, on obtient une réalisation de E_1, et donc de T_1, par un *Σ-graphe de contrôle.* Par exemple :

G_1

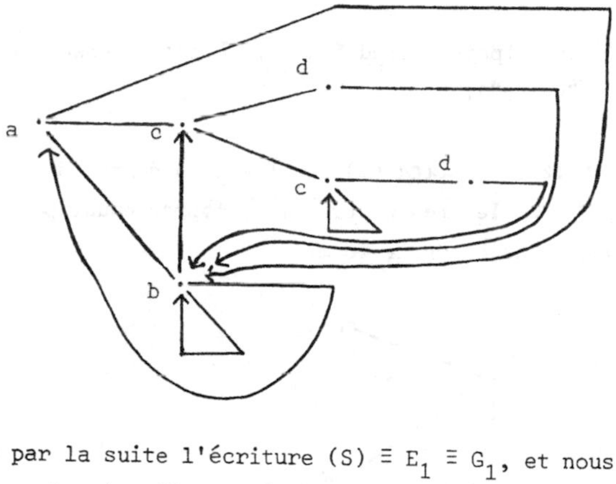

Nous admettrons par la suite l'écriture $(S) \equiv E_1 \equiv G_1$, et nous dirons que ce sont des représentations structurellement équivalentes du même arbre infini T_1.

Expansions stratifiées

Une expansion de (S) est dite *stratifiée* si chaque feuille de valeur T_i admet pour ancêtre au moins un sommet de référence i.

Comme T_1 est un Σ-arbre régulier, il admet au moins une expansion stratifiée. Il admet même une et une seule expansion stratifiée minimale. Nous donnons ici l'expansion stratifiée minimale de l'arbre infini T :

(F$_1$)

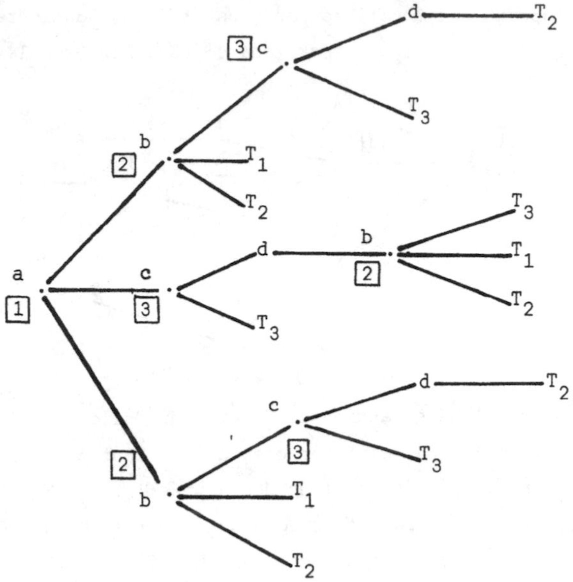

<u>**Reprenons cela plus formellement**</u> :

<u>Systèmes reguliers sur Σ</u>

Soit (Σ, δ) un alphabet gradué *fini*. (Σ est un ensemble fini, et $\delta : \Sigma \longrightarrow \mathbb{N}$ est l'application "arité").

<u>*Définition*</u> : Un *système régulier* (*T*) sur Σ est la donnée d'un vecteur $\vec{T} = (T_1, T_2, \ldots, T_n)$ de lettres distinctes (n'appartenant pas à Σ), et, pour chaque i ($1 \leq i \leq n$), d'une équation de la forme

(*T*)

soit T_i =

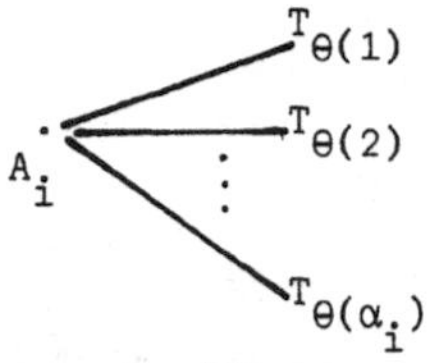

avec $A_i \in \Sigma$, $\delta(A_i) = \alpha_i$,

et $(\forall j \in [1, \alpha_i]), (\theta(j) \in [1, n])$

soit $T_i = B_i$

avec $B_i \in \Sigma$ et $\delta(B_i) = 0$

et où la lettre T_1 a été distinguée

Un tel système admet (voir G. Cousineau, Courcelle) un et un seul arbre (infini) solution, noté $A(\mathcal{T})$.

Pour définir les expressions de $(\mathcal{T})$, nous introduisons la grammaire d'arbre $G(\mathcal{T})$ suivante :

Pour chaque $i \in [1, \alpha_i]$, on a, suivant l'équation de T_i dans $(\mathcal{T})$, la règle de réécriture

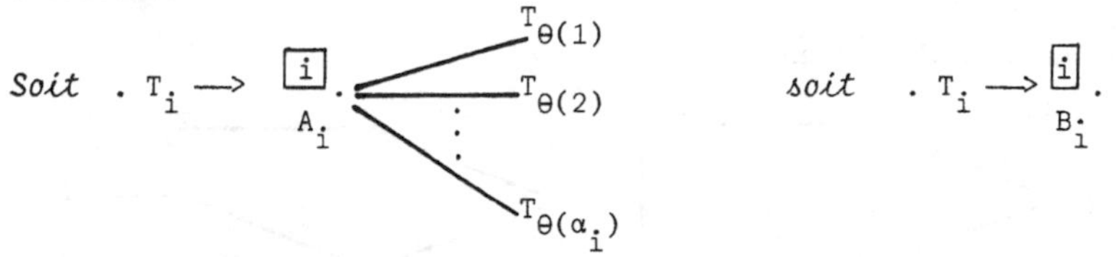

L'axiome est l'arbre réduit à un sommet

T_1

Tout arbre engendré par $G(\mathcal{T})$ a deux types de sommets :

- les "sommets d'appel" qui sont les feuilles étiquetées par une lettre T_i
- les "sommets de calcul" dont chacun
  - a une *référence* $\boxed{i}$ (avec $1 \leq i \leq n$)
  - a pour étiquette la lettre A_i (resp. B_i) correspondante dans le système $\mathcal{T}$,
  - et pour arité α_i (resp. 0).

Définition : Nous appelons *expansion* de $\mathcal{T}$ tout arbre E de $G(\mathcal{T})$ vérifiant

- si T_i est étiquette d'une feuille de E, alors $\boxed{i}$ est référence d'au moins un sommet de E.

Nous appelons *expansion stratifiée* de $\mathcal{T}$ tout arbre E de $G(\mathcal{T})$ vérifiant

- si T_i est étiquette d'une feuille ϕ de E, alors $\boxed{i}$ est référence d'au moins un ancêtre de ϕ.

II. EXPRESSIONS ITERATIVES SCALAIRES

L'itération scalaire d'une expression E s'écrit $\{E\}$. Nous montrons, en construisant une expression itérative scalaire structurellement équivalente à l'expansion F_1, comment apparaîssent les sorties indiciées ! j.

Considérons d'abord les deux expansions A et B extraites de F_1 :

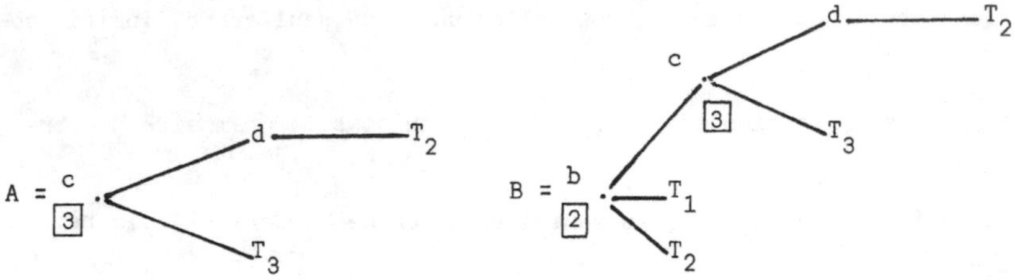

L'expansion A s'écrit encore :

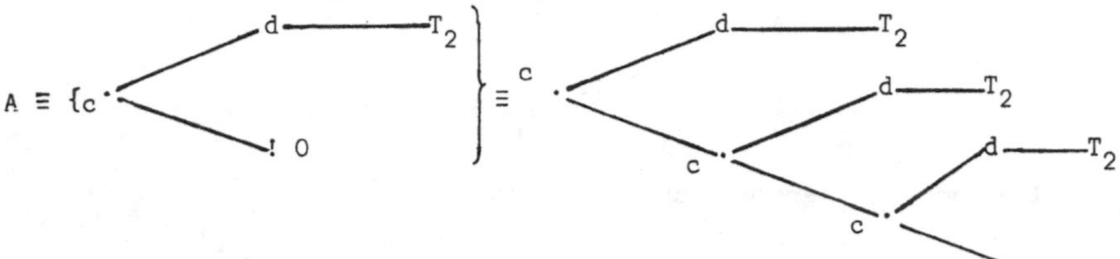

L'expansion B s'écrit :

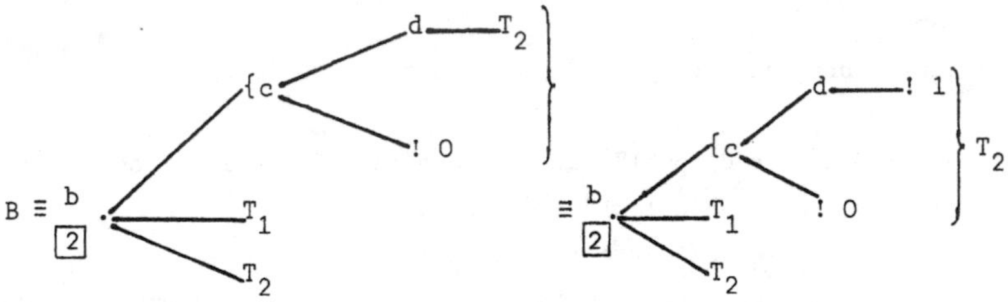

D'où l'écriture de B avec une nouvelle itération :

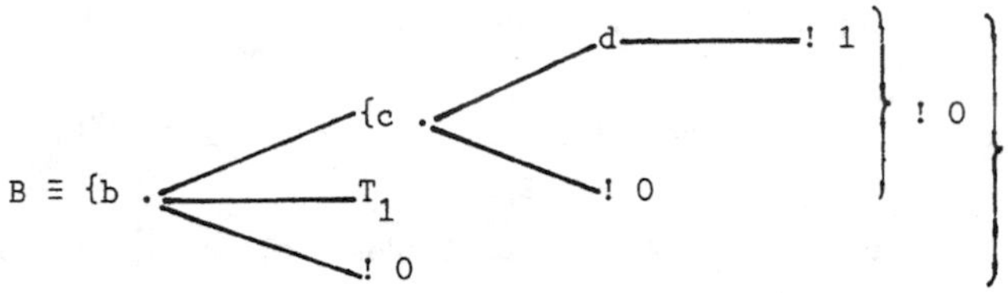

De même, on peut calculer :

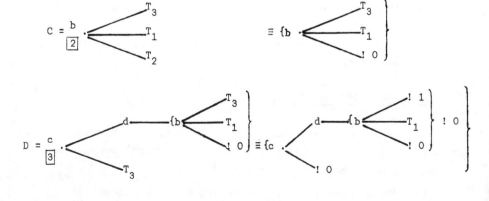

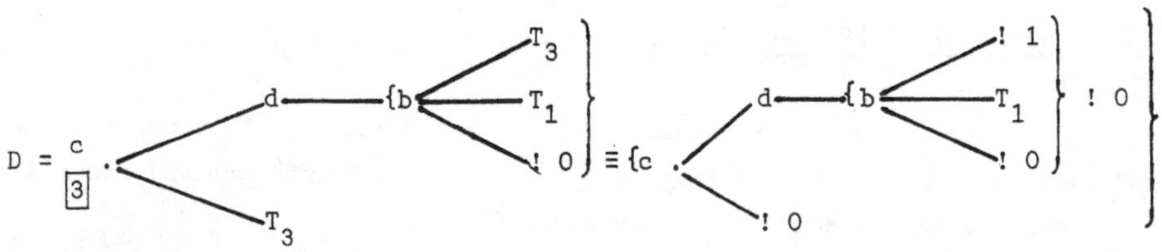

Et finalement, on obtient une expression itérative f_1 de F_1, que respecte la structure d'arborescence de cette expansion :

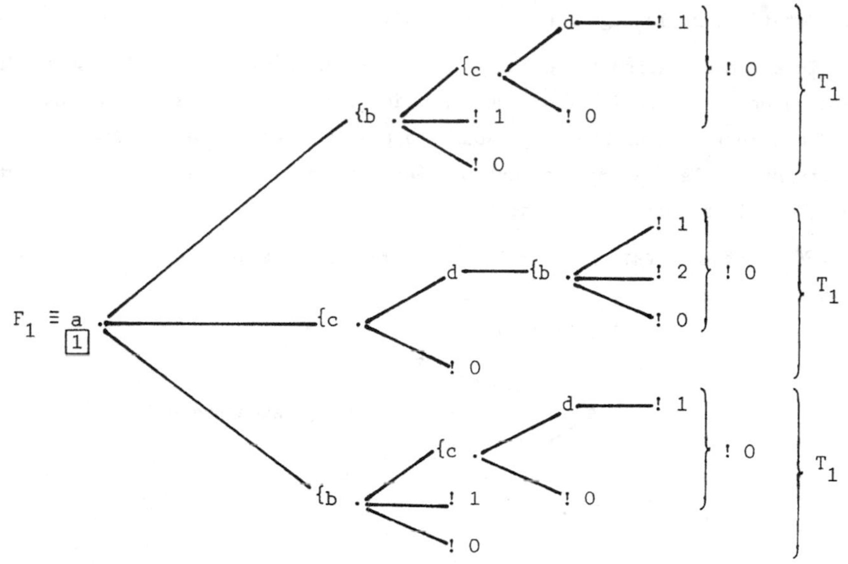

Nous écrivons finalement l'expression f_1 sous les deux formes, arborescentes et linéaires.

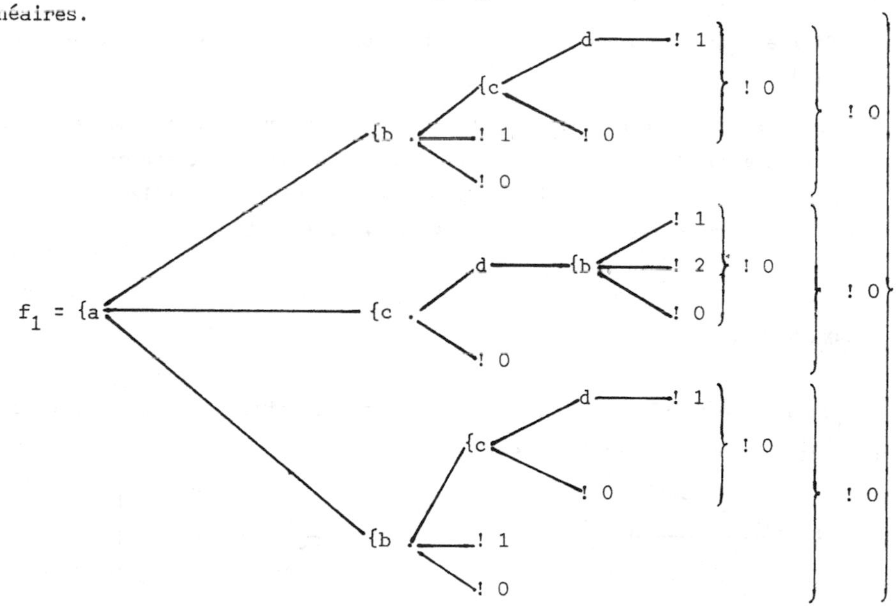

Et en représentation parenthésée des successeurs, pris "de haut en bas", on peut linéariser : f_1 = {a({b({c(d(! 1), ! 0)} ! 0, ! 1, ! 0)} ! 0,

{c(d({b(! 1, ! 2, ! 0)} ! 0), ! 0} ! 0,

{b({c(d(! 1), ! 0)} ! 0, ! 1, ! 0} ! 0}.

f_1 est une expression itérative scalaire *stratifiée* structurellement équivalente à l'arbre T. On dira encore que f_1 est une représentation itérative scalaire stratifiée de T. (Notations du langage EXEL : voir [1]).

Inversement, lorsqu'on donne une expression itérative scalaire stratifiée g, on peut en déduire canoniquement une expansion stratifiée qui est l'une des expansions structurellement équivalentes à g. Nous l'appelerons *expansion support* de g. Dans une telle expansion, les références sont portées par les sommets qui étaient débuts d'itérations, et elles sont toutes distinctes.

Plus formellement, l'ensemble FS des formules stratifiées sur Σ est défini par la grammaire d'axiome FS, et de règles

$FS \rightarrow\ !\ i \ (i \in \mathbb{N})$

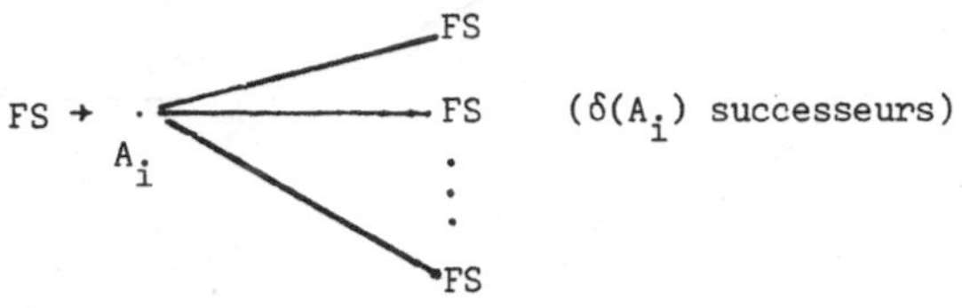

$FS \rightarrow \{FS\}$

Le signe de sortie ! i indique une sortie de i itérations emboîtées. Les accolades {...} peuvent se lire REPEAT ..... END.

Si f est une formule stratifiée, elle s'écrit comme une arborescence parenthésée par les { }.

En attribuant une référence différente $\boxed{j}$ à chaque sommet, chaque signe ! i est une feuille de l'arborescence, et elle renvoie à un sommet, par exemple de référence $\boxed{j}$. On remplace alors ! i par T_j, et les { } devenus inutiles peuvent être supprimés : on obtient ainsi l'*expansion stratifiée support* de f.

IV. RANG STRICT

On appelle *degré* d'une itération le nombre de "niveaux de sorties" de cette itération. Par exemple, les itérations

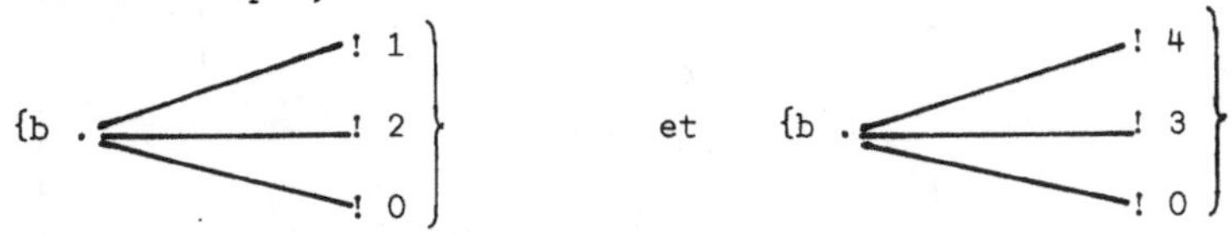

sont de degré 2. Par contre l'itération

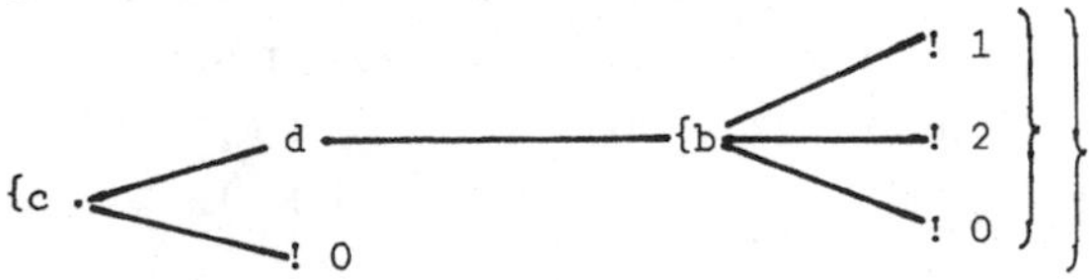

est de degré 1. Elle décrit un Σ-arbre régulier dont certaines feuilles ont pour valeur ! 0.

Tout ceci se formalise en introduisant la définition de "*valeur absolue*" d'une instruction de sortie, et la notion de "profondeur dans g d'une sous-formule de g" (Voir J. Arsac, G. cousineau).

On appelle *rang d'une expression* itérative stratifiée f le plus grand des degrés des itérations qui sont sous-expressions de f. (C'est donc bien la taille à prévoir pour un tableau gérant les sorties d'itération, et alloué dynamiquement à chaque entrée dans une itération).

Ainsi l'expression (ou formule) f_1 est de rang 2.

On appelle *rang strict* d'un Σ-arbre régulier infini T le plus petit des rangs des expressions itératives scalaires stratifiées représentant T (i.e. structurellement équivalentes à T).

V. CALCUL DU RANG STRICT

Il semble y avoir un lien certain entre le rang strict d'un Σ-arbre régulier infini, et certaines propriétés de ses graphes de contrôle, exprimables en termes de complexité de la structure de ses circuits (G. Jacob [19]). Cependant, ce lien n'a pas encore conduit à une caractérisation simple. (On peut s'en convaincre en calculant les "plus petits" Σ-graphes de contrôle de rang ≥ 3 sur des arbres binaires).

Notre but est ici la présentation d'un algorithme permettant le calcul du rang strict d'un Σ-arbre donné. Nous le mettrons en oeuvre sur l'exemple $T = T_1$, en montrant que cet arbre est en fait de rang strict égal à 1.

Notre algorithme part d'un système minimal d'équations régulières telles que (S). Nous allons progressivement adjoindre à (S) de nouvelles équations, dont les membres droits seront des itérations de rang d(sur l'exemple : d = 1).

Nous allons ainsi progressivement enrichir (S) en construisant des stades d'évolution (S | 1), (S | 2), etc... Pour passer de (S | n) à (S | n + 1), il faudra découvrir un sommet ($\boxed{1}$ ou $\boxed{2}$ ou $\boxed{3}$) qui soit "de degré strict au plus d pour le stade (S | n)".

Le stade (S | n) est formé de (Σ), et d'un certain nombre d'équations de la forme

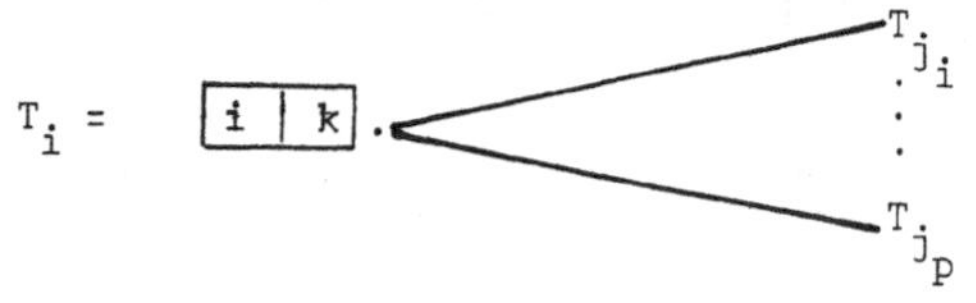

où $\boxed{i \mid k}$ est un "sommet itératif" ayant au plus d successeurs.

Un sommet $\boxed{i}$ de (S) est "de degré strict au plus d pour le stade (S | n)" si et seulement si on peut trouver une expansion de T_i dans (S | n).

- *de degré au plus* d, i.e. ayant au plus d valeurs de feuilles non constantes, dans tenir compte de la valeur T_i.
- telle que les sommets itératifs n'aient que des feuilles pour successeur.

En résumé, T_i admet "une bonne expansion de degré au plus d". On passe alors au stade (S | n + 1) par adjonction d'un sommet itératif $\boxed{i \mid n+1}$ et d'une équation correspondante.

Passons à l'exemple de l'arbre T donné par le système (S).

Le sommet $\boxed{3}$ est de degré 1, comme le montre l'expansion :

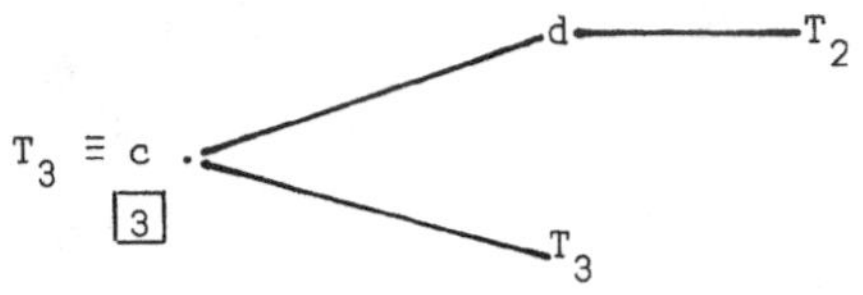

et le stade (S | 1) est obtenu en adjoignant le sommet itératif $\boxed{3 \mid 1}$ et l'équation :

$$T_3 \equiv \boxed{3 \mid 1} \text{———} T_2$$

Le sommet $\boxed{2}$ est de degré 1 pour le stade (S | 1), car on a la "bonne" expansion

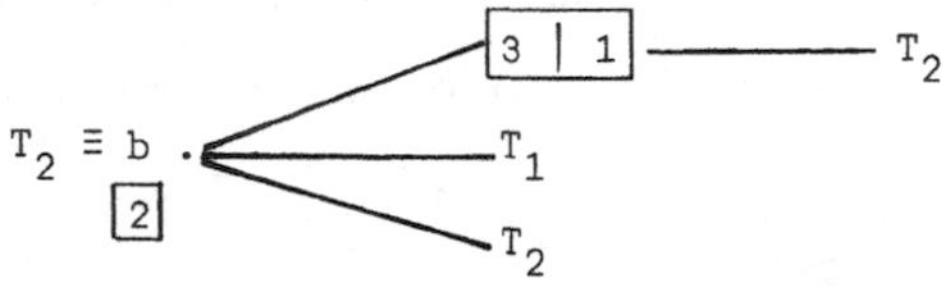

D'où l'adjonction du sommet itératif $\boxed{2 \mid 2}$ et de l'équation :

$$T_2 \equiv \boxed{2 \mid 2} \text{———} T_1$$

Nous sommes au stade (S | 2), et maintenant le sommet $\boxed{3}$ est de degré strict 1, grâce à la bonne expansion :

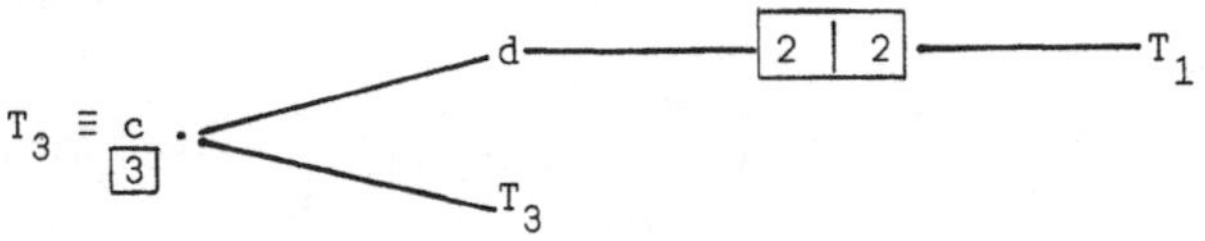

D'où le stade (S | 3), en adjoignant $\boxed{3 \mid 3}$ et l'équation :

$$T_3 \equiv \boxed{3 \mid 3} \text{———} T_1$$

A ce stade, T_1 devient un sommet de degré strict 0, grâce à la bonne expansion

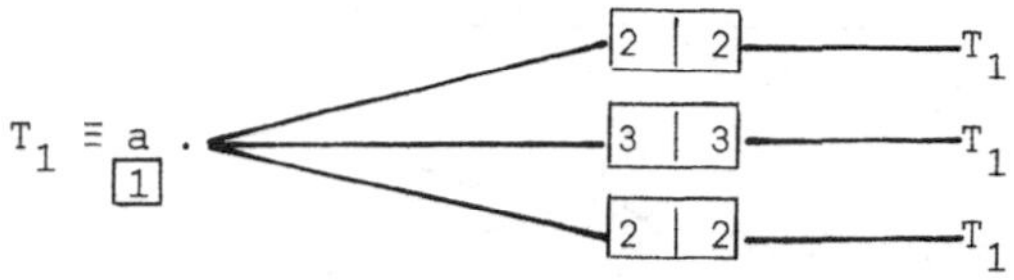

D'où le stade (S | 4), grâce au sommet [1 | 4] et à l'équation

$T_1 \equiv$ [1 | 4]

Au stade (S | 4), T_1 est donc réécrit comme un "sommet itératif" sans successeur (et donc sans "feuilles non constantes").

Nous en déduisons que T est un Σ-arbre de rang strict égal à 1, et que les constructions de notre algorithme sont en fait un codage d'une expression itérative stricte décrivant T.

Elle correspond en effet à l'expansion :

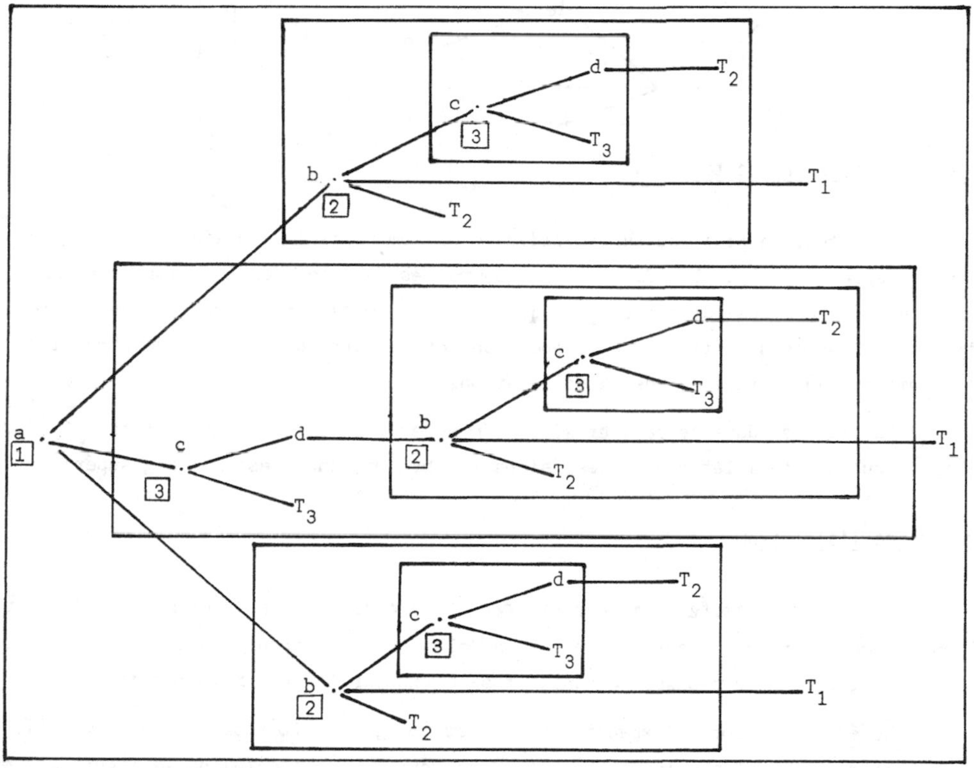

et fournit l'expression rationnelle de rang 1 :

g_1 = {a({b({c(d(! 1), ! 0)} ! 0, ! 1, ! 0)} ! 0,
{c(d({b({c(d(! 1), ! 0)} ! 0, ! 2, !0)} ! 0, ! 0)} ! 0,
{b({c(d(! 1), ! 0)} ! 0, ! 1, ! 0)} ! 0}.

Ce qui s'écrit en notation arborescente :

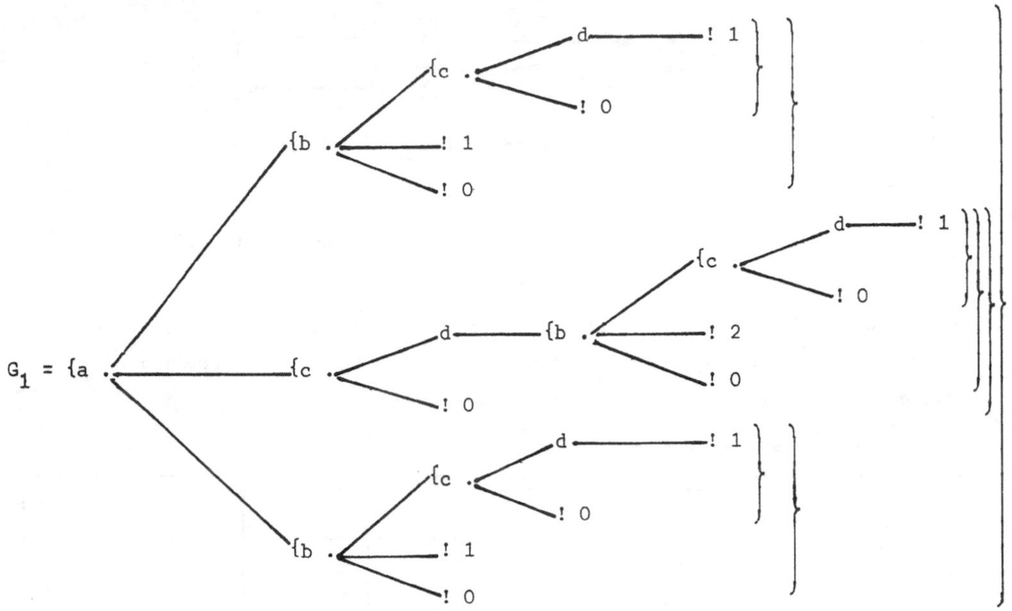

L'algorithme s'arrête toujours

En effet, on peut toujours décider en un nombre fini d'essais si un sommet donné est de degré strict au plus d. En outre, comme les feuilles des équations surajoutées sont indexées par les références $\boxed{i}$, qui sont en nombre fixé, après un certain nombre de pas, les seuls sommets itératifs que l'on pourra introduire seront des "rééditions" de sommets déjà décrits, et de leurs équations.

Si l'on est dans ce cas, et si l'on n'a pas trouvé de sommet itératif $\boxed{1 \mid m}$ n'ayant que des feuilles constantes, alors l'arbre rationnel est de rang supérieur à d.

Preuve de l'algorithme

Soit f *une formule itérative stratifiée de rang* d. On peut d'abord la ramener à une forme canonique, c'est-à-dire n'ayant plus de sous-formule de la forme {g} ! i, ou {{g}} et ceci sans changer le rang, et sans changer l'arbre régulier défini.

En écrivant alors l'expansion F support de f, on voit alors que les itérations découpent dans F un "emboîtement de blocs" du même type que celui de la figure, chaque bloc ayant au plus d feuilles T_i distinctes (dont le calcul est à chercher hors de bloc). On peut alors construire l'algorithme (pour le rang d) en remplaçant une feuille de profondeur maximale par un sommet itératif ayant au plus d successeurs ... et on continue avec ce nouveau sommet. Le calcul se terminera avec une formule réduite à un seul sommet, et dont les feuilles sont des symboles constants.

Inversement, si l'algorithme s'applique, on peut en partant de la définition successive des sommets itératifs, reconstruire une formule de rang d qui définit le même arbre régulier. De façon plus précise, on considère chaque bonne expansion de degré au plus d (au stade n) comme définissant une équation, dont la solution est une itération scalaire de degré au plus d. Comme dans une "bonne expansion" les sommets itératifs n'ont que des feuilles pour successeurs, on est ainsi assuré que les appels ! i renverront toujours *au début* d'une itération englobante.

V. EXPRESSIONS AVEC CONCATENATION

Dans l'expansion précédente, on recopie à trois reprises une même sous-expansion. La *concaténation* permet d'éviter ces "redites", en écrivant par exemple

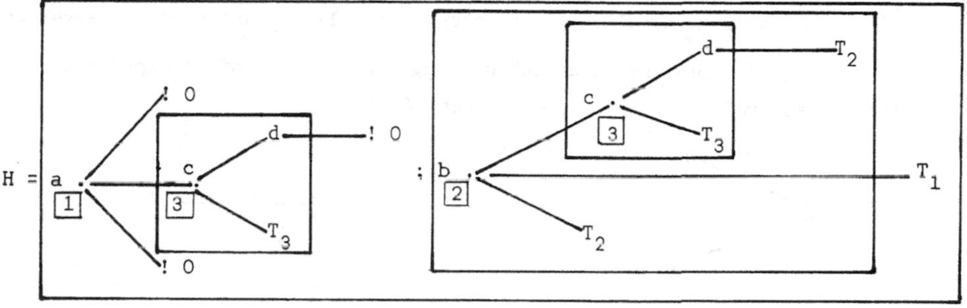

Soit encore

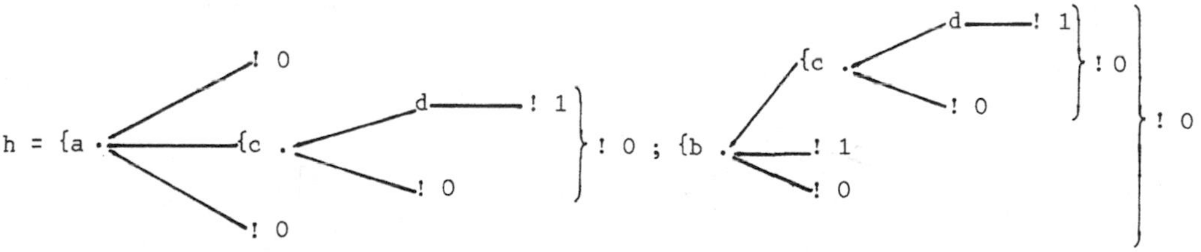

h = {a(! 0, {c(d(! 1), ! 0)} ! 0, ! 0) ; {b({c(d(! 1), ! 0} ! 0, ! 1, ! 0} ! 0}

Le *support* de h est le Σ-graphe de contrôle

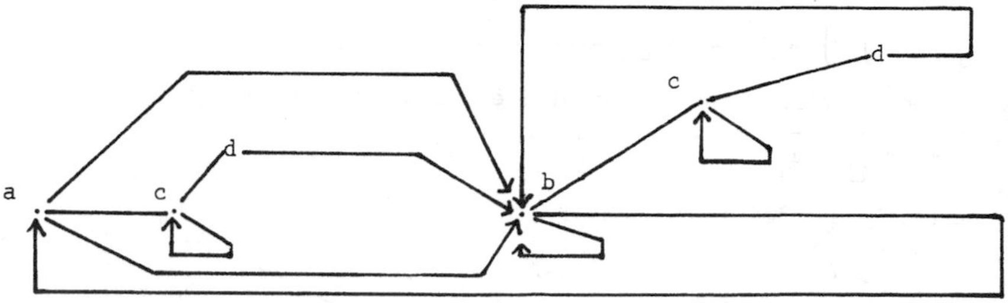

h = f ; g peut aussi être décrit canoniquement par une expansion stratifiée, mais cette *expansion canonique* peut nécessiter des recopies de g.

VI. LE CALCUL DU RANG

Nous considérons donc à présent les expressions itératives avec concaténation.

Le *degré* d'une itération est le nombre de ses niveaux de sortie. Le *rang* d'une expression itérative rationnelle est le plus grand degré de ses sous-expressions.

Le *rang d'un Σ-arbre régulier* est le plus petit rang des expressions itératives avec concaténation qui le décrivent.

Pour calculer le rang d'un Σ-arbre régulier infini, on reprend l'algorithme du calcul du rang strict en le modifiant comme suit :

pour passer du stade d'évolution $(S \mid n)$ au stade $(S \mid n + 1)$, il faut trouver un sommet quelconque de $(S \mid n)$ de la forme $\boxed{i}$, ou $\boxed{i \mid k}$, admettant dans $(S \mid n)$ une expansion ayant au plus d valeurs distinctes de feuilles, sans compter la valeur T_i.

Par exemple, l'algorithme de calcul du rang permet d'obtenir l'expression H structurellement équivalente à $T = T_1$, et de rang égal à 1 :

$$T_3 \equiv c \cdot \boxed{3} \quad \begin{array}{l} d \longrightarrow T_2 \\ \\ T_3 \end{array} \qquad\qquad T_3 \equiv \boxed{3 \mid 1} \longrightarrow T_2$$

$$T_2 \equiv b \cdot \boxed{2} \quad \begin{array}{l} \boxed{3 \mid 1} \longrightarrow T_2 \\ T_1 \\ T_2 \end{array} \qquad\qquad T_2 \equiv \boxed{2 \mid 2} \longrightarrow T_1$$

$$T_1 \equiv a \cdot \boxed{1} \quad \begin{array}{l} T_2 \\ \boxed{3 \mid 1} \longrightarrow T_2 \\ T_2 \end{array} \qquad\qquad T_1 \equiv \boxed{1 \mid 3} \longrightarrow T_2$$

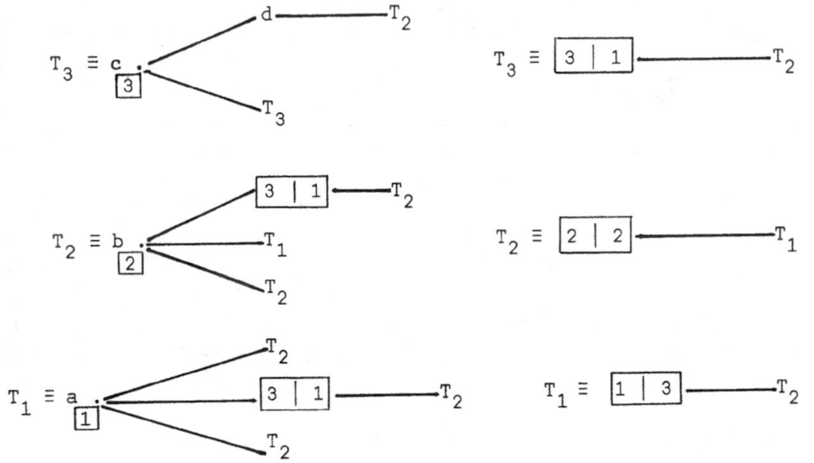

Tout ceci restait dans le cadre de l'algorithme précédent, mais on écrit à présent :

$$T_1 \equiv \boxed{1 \mid 3} \longrightarrow \boxed{2 \mid 2} \longrightarrow T_1$$

expansion de degré 0, introduisant le sommet itératif $\boxed{1 \mid 4}$ et l'équation

$$T_1 \equiv \boxed{1 \mid 4}$$

Par contre, le système d'équations :

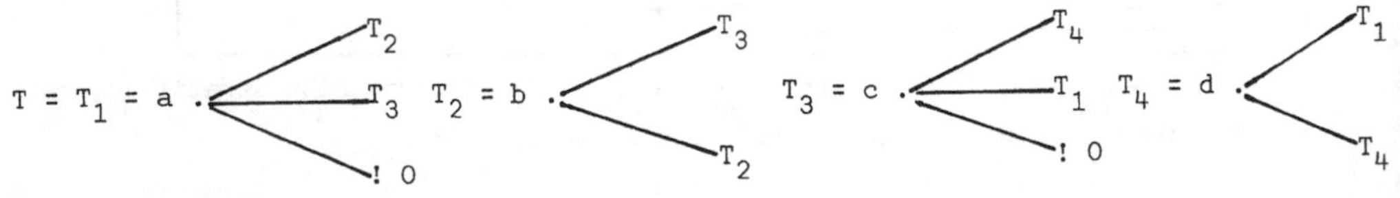

définit un Σ-arbre régulier infini de degré 1 (en raison des feuilles de valeur ! 0) dont le *rang strict* est 2, et dont *le rang* est 1, comme le montre l'expression itérative avec concaténation structurellement équivalente :

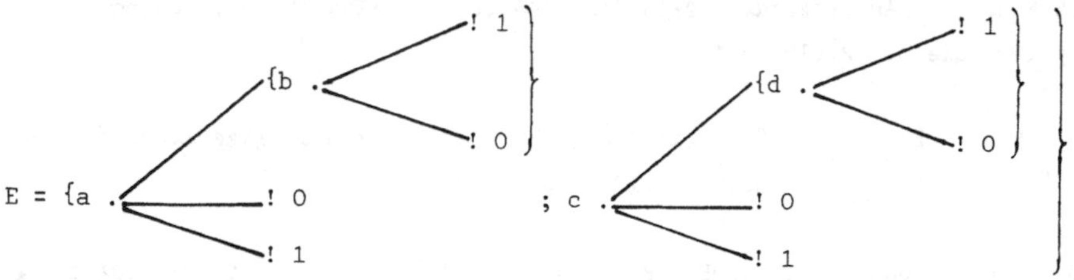

On peut donner aussi un Σ-arbre régulier de degré 0 de rang strict égal à 2 et de rang égal à 1.

Par exemple, celui donné par le système d'équations

$$R = R_1 = a \cdot \begin{cases} R_2 \\ R_3 \\ R_5 \end{cases} \quad R_2 = b \cdot \begin{cases} R_3 \\ R_2 \end{cases} \quad R_3 = c \cdot \begin{cases} R_4 \\ R_1 \\ R_5 \end{cases} \quad R_4 \ d \cdot \begin{cases} R \\ R_1 \end{cases}$$

$$R_5 = a' \cdot \begin{cases} R_6 \\ R_7 \\ R_1 \end{cases} \quad R_6 = b' \cdot \begin{cases} R_7 \\ R_6 \end{cases} \quad R_7 = c' \cdot \begin{cases} R_8 \\ R_5 \\ R_1 \end{cases} \quad R_8 = d' \cdot \begin{cases} R_1 \\ R_1 \end{cases}$$

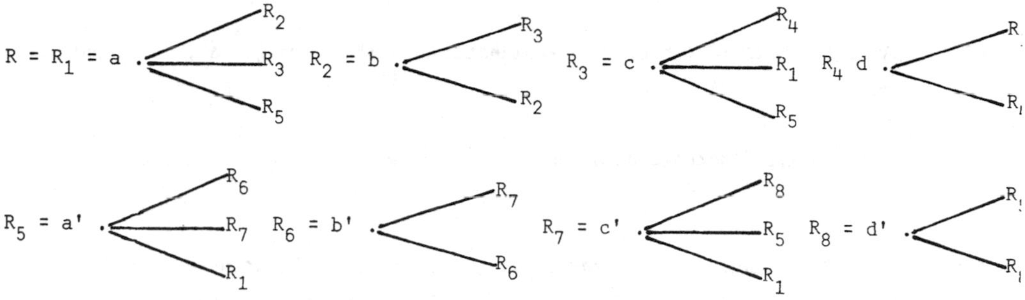

L'expression itérative F avec concaténation *de rang 1* décrivant R est obtenue comme suit : Notons E' une copie de l'expression E où l'on a "primé" toutes les lettres graduées. On obtient alors

$$F = \{E \ ; \ E'\}$$

Preuve : Elle se fait comme celle du calcul du rang strict, (par induction sur la hauteur d'itération).

Bibliographie

[1]   ARSAC J., NOLIN L., RUGGIU G., VASSEUR J.P., *"Le système de Programmation EXEL"*, Revue technique Thompson-CSF, vol. 6, 3 (1974).

[2]   COUSINEAU G., *"Transformations de programmes itératifs"*, in Programmation, Proc. of the 2nd international symposium on Programming, B. Robinet Ed., Paris (1976-DUNOD) 53-74.

[3] COUSINEAU G., "*Arbres à feuilles indiciées et transformations de programmes*", Thèse es Sci-Mathématiques, Université de Paris VII (1977).

[4] COUSINEAU G., "*An algebraic definition for control structures*", Theoretical computer Science 12 (1980) 175-192.

[5] BLOOM S.L., ELGOT C.C., "*The existence and Construction of free iterative theories*", J. Comput. Syst. Sci. 12 (1976) 305-318.

[6] ELGOT C.C., BLOOM S.L., TINDELL R., "*On the algebraic structure of Rooted Trees*", J. Comput. Syst. Sci. 16 (1978) 362-399.

[7] IANOV I.I., "*The logical schemes of algorithms*", Problemy Kibernet., 1 (1960) 82-140.

[8] KASAI T., "*Translatability of flowcharts into WHILE programs*", J. Comput. Syst. Sci. 9 (1974) 177-195.

[9] KOSARAJU R., "*Analysis of structured programs*", J. Comput. Syst. Sci., 9 (1974) 232-255.

[10] JACOB G. "*Structural Invariants for some classes of structured programs*", MFCS 78, Zakopane (Poland) ; Lect-Notes in Comput. Sci. n° 64.

A class of tree-like UNION-FIND data structures and the nonlinearity

Marek J. Lao

Institute of Informatics
Warsaw University
PKiN VIII p. skr. poczt.1210
00-901 Warszawa, Poland

Key words. computational complexity, data structures, set union, trees, UNION-FIND

0. Abstract

This paper defines a class of tree-like data structures for the UNION-FIND problem. A structure from this class is injectable in another if each tree in the latter one can be obtained as a result of some program in the former as well. By means of injection of structures the nonlinearity in this class is proved.

1. Introduction

UNION-FIND data structures should represent a set partitioned into disjoint classes and make it possible to execute operations of the following types:

-FIND(x) computes the unique class which contains the element x,

-UNION(A,B) combines the classes A and B into a new class.

At first UNION-FIND data structures were used in FORTRAN compilers for the EQUIVALENCE declaration [2]. They have also many applications in set and graph processing [2,3], finding minimum spanning trees [2,4] and a few others. There exist a number of algorithms [1,3,6] which use lists or trees to represent classes of the partitioned set. Those using trees compute an UNION-FIND program in a nearly linear time (with respect to the number of UNIONs and FINDs in the program).

For many years it was an open question if there was no linear time algorithm for the UNION-FIND problem. Recently Tarjan in [8] has proved that every UNION-FIND algorithm which can be executed on a pointer machine requires nonlinear time in worst case. That class encompasses all known UNION FIND algorithms which use trees or lists.

This paper is an extension of the idea given in [6]. I conjectured there that every UNION-FIND data structures is a special kind of very low trees and there -

fore in order to prove the nonlinearity of the problem it should be sufficient to consider only this class of algorithms. This paper defines a generalization of very low trees and proves the nonlinearity in this class without taking account of any special models of computations. However Tarjan's method is more general and because of his result this paper describes only one algorithm with various data structures.

2. The structure S_F and the algorithms

Each class of elements of the partitioned set is represented by a tree. The nodes correspond to the elements of the class. The root identifies the class. Any function $F: N \to N$ such that $F(0)=1$ and $F(h) \geq 2$ (for $h \geq 1$) defines one of the UNION-FIND data structures. We denote this structure by S_F. The function F defines a minimal number of trees required in order to construct a taller tree.

In structures of this type, each node contains a father edge. Passing those edges from the node to the root we can execute the simplest version of FIND Each root has some additional information such as:

- the level of the tree *),
- the list of the tallest subtrees,
- the length of any single element tree is equal to 0.

In order to reduce the costs of UNION -FIND programs the following rules are often used [1,3,6,7] :

Weighted union rule for lists. The shorter list is merged to the longer one.

Weighted union rule for trees. The lower tree is joined to the taller one as a son of its root.

Collapsing rule. After a FIND operation all nodes on the path become sons of the root (i.e. the path is compressed).

Trees in S_F are built by the following algorithm of UNION:

```
procedure UNION_F (A,B) ;
  if the level of A and the level of B are not equal
    then {weighted union rule for trees}
      wlg the level of A is greater than the level of B
        otherwise change A and B in
        make all the subtrees in B's list sons of A {Fig. 1}
    else
```

---

*) This attribute is equal to the height of the tree if during FIND operations the path compression is not made.

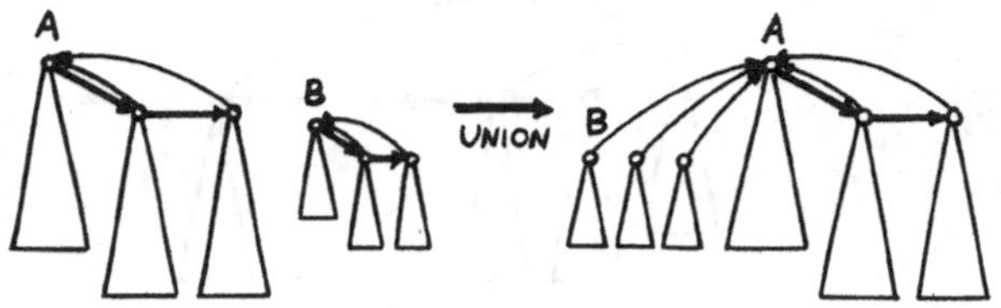

Figure 1.

UNION(A,B) in the case when trees are not of the same level. The subtrees of the lower tree are joined to the taller one.

```
wlg  A's list is longer than  B's
  otherwise  change  A  and  B  in
  begin
    {weighted union rule for lists}
    change all the 'father edges' in the roots of the subtrees in B's
    list;
    merge B's list to A's   {Fig. 2} ;
    if  the length of the new list is greater than or equal to F(A.level)
      then  {a new taller tree A  is being created }
        begin
          A.level := A.level+1;
          A.list := A;
          A.length := 1
        end
  end {of UNION} .
```

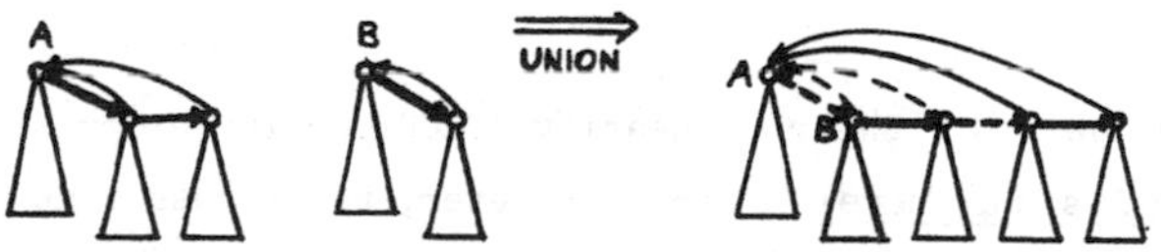

Figure 2.

UNION(A,B) when the levels of the trees are equal. The shorter list of subtrees is merged to the longer one.

We call a tree which contains a single element list a complete tree. Others are called incomplete {Fig. 3}.

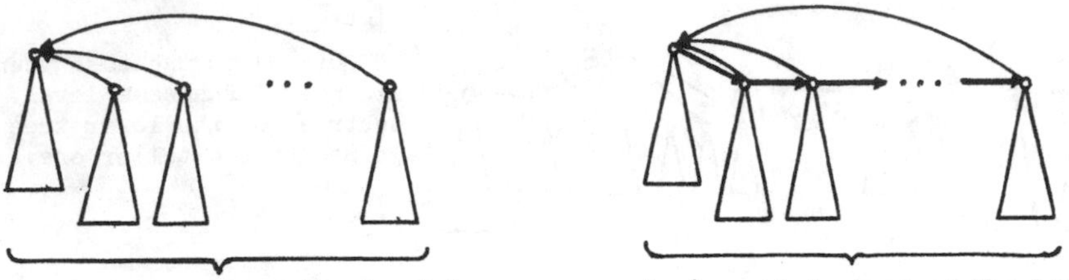

A complete tree of the level h
−F(h) subtrees, each of the level of h-1, at least

An incomplete tree of the level h-1
− not more than F(h)-1 subtrees of the levels of h-1

Figure 3.

A complete and an incomplete tree in S_F

In order to carry out FIND(x) we follow 'father edges' from the element x to the root of the tree which contains it. All the nodes reached during this FIND become sons of the root (Fig. 4). The path compression do not change any other attributes of nodes.

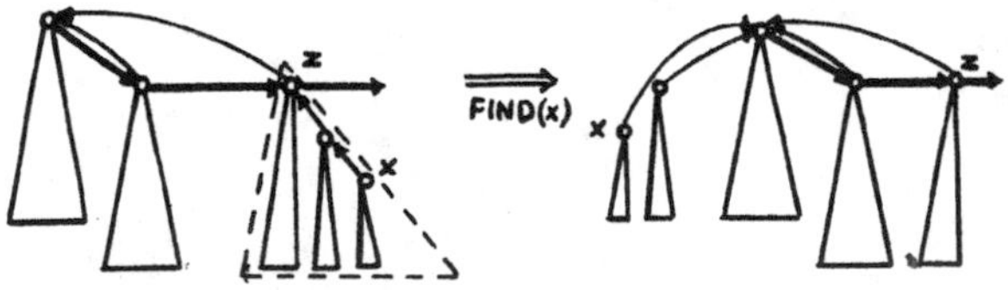

Figure 4.

Path compression during FIND(x)

3. Injection of structures

For every function F we prove the nonlinearity of UNION-FIND programs in S_F by means of injection of structures . First, however, we discuss some facts about the cost of UNIONs. In much the same way as in [6] we can prove that in S_F each node of the level of h can be joined by UNIONs $\log F(h+1)$ times at most. In any n-element tree the number of nodes of the level not less than h is not greater than $n/(F(0)\dots F(h))$. Therefore, the upper bound of the cost of UNIONs which build an n-element tree is of the range of

$$n \sum_{h=0}^{l} \frac{\log F(h+1)}{F(0)\dots F(h)}$$

where l is the level of the tallest tree consisting of n elements.

The lower bound is of the same range. The proof can be also done in the way like in [6] . First we prove that p elements may be merged into a list by p-1 UNIONs which change 'father edges' at least $p \lfloor \log p \rfloor / 4$ times. Then by induction on the height h we obtain the result that in each S_F a complete tree of this height may be created by a program of $F(0)\ldots F(h)-1$ UNIONs which build

$$\frac{1}{4} F(0)\ldots F(h) \sum_{i=0}^{h-1} \frac{\lfloor \log F(i+1) \rfloor}{F(0)\ \ldots\ F(i)}$$

'father edges'. As a corollary follows the lower bound of the range given above.

Now we introduce the notion of injection.

Definition 1.

S_F is injectable in S_G iff for every tree T_F from S_F there exists a tree T_G from S_G such that:

(i) the level of T_G is equal to the level of T_F,

(ii) T_F is a subtree of T_G (i.e. T_F is injected in T_G).

Example 1.

Let $F(h)=2$ and $H(h)=3$ for every $h \geq 1$. Then S_F is injectable in S_H (Fig. 5).

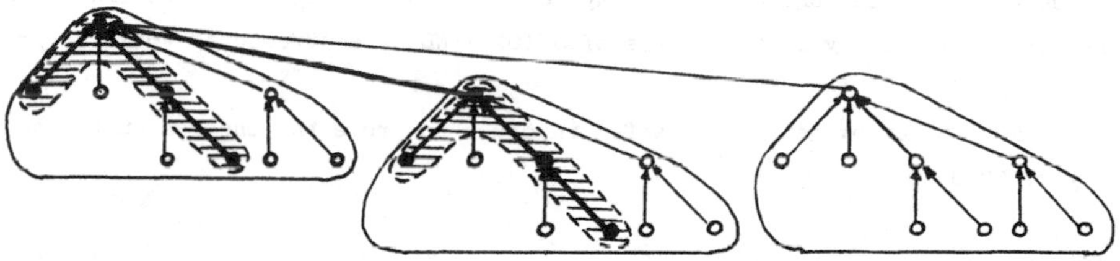

Figure 5.
The injection of structures (Example 1). The injected tree T_F is tinted.

From the definition follow properties given bellow.

Property 1.

S_F is injectable in S_G iff $2 \leq F(h) \leq G(h)$ for every $h \geq 1$.

Property 2.

If S_F is injectable in S_G then for every sequence of UNIONs which constructs T_F there exists a sequence of UNIONs such that

(i) it constructs corresponding tree T_G in S_G,

(ii) the sequence constructing T_F is a subsequence of it.

Property 3.

S_F is injectable in S_G iff every complete tree in S_G can be obtained as a result of some program using $UNION_F$ operations as well as $UNION_G$.

Property 3 can be proved from Property 1 by induction on the level of the tree in S_G.

Now we are able to formulate the theorem on the nonlinearity of FINDs.

Theorem 1. (nonlinearity of FINDs and injection of structures)

Let S_F be injectable in S_G and let the cost of FINDs in S_G be not a linear function of the number of elements and FINDs.

Then in S_F the cost of FINDs is not linear, either.

This theorem follows from Property 3 and from the fact that the algorithm of FIND is not related to any function. The theorem can be formulated stronger. The lower bound of FINDs' cost in S_F is not less than in S_G. From this theorem we can prove the nonlinearity of the UNION-FIND problem finding a structure S_G in which every structure of linear UNIONs' cost may be injected. If S_G had a nonlinear cost of FINDs, the nonlinearity in this class of UNION-FIND data structures would be proved.

The following notion is useful in order to prove the theorem on the nonlinearity of UNIONs.

Definition 2.

S_F is strongly injectable in S_G iff

$$\frac{\log F(h+1)}{F(0)\ \dots\ F(h)} \leqslant \frac{\log G(h+1)}{G(0)\ \dots\ G(h)} \qquad \text{for every } h .$$

By induction on h we can prove that $F(h) \leqslant G(h)$ follows from the condition of strong injection. Thus every strongly injectable structure is also an injectable one.

Theorem 2. (nonlinearity of UNIONs and injection of structures)

Let S_F be strongly injectable in S_G and let the cost of UNIONs in S_F be not a linear function of the number of UNIONs.

Then in S_G the cost of UNIONs is not linear, either.

Proof. (ad absurdum)

Let us assume that the cost of UNIONs in S_F is not linear and it is linear in S_G. Let the upper bound of the cost of $n-1$ UNIONs in S_G be equal to cn. Then

$$\sum_{h=0}^{l} \frac{\log G(h+1)}{G(0)\ \dots\ G(h)} \leqslant c \qquad \text{for every } l .$$

From the nonlinearity of UNIONs in S_F it follows that there exists a program consisting of k UNIONs such that its cost is greater than ck. Because of the strong injection, the cost of this program is limited by

$$k \sum_{h=0}^{l} \frac{\log F(h+1)}{F(0)\ \dots\ F(h)} \leqslant k \sum_{h=0}^{l} \frac{\log G(h+1)}{G(0)\ \dots\ G(h)} \leqslant c\cdot k$$

where l is the height of the tree created by this program in S_F.
It contradicts the assumption of the nonlinearity in S_F.

□

The following example shows that the property of injection is not sufficient to prove the nonlinearity.

Example 2.

Let $F: N \to N$ be defined as follows

$$\begin{cases} F(0) = 1 \\ F(i) = 2^{F(i-1)} & \text{for } i \geqslant 1 . \end{cases}$$

It is easy to show that in S_F the cost of UNIONs is linear.

Let $f: N \to N$ be

$$\begin{cases} f(0) = 1 \\ f(i) = \dfrac{F(i)}{F(i-1)} & \text{for } i \geqslant 1 . \end{cases}$$

It has been proved in [6] that the cost of UNIONs in S_f is not linear.
Because $f(i) \leqslant F(i)$, S_f is injectable in S_F. It is not strongly injectable, though.

4. The subclass of structures which have linear costs of UNIONs

First we define function $G: N \to N$ in the following way:

$$\begin{cases} G(0) = 1 \\ G(1) = 2 \\ G(h) = G(h-1)^{G(h-1)} & \text{for } h \geqslant 2 . \end{cases}$$

Let G_j denote the j-th shift of G, i.e.

$$\begin{cases} G_j(0) = 1 \\ G_j(h) = G(j+h-1) \qquad \text{for } h \geq 1,\ j \geq 1 . \end{cases}$$

Therefore, S_G is injectable in each of S_{G_j}-s.

Now let us consider a structure S_F with a linear cost of UNIONs. The tight bound of UNIONs' cost is of the range of

$$n \sum_{h=0}^{l} \frac{\log F(h+1)}{F(0) \ldots F(h)} \qquad l \text{ - the level of the tallest n-elément tree in } S_F$$

This series must be convergent in order to reach the linear cost. By d'Alemert's criterion, for some K and every $h \geq K$ the following condition must hold

$$\frac{\log F(h+2)}{F(0) \ldots F(h)F(h+1)} \leq \frac{\log F(h+1)}{F(0) \ldots F(h)}$$

So

$$\log F(h+2) \leq F(h+1) \cdot \log F(h+1)$$

$$F(h+2) \leq F(h+1)^{F(h+1)}$$

There exists j such that

$$G(j) \geq \max \{ F(i) : i \leq K \}$$

Thus $G_j(i) \geq F(i)$ for every i, so S_F is injectable in S_{G_j}. The nonlinearity of FINDs in each of S_{G_j}-s is proved in Appendix. Therefore, by Theorem 1 we complete the proof of the nonlinearity of the UNION-FIND problem in this class of structures.

5. An extension of structures S_F

We consider a class of structures with the following properties:

- the function defining the number of subtrees required to create a taller tree may be related to the number of UNIONs and FINDs to be executed,
- the function is fixed at the first step of the program and cannot be changed.

So by a structure we call a set of functions $\{F_{n,m}\}$ now. Notice that $F_{n,m}$ does not have to be defined for each argument. Let us assume its value is equal to 0 in that case.

At first the UNION-FIND program chooses the function with respect to $n-1$ the number of UNIONs, and m, the number of FINDs. Then UNIONs and FINDs are being executed in $S_{F_{n,m}}$

Suppose that in this class there exists a structure realizing the linear cost of UNION-FIND programs. That means for every n and m the cost is less than $c(n+m)$ for some $c>0$. From the upper bound of the cost of UNIONs it must hold that for every n,m and i

$$\frac{\log F_{n,m}(i+1)}{F_{n,m}(0)\ \dots\ F_{n,m}(i)} \leqslant c(1+\frac{m}{n})$$

Let us choose an arbitrary value $w>1/4$ and let L_n denote $F_{n,w\cdot n}$. Then the above condition is equivalent to

$$\frac{\log L_n(i+1)}{L_n(0)\ \dots\ L_n(i)} \leqslant c\ (1+w)$$

Then in order to have a linear cost of UNIONs the function L_n must satisfy the following conditions:

$$L_n(1) \leqslant 2^{c(1+w)}$$

$$L_n(2) \leqslant 2^{c(1+w)\ L_n(1)}$$

$$\vdots$$

$$L_n(i) \leqslant 2^{c(1+w)\cdot L_n(1)\ \dots\ L\ (i-1)}$$

$$\vdots$$

These conditions taken with equalities give us the function:

$$\begin{cases} H(0) = 1 \\ H(1) = 2^{c(1+w)} \\ H(i) = H(i-1)^{H(i-1)} & \text{for} \quad i\geqslant 2\ . \end{cases}$$

For each n, S_{L_n} is <u>weakly injectable</u> in S_H. That means, for every i $L_n(i) \leqslant H(i)$, but some values of L_n may be equal to zero. So we are not able to obtain the definability of trees in S_H by a program in S_{L_n}.

<u>Property 4.</u>

If S is weakly injectable in S' then the maximal height of an n-element tree in S is not less than the maximal height of an n-element tree in S'.

From the weak injection of S_{L_n}-s and Property 4 follows that for each value of n there exists an n-element tree in S_{L_n}, not lower than any n-element tree in S_H. FINDs are not linear in S_H. Therefore by Theorem 1, there exists such n that the cost of some sequence of $n/4$ FINDs is not less than $n/4 + c(1+w)$ (see Appendix). Adding FINDs, each of the cost of 1, we obtain the total cost of FINDs greater than

$$m+n\cdot c(1+w) = m + c(n+m)\ .$$

So the assumption that the upper bound of the cost in S_{L_n} can be $c(n+m)$ does not hold. This way we have proved the nonlinearity of the UNION-FIND problem in any of structures $S_{F_{n,m}}$.

6. Final remarks

In the same way as in the previous section we can prove the nonlinearity of the involved algorithm in the case of a function F which is defined (deterministically or not) with respect to the history of the program being executed. We can also prove the nonlinearity if the structure depends on some other attributes (like the size of the tree), not only on the level of the tree. The algorithm of FIND may also change values of this attribute.

Because of Tarjan's result which is more general in use than the presented above, the properties of such structures are not very interesting however. Taking account of Tarjan's result we can show that every structure defined in Section 2 has the same range of FINDs' cost, equal to $\Theta(m \cdot \alpha(m,n))$, where α is a very slowly growing function [7,8] . Because there are many structures with linear costs of UNIONs it could be possible to determine the optimal structure in the way this paper has described.

References.

[1] A.V. Aho, J.E. Hopcroft, J.D. Ullman, "The design and analysis of computer algorithms", Addison-Wesley, Reading, MA, 2-nd printing, 1975, 124-139

[2] B.W. Arden, B.A. Galler, R.M. Graham, An algorithm for equivalence declarations, Comm. ACM 4 (7) (1961) 310-344

[3] J.E. Hopcroft, J.D. Ullman, Set merging algorithms, SIAM J. Comput. 2 (1973) 294-303

[4] A. Kirschenbaum, R. van Slyke, Computing minimum spanning trees efficiently, Proc. 25th Annual Conf. of the ACM (1972) 518-527

[5] D.E. Knuth, A. Schönhage, The expected linearity of a simple equivalence algorithm, Theoret. Comput. Sci. 6 (3) (1978) 281-315

[6] M.J. Lao, A new data structure for the UNION-FIND problem, Inform. Processing Lett. 9 (1979) 39-45

[7] R.E. Tarjan, Efficiency of a good but not linear set union algorithm, J. ACM 22 (2) (1975) 215-225

[8] R.E. Tarjan, A class of algorithms which require nonlinear time to maintain disjoint sets, J. Comp. System Sci. 2 (1979) 110-127

Appendix

<u>The nonlinearity of FINDs in structures S_{G_j}</u>

We consider the function $G: N \rightarrow N$ defined by

$$\begin{cases} G(0) = 1 \\ G(1) = 2 \\ G(i) = G(i-1)^{G(i-1)} & \text{for } i \leqslant 2 \end{cases}$$

Let T be any tree (not necessary from S_G). Let $\mathcal{T}(T,h,l)$, where $h \geqslant 1$ and $l \geqslant 1$, denote a tree constructed in the following way:

- $\mathcal{T}(T,h,1)$ from $G(h)$ copies of T (Fig. 6),
- $\mathcal{T}(T,h,i)$ as
  $\mathcal{T}(\mathcal{T}(T,h,1),h+1,i-1)$ for $i \geqslant 2$.

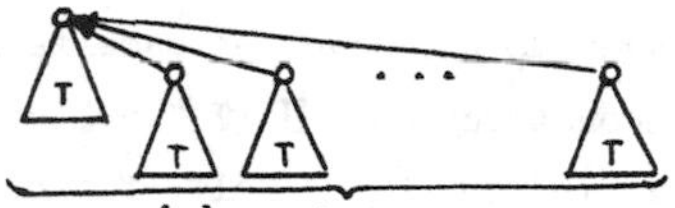

G(h) copies of T

<u>Figure 6.</u>
A tree $\mathcal{T}(T,h,1)$

That means, trees of the type of look like trees constructed in S_G with leaves which are of the form of T.

Now we can formulate the theorem on nonlinearity.

<u>Lemma.</u>

Let T be any tree containing at least two nodes - the leaf and the root. Then

for every $h \geqslant 3$ and $k \geqslant 1$ there exists $l \geqslant 1$ such that constructing a tree of the type $\mathcal{T}(T,h,l)$ we can perform FINDs, each of the cost not less than k, on at least half the leaves.

<u>Proof.</u>

Let s be the number of leaves in T. The proof is similar to Tarjan's method [7,8] and will be done by double induction on k and s.

(1) $k=1$ and s is arbitrary. In the tree $\mathcal{T}(T,h,1)$ we can execute FINDs for all leaves. Of course, each FIND requires a cost not less than 1.

(2) Let us assume that the thesis holds for all $k' < k$ and arbitrary s'. We prove it holds for k and $s=1$. Tree $\mathcal{T}(T,h,1)$ is of the form shown by Fig. 7. Let T' denote the tree consisting only of the roots of the subtrees T in $\mathcal{T}(T,h,1)$. From the induction hypothesis there exists such l' that constructing $\mathcal{T}(T',h+1,l')$ we are able to carry out FINDs of the costs not less than $k-1$ for

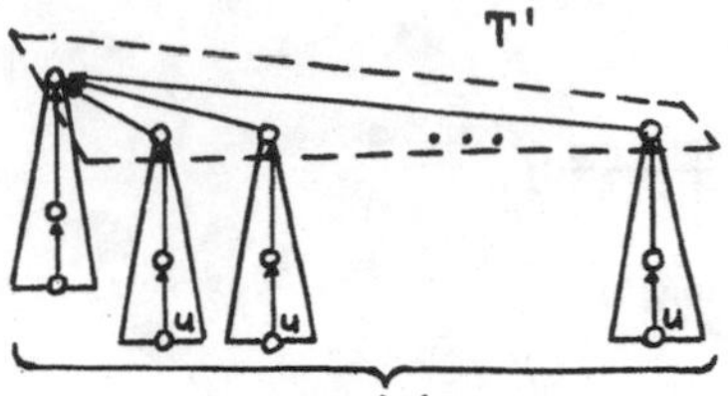

Figure 7.

An illustration to the proof of the theorem on nonlinearity of FINDs. U-leaves and the tree T' are shown.

for half the leaves of T'-s. So constructing $\mathcal{T}$ (T,h,l'+1) we can execute FINDs for a half of u-leaves at least and each FIND is of the cost not less than k .

Now let us consider a tree T" formed from $\mathcal{T}$ (T,h,l'+1) consisting only of the paths from the fathers of remaining u-leaves to the root. Notice that two u-leaves do not have the same father. From the hypothesis there exists l" such that constructing $\mathcal{T}$ (T",h+l'+1,l") FINDs of the costs not less than k-1 may be carried out for a half of the leaves at least. Therefore constructing $\mathcal{T}$ (T,h,l'+l"+1) we execute FINDs of the costs not less than k. They may be performed on at least three-fourth of all the u-leaves. Because u-leaves amount 63/64 of all the leaves, the thesis holds.

(3) Now suppose that the thesis holds for each couple of k' and s' such that

-either $k' < k$ and s' is arbitrary

-or $k'=k$ and $s' < s$.

We prove the thesis for k and s in much the same way as previously.

First we remove one leaf from the tree T (removing also all nodes on the path going only to this leaf). Denote this tree by T'. Then from the induction hypothesis there exists l' such that in $\mathcal{T}$ (T',h,l') = T" we can execute FINDs of the costs not less than k for half the leaves at least.

Now let us consider only the leaves previously removed. These of the depth in T" not less than 2 will be called u-leaves. They amount at least $\frac{G(3)-1}{G(3)} = \frac{63}{64}$ of all these leaves. No one of their fathers is another leaf's predecessor. Using an argument like in (2) we prove that for some l" constructing $\mathcal{T}$ (T",h+l',l") FINDs of the required costs may be done on at least half the u-leaves. Denote the tree by T‴.

Notice that remaining u-leaves are still of the depth not less than 2 and they have no common fathers. Therefore in the same way, for some l‴ we can construct $\mathcal{T}$ (T‴,h+l'+l",l‴) obtaining required FINDs for a half of these u-leaves. Finally constructing $\mathcal{T}$ (T",h+l',l"+l‴) on three-fourth of u-leaves we can execute FINDs of the cost not less than k.

Thus for $\frac{3}{4}\cdot\frac{63}{64} > \frac{1}{2}$ of the leaves not considered at the first step, the required FINDs may be carried out. This way we obtain the thesis.

Corollary. (nonlinearity of FINDs)

For every $k \geq 1$ there exists $n \geq 1$ such that for every $m \geq n/4$ there exists a sequence of $n-1$ UNIONs and m FINDs and in S_G the cost of these FINDs is at least $m + kn$.

Corollary holds because the leaves amount a half of the nodes in any tree from S_G structure.

We remind of the definition of functions G_j .

$G_j : N \rightarrow N$ is defined as

$$\begin{cases} G_j(0) = 1 \\ G_j(1) = G(j) \\ G_j(i) = G_j(i-1)^{G\ (i-1)} \quad \text{for } i \geq 2 \ . \end{cases}$$

Notice that the cost of FINDs in S_{G_j} can be less than in S_G only of the amount of nj . It collects the proof of the nonlinearity of FINDs in structures S_{G_j} .

GRAMMARS WITHOUT ERASING RULES. THE OI CASE

Bernard LEGUY - Université de Lille I

ABSTRACT

The problem of ε-rules in context-free languages is generalized to the tree-case. For context-free tree grammars, we distinguish three classes of erasing rules : incomplete rules, ε-rules and monadic ε-rules (i.e. rules like $X(x_2) \to x_1$). For grammars with erasing rules of just the third class, erasing-free grammars can be obtained and a construction is provided. Other results are negative and we prove that generally erasing rules cannot be avoided.

I. INTRODUCTION

Context-free tree grammars may be interpreted as non deterministic recursive program schemes [4], but, in this paper, they are viewed as a generalization of the word case. This generalization is well known and has been studied a long time [1, 5, 6, 7, 8, 11].

In language theory, for technical reasons, it is often desirable to have grammars without useless symbols or rules and without erasing rules, so is it in the tree case. Furthermore when tree grammars are viewed as program schemes, an approach to obtain efficient programs consists in avoiding useless things and use of things that may be erased later. Such an approach has been achieved in the case of deterministic program schemes [9].

In the non-deterministic case, algorithms for the Ø-reduction (i.e. carrying off symbols that generate empty sets) were given in [1, 5, 6] and the problem of the useless arities was studied in [5]. A notion of reduced tree-grammar which is a wide generalization of the well known one for words, and contains the two notions of reduction given above, is defined in this paper.

Up to now there were few results about ε-free reduction in the non-deterministic case. Sometimes people wrote this reduction could always be achieved as in the word-case, unfortunately, this paper exhibits a counter-example. Furthermore we define and study the three types of erasing rules we call incomplete rules, ε-rules and monadic ε-rules.

Finally, denoting by Alg (resp. S, C, CS) the class of sets of trees generated by context-free grammars without restriction (resp. without ε-rules, without incomplete rules, without ε-rules or incomplete rules), we obtain the result

Alg $\supsetneq$ S $\supsetneq$ C $\supsetneq$ CS.

This paper investigates the OI-case only. In the IO-case, all the results are positive and can be found in [10].

Section II introduces some notations that we need for the proofs. These notations are the ones of the Theory of the Magmoïds which is precisely defined in [3].

Section III defines context-free grammars and the sets of trees generated by them.

Section IV is devoted to the notion of reduced grammar.

In section V a method for carrying off the monadic ε-rules is given.

Section VI defines a special class of sets of trees built with balanced trees. The counter-examples will be choosed in this class and a very usefull lemma is given.

Section VII contains the proof that the ε-reduction cannot always be achieved.

Section VIII contains the other results and finally proves the hierarchy Alg $\supsetneq$ S $\supsetneq$ C $\supsetneq$ CS.

II. PRELIMINARIES

A *ranked alphabet* is a pair (Σ, d) where Σ is a finite set of symbols and d a function of Σ into $\mathbb{N}$. For any α in Σ we call d(α) the *degree* or *arity* of α. For any integer n we denote by Σ_n the set $d^{-1}(n)$.

T_Σ is the *set of trees over* Σ. It is the smallest set such that

$T_\Sigma = \Sigma_0 \cup \{\sigma(t_1,\dots, t_n) \mid \sigma \in \Sigma_n$ and $t_1,\dots, t_n \in T_\Sigma\}$. So, if $\Sigma_0 = \{\#\}$, $\Sigma_1 = \{a, \bar{a}\}$, $\Sigma_2 = \{b\}$, $\Sigma_4 = \beta$ then $\bar{t} = b(\beta(a(\#), \#, \bar{a}(\#), b(\#, \#)), b(b(\#, \#), a(\#)))$ is a tree in T_Σ. Generally it will be usefull to join a denumerable set of *variables* $\chi = \{x_1,\dots, x_n,\dots\}$ to Σ. The degree of a variable is zero. For any integer n, the set $\{x_1,\dots, x_n\}$ ist denoted by χ_n and $T(\Sigma)^1_n$ is the *set of trees over* Σ *indexed by* χ_n that may be defined as $T_{\Sigma\cup\chi_n}$. So, the tree $\tau = b(\beta(x_1, \#, x_2, x_3), b(x_3, x_1))$ is in $T(\Sigma)^1_3$ and $\tau_1 = a(\#)$, $\tau_2 = \bar{a}(\#)$ and $\tau_3 = b(\#, \#)$ are trees in $T(\Sigma)^1_0$. We will say the vector $[\tau_1, \tau_2, \tau_3]$ is in $T(\Sigma)^3_0$ and more generally if $u_1 \in T(\Sigma)^1_p,\dots, u_n \in T(\Sigma)^1_p$ then we will say $[u_1, u_2,\dots, u_n]$ is in $T(\Sigma)^n_p$. When we say that $\vec{u} \in T(\Sigma)^n_p$ we mean $\vec{u}$ is composed of n trees and any variable of χ_p may be in $\vec{u}$. So $x_2 \in T(\Sigma)^1_3$. It would be better to denote any vector $[t_1,\dots, t_p]$ in $T(\Sigma)^p_q$ by the pair $<q ; [t_1,\dots, t_p]>$ because two vectors that are defined over different sets of variables are considered as different vectors by the two Magmoïd-operations. In this paper we will use only $[t_1,\dots, t_p]$ because it is shorter.

The set $T(\Sigma) = \bigcup_{n,p\in\mathbb{N}} T(\Sigma)^n_p$ is called a *Magmoïd* when the two following operations are given. The *composition-product* of $\vec{u} = [u_1,\dots, u_n] \in T(\Sigma)^n_p$ by $\vec{v} = [v_1,\dots, v_p] \in T(\Sigma)^p_q$ is denoted by $\vec{u} \cdot \vec{v}$ and defined by

$\vec{u} \cdot \vec{v} = [u_1 \cdot \vec{v}, u_2 \cdot \vec{v},\ldots, u_n \cdot \vec{v}]$, where $u_i \cdot \vec{v}$ is defined by : if $u_i = x_j \in X_p$ then $x_j \cdot \vec{v} = v_j$, if $u_i = \alpha \in \Sigma_0$ then $\alpha \cdot \vec{v} = \alpha$ and if $u_i = \sigma(t_1,\ldots,t_r)$, $\sigma \in \Sigma_r$ and $t_1,\ldots, t_r \in T(\Sigma)^1_p$ then $u_i \cdot \vec{v} = \sigma(t_1 \cdot \vec{v},\ldots, t_r \cdot \vec{v})$. This composition-product is associative and $\vec{u} \cdot \vec{v} \in T(\Sigma)^n_q$. The equality $\bar{t} = \tau \cdot [\tau_1, \tau_2, \tau_3]$ where $\bar{t}$, τ, τ_1, τ_2 and τ_3 are defined above, can be checked easely.

The second operation is the *tensorial-product*. If $\vec{u} = [u_1,\ldots, u_p] \in T(\Sigma)^p_q$ and $\vec{v} = [v_1,\ldots, v_{p'}] \in T(\Sigma)^{p'}_{q'}$, then the tensorial-product of $\vec{u}$ and $\vec{v}$ is denoted by $\vec{u} \otimes \vec{v}$ and defined by

$\vec{u} \otimes \vec{v} = [u_1, u_2,\ldots, u_p, v_1 \cdot [x_{q+1},\ldots, x_{q+q'}],\ldots, v_{p'} \cdot [x_{q+1},\ldots, x_{q+q'}]]$. This tensorial-product is always defined and is also associative. It is obvious that $\vec{u} \otimes \vec{v} \in T(\Sigma)^{p+q'}_{q+q'}$.

Any subset of $T(\Sigma)$ is called a *forest*. For any tree t in $T(\Sigma)^1$, if $t = \sigma(t_1,\ldots, t_n)$ where $\sigma \in \Sigma_n$ or if $t = \sigma$ with $\sigma \in \Sigma_0$, then *root* (t) = σ. In any $\vec{u} \in T(\Sigma)$, each symbol that is in $\Sigma_0 \cup X$ (i.e. each zero degree symbol), is a *leaf*. The *yield* of $\vec{u}$ is the word over $\Sigma_0 \cup X$ defined by :

- For any $\alpha \in \Sigma_0 \cup X$, yield $(\alpha) = \alpha$,
- if $\sigma(t_1,\ldots, t_n) \in T(\Sigma)^1$ and $\sigma \in \Sigma$ then
  yield $(\sigma(t_1,\ldots, t_n))$ = yield (t_1) yield $(t_2),\ldots,$ yield (t_n),
- if $\vec{u} = [u_1,\ldots, u_p]$ then yield $(\vec{u})$ = yield (u_1) yield $(u_2),\ldots,$ yield (u_n).

The *depth* of $\vec{u}$ is an integer denoted $|\vec{u}|$ and defined by :

- For any $x_i \in X$, $|x_i| = 0$,
- For any $\alpha \in \Sigma_0$, $|\alpha| = 1$,
- if $\sigma(t_1,\ldots, t_n) \in T(\Sigma)^1$ and $\sigma \in \Sigma$ then
  $|\sigma(t_1,\ldots, t_n)| = 1 + \mathrm{MAX}(|t_1|,\ldots, |t_n|)$,
- if $\vec{u} = [u_1,\ldots, u_p]$ then $|\vec{u}| = \mathrm{MAX}(|u_1|,\ldots, |u_p|)$.

The set of variables that are actually in $\vec{u}$, is denoted by
var$(\vec{u}) = \{x_i \mid \exists f_1, f_2 \in (\Sigma_0 \cup V)^*$ such that yield $(\vec{u}) = f_1\, x_i\, f_2\}$.
A vector $\vec{v}$ in $T(\Sigma)^n_p$ is said *initial* in $T(\Sigma)^n_p$ if and only if yield $(\vec{v}) = u_0\, x_1\, u_1\, x_2\, u_2,\ldots, x_p\, u_p$ for $u_0, u_1,\ldots, u_p$ in $(\Sigma_0)^*$. We denote by $\tilde{T}(\Sigma)^n_p$ the set of initial vectors in $T(\Sigma)^n_p$ and in this paper any tree with the mark $\sim$ is initial. For any symbol σ in Σ_n it is convenient to denote by σ the initial tree $\sigma(x_1,\ldots, x_n)$. For any symbol a in Σ_1 we write $a^0 = x_1$, for any tree t, a t = a(t) and for any $i > 0$, $a^i = a\, a^{i-1}$. For any b in Σ_2 we write $\tilde{b}^0 = x_1$ and for any $i > 0$, $\tilde{b}^i = b(\tilde{b}^{i-1} \otimes \tilde{b}^{i-1})$. So $\tilde{b}^2 = b(b(x_1, x_2), b(x_3, x_4))$ is an initial balanced tree and its depth is 2.

III. CONTEXT-FREE GRAMMARS ON TREES

i) Définition

$G = \langle V, \Sigma, X_0, R\rangle$ is a grammar iff :

- V is a ranked alphabet of symbols which are called *non-terminals*,
- Σ is a ranked alphabet of *terminal* symbols and $\Sigma \cap V = \emptyset$,
- X_0 is an element of V_0. It is called the *axiom*,
- R is a finite set of *rules*. Rules are of the form $X \to \tau$ with X in V and τ in $T(\Sigma \cup V)^1_{d(X)}$. If $n = d(X)$, we can also write $X(x_1,\ldots, x_n) \to \tau$.

Let be t in $T(\Sigma \cup V)^1_p$ and X in V_n. We say X *occurs* in t when there exist

$\tilde{u} \in \tilde{T}(\Sigma \cup V)^1_q$, $\vec{v}_1 \in T(\Sigma \cup V)^{q_1}_{r_1}$, $\vec{v}_2 \in T(\Sigma \cup V)^{q_2}_{r_2}$, $\vec{w} \in T(\Sigma \cup V)^r_p$ where $q = q_1 + 1 + q_2$

and $r = r_1 + n + r_2$, such that $t = \vec{u} \cdot (\vec{v}_1 \otimes X \otimes \vec{v}_2) \cdot \vec{w}$. We say the rule $X \to \tau$ is applied to this *occurrence* of X in t, giving t' if $t' = \tilde{u} \cdot (\vec{v}_1 \otimes \tau \otimes \vec{v}_2) \cdot \vec{w}$. We write $t \underset{G}{\Longrightarrow} t'$ if t' results of the application of one rule of G to one occurrence of non-terminal in t. Furthermore we say the derivation $t = \tilde{u} \cdot (\vec{v}_1 \otimes X \otimes \vec{v}_2) \cdot \vec{w} \underset{G}{\Longrightarrow} t' = \tilde{u} \cdot (\vec{v}_1 \otimes \tau \otimes \vec{v}_2) \cdot w$ is *top-down* or OI and we write $t \ D\!\underset{G}{\Longrightarrow} t'$ if and only if $\tilde{u} \in \tilde{T}(\Sigma)$. We write $t \overset{*}{\underset{G}{\Longrightarrow}} t'$ (resp. $t \ D\!\overset{*}{\underset{G}{\Longrightarrow}} t'$) if there exist $t_0, t_1,\ldots, t_j$ such that $t = t_0$, $t' = t_j$ and for each i in $[0, j-1]$, we have $t_i \underset{G}{\Longrightarrow} t_{i+1}$ (resp. $t_i \ D\!\underset{G}{\Longrightarrow} t_{i+1}$). In this case the *length* of the derivation is j (note that j can be equal to 0). A top-down derivation $X_0 \ D\!\Longrightarrow t_1 \ D\!\Longrightarrow t_2 \ D\!\Longrightarrow,\ldots, D\!\Longrightarrow t_{n-1} \ D\!\Longrightarrow t_n$ is *initial* iff the roots of $t_1, t_2,\ldots, t_{n-1}$ are in V.

ii) Properties of rules

A rule $X \to \tau$ is *strict* iff $|\tau| > 0$. It is *complete* if $var(\tau) = var(X)$. If the rule $X \to \tau$ is complete but not ε-free, this implies that $d(X) = 1$ and $\tau = x_1$, then the rule is $X(x_1) \to x_1$ and is said a *monadic* ε-*rule* (a not strict rule is said an ε-*rule*).

Grammars without ε-rules (resp. incomplete rules) are said ε-free (resp. *complete*). Grammars such that the only ε-rules are monadic ε-rules, are said *monadicly unstrict*.

iii) Context-free forests

Given the grammar $G = \langle V, \Sigma, X_0, R\rangle$, for any τ in $T(\Sigma \cup V)$ we denote by $F(G, \tau)$ the set $\{t \in T(\Sigma) \mid \tau \overset{*}{\underset{G}{\Longrightarrow}} t\}$ and by $F_{OI}(G, \tau)$ the set $\{t \in T(\Sigma) \mid \tau \ D\!\overset{*}{\underset{G}{\Longrightarrow}} t\}$. It is well known that $F(G, \tau) = F_{OI}(G, \tau)$, so we will often be concerned by the top-down derivations only. We say the grammar G generated the forest $F(G) = F(G, X_0)$.

IV. REDUCED GRAMMARS

A grammar $G = <V, \Sigma, X_0, R>$ is reduced (it would be better to say OI-reduced) iff :

i) for each X in V, one can find t such that $X_0 \overset{*}{\underset{G}{\Longrightarrow}} t$ and X occurs in t,

ii) for each X in V, the set F(G, X) is not empty,

iii) for each X in V and x_i in var(X), one can find t in F(G, X) such that $x_i \in var(t)$. We say that x_i is usefull in X.

Theorem 1 : For any grammar G such that $F(G) \neq \emptyset$ one can find a reduced grammar G' that generates the same forest as G and such that ε-rules or incomplete rules are never applied in any initial derivation. Furthermore, if G is ε-free or complete, so is G'.

Algorithms and proofs are too long for being given here. They are wholly given in [10].

V. REDUCTION OF THE MONADIC UNSTRICTRESS

Let be $G = <V, \Sigma, X_0, R>$, a monadicly unstrict grammar. For each X in V, for each x_i in var(X), one can decide if x_i is in F(G, X). So we can build $E = \{Z \in V_1 \mid x_1 \in F(G, Z)\}$. E is the set of erasable non-terminals. Without change in F(G), we used in [10] a long sequence of simple transformations to get these properties for G :

a) $E = \{Z_1, Z_2, \dots, Z_k\}$; for each Z_i in E, there exists $Y_i \in V_1 \backslash E$, and the only rules that can be applied to Z_i are $Z_i \to x_1$ and $Z_i \to Y_i$. So, it is obvious that $F(G, Y_i) = F(G, Z_i) \backslash \{x_1\}$.

b) There is no symbol of E above any symbol in any rule in R : any rule that contains an occurrence of a symbol Z of E in its right-part, is of the form $X \to \tilde{u} \cdot [\vec{v}_1, Z(x_i), \vec{v}_2]$ where x_i is in var(X).

We suppose now that G has got the properties a and b, then we can explain the main part of the reduction. For each X in $\Sigma \cup V \backslash E$ we build a new symbol $\bar{X}$ such that $d(\bar{X}) = (k + 1)\, d(X)$ and we denote by W the set $\{\bar{X} \mid X \in \Sigma \cup V \backslash E\}$. For any tree t in $T(\Sigma \cup V)$, we define k + 1 trees in T(W). These trees can be considered as the results of k + 1 functions denoted by $H_0, H_1, \dots, H_k$. We define theses functions as follows :

- if $t = x_j$, for each i from 0 to k, we have $H_i(x_j) = x_{(j-1)(k+1)+i+1}$

- if $t = Z_j(t')$, where Z_j is in E, then $H_j(t) = Y_j(H_0(t'),\ldots, H_k(t'))$ and for each i in $\{0,\ldots, j-1, j+1,\ldots, k\}$, we have $H_i(t) = H_i(t')$

- if $t = X(t_1,\ldots, t_p)$, where $X \notin E$, then $H_0(t) = H_1(t) =\ldots= H_k(t)$ and the result is $H(t) = \bar{X}(H_0(t_1),\ldots, H_k(t_1), H_0(t_2),\ldots, H_k(t_p))$.

For strings that contain erasable symbols, we must choose to apply ε-rules or strict-rules to each occurrence of erasable symbol. The function H replaces such strings by choosing-trees which give terminal strings by using incomplete rules.

We define $G' = \langle W, \Sigma, \bar{X}_0, R_1 \cup R_2\rangle$ where $R_1 = \{(\bar{X} \to H(t)) \mid (x \to t) \in R$ and $X \in V\backslash E\}$ and $R_2 = \{(\bar{\alpha}(x_1,\ldots, x_{n(k+1)}) \to \alpha(x_{i_1}, x_{i_2},\ldots, x_{i_n})) \mid n = d(\alpha), \alpha \in \Sigma$ and for each j from 1 to n we have $(j-1)(k+1) < i_j \le j(k+1)\}$.

Theorem 2 : For any monadicly unstrict grammar G *one can find an ε-free grammar* G' *such that* $F(G) = F(G')$.

Proof : For any rule $X \to t$ in R such that $X \notin E$, we have $|t| > 0$ and the property b implies $root(t) \notin E$, then $|H(t)| > 0$ and it becomes obvious that G' is ε-free. The proof of $F(G) = F(G')$ is too long and can be found in [10]. □

Most of the difficulties that relate to this reduction can be discovered in the following example : $X_0 \to X(\#, \#)$; $X(x_1, x_2) \to X(Z_1(x_1), a(x_2))$; $X(x_1, x_2) \to X(Z_2(x_1), \alpha(x_2))$; $X(x_1, x_2) \to D(\beta(x_1, x_2))$; $D(x_1) \to D(b(x_1, x_1))$; $D(x_1) \to b(x_1, x_1)$; $Z_1(x_1) \to a(x_1)$; $Z_1(x_1) \to x_1$; $Z_2(x_1) \to \alpha(x_1)$; $Z_2(x_1) \to x_1$.

Corollary 3 : For any complete grammar G, *one can find an ε-free grammar* G' *such that* $F(G) = F(G')$.

Proof : If G is not ε-free then it is monadicly unstrict. So the theorem 2 gives the result. It must be noted that generally G' is not complete. □

VI. FORESTS WITH WIDE BALANCED TREES

For any ranked alphabet Σ that contains a symbol b of degree 2 and for any set $\mathcal{F}$ of forests in $T(\Sigma\backslash\{b\})_0^1$, we define the new forest :

$B\mathcal{F} = \{\tilde{b}^q \cdot [t_1,\ldots, t_m] \mid m = 2^q$ and $\exists F \in \mathcal{F}$ such that $t_1,\ldots, t_m \in F\}$.

We are studying here forests of the $B\mathcal{F}$ shape that can be generated by grammars. Then, the theorem 1 involves we may assume G is reduced and does not use ε-rules or incomplete rules in initial derivations. Let be $G = \langle V, \Sigma, X_0, R\rangle$ such a grammar that generates $B\mathcal{F}$.

For any τ in $T(\Sigma \cup V)_0^1$, we say the b-*level* of τ is defined with the value q iff for each t in $F(G, \tau)$ the shape of t is $\tilde{b}^q \cdot [t_1, \ldots, t_m]$ where $t_1, \ldots, t_m$ are in $T(\Sigma \setminus \{b\})$.

<u>Lemma 4</u> : *For* G *we can find two constants* h *and* k *such that for each* t *in* BF *there exists a derivation* $X_0 \overset{*}{\underset{G}{D\Longrightarrow}} u \cdot \vec{v} \overset{*}{\underset{G}{D\Longrightarrow}} t$ *where* $u \in T(\{b\})^1_p$, $\vec{v} \in T(\Sigma \cup V)^p_0$, $p \leq k$ *and all the* p *components of* $\vec{v}$ *have the same* b-*level* $q \leq h$.

The most important property used for proving this lemma is that for any derivation $X_0 \overset{*}{\underset{G}{D\Longrightarrow}} X \cdot \vec{\omega}$ where $X \in V$ and $F(G, X) \cap T(\{b\}) = \emptyset$, the b-level of each compc nent of $\vec{\omega}$ is defined and bounded (do not forget that G has the properties of theorem 1). Furthermore these b-levels are not changed by derivations that may be appliec later. The proof which is long is given in [10] pp. 150-161 and this lemma has numbeı IV 33.

We define now a notation for examining the power of generation of any tree t in $T(\Sigma \cup V)^1_0$. For any tree t in $T(\Sigma)^1_0$, if root(t) is not b then we write $\sigma(t) = \{t\}$ else we have $t = b(t_1, t_2)$ and $\sigma(t) = \sigma(t_1) \cup \sigma(t_2)$. Then for any tree t in $T(\Sigma \cup V)^1_($ we define $\sigma(t)$ as the set $\{\tau \in \sigma(t') \mid t' \in F(G, t)\}$. Any tree π of $T(\Sigma \cup V)^1_0$ such tł $\sigma(\pi) \subset F \in \mathcal{F}$ is said a *prototype* for F.

The counter-examples that will be used in this paper, are built by using families $\mathcal{F}$ such that for any F and F' in $\mathcal{F}$ we have always $F \cap F' = \emptyset$. So the lemma 5 thai can be considered as a weak form of lemma 4, will be usefull.

<u>Lemma 5</u> : *Let be* $\mathcal{F}$ *such that for any* F *and* F' *in* $\mathcal{F}$ *we have* $F \cap F' = \emptyset$ *and let be* $G = \langle V, \Sigma, X_0, R\rangle$ *that generates* BF *and verifies the property of theorem 1. There exist two constants* h *and* k *such that for any finite part* E *of any forest* F *in* $\mathcal{F}$, *it must exist a prototype* π *for* F *such that the* b-*level of* π *is defined with a value* $q \leq h$ *and* $\text{Card}(\sigma(\pi) \cap E) > \text{Card}(E)/k$. *(Where* Card *is the number of elements).*

<u>Proof</u> : Let be r an integer such that $r > h$ and $2^r > \text{Card}(E)$. One can find t in BF such that b-level (t) = r and $\sigma(t) = E$. Let us examine the derivation of lemma 4 : $X_0 \overset{*}{D\Longrightarrow} u \cdot \vec{v} \overset{*}{D\Longrightarrow} t$. There exists only one vector $\theta = [x_{i_1}, \ldots, x_{i_n}]$ such that $\tilde{u} = u \cdot \theta$ is initial (see [3]). Then $u \cdot \vec{v} = \tilde{u} \cdot \theta \cdot \vec{v} = \tilde{u} \cdot [\nu_1, \ldots, \nu_n]$ and $\nu_1, \nu_2, \ldots, \nu_n$ are components of $\vec{v}$ (generally copied several times). Lemma 4 says there are not more than k components in $\vec{v}$. Let be $\pi_1, \pi_2, \ldots, \pi_\ell$ theses components and $\ell \leq k$. Lemma 4 says also that the b-level of each π_i is $q \leq h$.

We prove now that $\pi_1, \ldots, \pi_\ell$ are prototypes for the same forest $F \in \mathcal{F}$. From $r > h$ it is obvious that $|\tilde{u}| \geq 1$ and $n \geq 2$. Let's take two different integers i_1 and i_2 between 1 and n. There exist t_1 and t_2 in $T(\Sigma)$ such that $\nu_{i_1} \overset{*}{D\Longrightarrow} t_1$, $\nu_{i_2} \overset{*}{D\Longrightarrow} t_2$ $\sigma(t_1) \subset F$ and $\sigma(t_2) \subset F$ because $u \cdot \vec{v} \overset{*}{D\Longrightarrow} t$, $t \in$ BF and $\sigma(t) \subset E \subset F$. If it were t_1'

in $F(G, \nu_{i_1})$ such that $\sigma(t_1') \not\subset F$ then we would have t' in $F(G, u \cdot \vec{v})$ such that $\sigma(t') \not\subset F$ and $\sigma(t') \cap F \neq \emptyset$ then G would not generate $B\mathcal{F}$. (Remember $F \cap F' = \emptyset$ for any F and F' in $\mathcal{F}$). Then $\sigma(\nu_{i_1})$ must be a part of F and $\pi_1, \pi_2, \ldots, \pi_\ell$ are prototypes for F. To generate t, we must have $E \subset \bigcup_{i=1}^{\ell} \sigma(\pi_i)$. Since $\ell \leq k$, the result becomes obvious. □

VII. IRREDUCIDILITY OF THE UNSTRICTNESS

We are going to use a counter-example that has already been used for an other purpose in [2]. Let be $F_{1,n} = \{\gamma(a^n\ \#,\ \delta(a^j\ \#,\ a^j\ \#)) \mid j \leq n\}$ for any n in $\mathbb{N}$ and $\mathcal{F}_1 = \{F_{1,n} \mid n \in \mathbb{N}\}$. The reader can easily verify that $B\mathcal{F}_1$ is generated by the grammar, the rules of which are : $X_0 \to X(\#, \delta(\#, \#))$; $X(x_1, x_2) \to X(a\ x_1, Y(a\ x_1, x_2))$; $X(x_1, x_2) \to Z(\gamma(x_1, x_2))$; $Z(x_1) \to b(x_1, x_1)$; $Z(x_1) \to Z(b(x_1, x_1))$; $Y(x_1, x_2) \to \delta(x_1, x_1)$; $Y(x_1, x_2) \to x_2$.

We want to show that $B\mathcal{F}_1$ cannot be generated by an ε-free grammar. We begin to assume there exists $G = \langle V, \Sigma, X_0, R\rangle$ and ε-free grammar that has the properties of theorem 1 and generates $B\mathcal{F}_1$, then we prove this assumption goes counter to lemma 4.

For any tree t in $T(\Sigma \cup V)_0^1$ and any occurrence X of a symbol in t, i.e. $t = \tilde{u} \cdot [\vec{v}_1, X \cdot \vec{w}, \vec{v}_2]$, if the symbol δ occurs in no tree in F(G, X) we say *variety* (X) = ∅. The integer j is in variety (X) iff either there exists $\tilde{\nu} \cdot [\vec{\nu}_1, \delta(a^j\ \#, \nu'), \vec{\nu}_2]$ in F(G, X) or there exist together $\tilde{\nu} \cdot [\vec{\nu}_1, \delta(a^{j-\ell}\ x_i, \nu'), \vec{\nu}_2]$ in F(G, X) and $a^\ell\ \#$ in $F(G, x_i \cdot \vec{w})$. Now we call *weight* (X) the number of elements in variety (X). Note that variety and weight relate to the occurrence of the symbol and not to the symbol itself. It is obvious that $\mathrm{Card}(\sigma(t)) \leq \sum_{X \text{ in } t} \text{weight } (X)$.

Lemma 6 : If π is a prototype for $F_{1,n}$, then for any subtree τ of π, the tree $a^q\ \#$ is in F(G, τ) iff $F(G, \tau) = \{a^q\ \#\}$.

Shetch of proof : It is based on examining the three cases :
$\pi \overset{*}{\Longrightarrow} \tilde{b}^p \cdot [\vec{v}_1, \gamma(a^m\ \tau, \delta(a^i\ \#, a^i\ \#)), \vec{v}_2]$ then $\tau \overset{*}{\Longrightarrow} a^{n-m}\ \#$,
$\pi \overset{*}{\Longrightarrow} \tilde{b}^p \cdot [\vec{v}_1, \gamma(a^n\ \#, \delta(a^m\ \tau, a^i\ \#)), \vec{v}_2]$ then $\tau \overset{*}{\Longrightarrow} a^{i-m}\ \#$,
$\pi \overset{*}{\Longrightarrow} \tilde{b}^p \cdot [\vec{v}_1, \gamma(a^n\ \#, \delta(a^i\ \#, a^m\ \tau)), \vec{v}_2]$ then $\tau \overset{*}{\Longrightarrow} a^{i-m}\ \#$ and in each case the derivation given for τ is the only terminal one that can be made. □

Lemma 7 : For any integers n and p and any prototype π for $F_{1,n}$, the b-level of which is p, the weight of any occurrence X of a symbol in π is bounded by a constant N associated to G.

Proof : It is obvious that if $X \in X\backslash\{\delta\}$ then weight (X) = 0 and the lemma 6 says if X = δ then weight (X) = 1.

If X is in V, let's denote by $\Delta(X)$ the set of integer pairs defined by :
$(m, 0) \in \Delta(X)$ iff $\exists \tilde{\nu} \cdot [\vec{v}_1, \delta(a^m \#, \tau), \vec{v}_2]$ in F(G, X) and for any integer $i \neq 0$,
$(m, i) \in \Delta(X)$ iff $\exists \tilde{\nu} \cdot [v_1, \delta(a^m x_i, \tau), \vec{v}_2]$ in F(G, X). By the lemma 6,
weight (X) $\leq$ Card(Δ(X)).
π is a prototype for $F_{1,n}$ and b-level (π) = p then F(G, π) is finite and so is F(G, X) for any occurrence in π.
N_1 = Sup(Card(Δ(Z))) for any non-terminal Z that generate a finite forest, i.e. F(G, Z) is finite. It is easy to show N_1 exists. Then we have weitht (X) $\leq N_1$ and it is sufficient to choice N = Sup(N_1, 1). □

Lemma 8 : For any integers n *and* p *and any prototype* π *for* $F_{1,n}$, *the* b-*level of which is* p, Card($\sigma(\pi)$) $\leq N \times D^{p+2}$ *where* N *is the constant in lemma 7 and* D *is* Sup(d(X)) *for any* X *in* $\Sigma \cup V$.

Proof : One can find $\tilde{u}$, the smallest initial subtree of π such that $\pi = \tilde{u} \cdot \vec{w}$ and for any occurrence X in $\vec{w}$, weight (X) = 0. Any tree t in F(G, Π) has the shape $\tilde{b}^p \cdot [t_1, \ldots, t_{2^p}]$ where $t_1, \ldots, t_{2^p}$ are in $F_{1,n}$. So any occurrence of δ in t is at a depth p + 2, then, since G is ε-free, $|\tilde{u}| \leq p + 2$. The number of symbols in $\tilde{u}$ is bounded by D^{p+2}, so the result becomes obvious. □

Proposition 9 : BF_1 *cannot be generated by an* ε-*free grammar.*

Proof : Since $F_{1,n} \cap F_{1,m} = \emptyset$ for any $n \neq m$, we may use the lemma 5.
We choose $n > k \times N \times D^{h+2}$. It must exist a prototype π for $F_{1,n}$ such that b-level (π) = q $\leq$ h and Card($\sigma(\pi)$) $\geq$ Card($F_{1,n}$)/k = (n + 1)/k > n/k > $N \times D^{h+2}$. Then the lemma 8 is denied. □

VIII. OTHER ERASING RULES

Let be $F_{2,n} = \{\alpha^i \gamma a^j \# \mid i + j = n\}$ for any integer n and $F_2 = \{F_{2,n} \mid n \in \mathbb{N}\}$. The forest BF_2 is generated by $X_0 \to X(\gamma \#, \#)$; $X_0 \to Z \gamma \#$; $X(x_1, x_2) \to Z\ Y(x_1, a\ x_2)$; $X(x_1, x_2) \to X(Y(X_1, a\ x_2), a\ x_2)$; $Z(x_1) \to Z\ b(x_1, x_1)$; $Z(x_1) \to b(x_1, x_1)$; $Y(x_1, x_2) \to \gamma\ x_2$; $Y(x_1, x_2) \to \alpha\ x_1$.

Proposition 10 : BF_2 *cannot be generated by a complete grammar.*

Proof : See [10]. □

Let be $F_{3,n} = \{\delta\ (a^j \#, a^n \#) \mid j \leq n\}$ for any integer n and $F_3 = \{F_{3,n} \mid n \in \mathbb{N}\}$. Let be the set of rules R = $\{X_0 \to X(\#, \#)$; $X(x_1, x_2) \to Z\ \delta(x_1, x_2)$;

$X(x_1, x_2) \to X(Y\ x_1, a\ x_2)$; $Z(x_1) \to b(x_1, x_1)$; $Z(x_1) \to Z\ b(x_1, x_1)$; $Y(x_1) \to a(x_1)\}$. The grammar, the rules of which are in $R \cup \{Y(x_1) \to x_1\}$ is (monadicly) unstrict but is complete and generates BF_3. Furthermore if we take the set $R \cup \{Y(x_1) \to \#\}$, the new grammar is not complete but it is ε-free and generates BF_3.

Proposition 11 : BF_3 *cannot be generated by a complete and ε-free grammar.*

Proof : See [10]. □

We denote by Alg (resp. S, C, CS) the class of forest that are generated by context-free grammars without restriction (resp. ε-free grammars, complete grammars, ε-free and complete grammars).

Theorem 12 : $CS \subsetneq C \subsetneq S \subsetneq Alg$.

Proof : Definitions give $CS \subseteq C$ and $S \subseteq Alg$. The proposition 11 says $CS \subsetneq C$. The corollary 3 says $C \subseteq S$ and the proposition 10, $C \subsetneq S$. Finally proposition 9 says $S \subsetneq Alg$. □

APPENDIX

You can find here an application of the construction that has been given in Section V. From the grammar that has been given like an example, we obtain easily a grammar that has the properties a and b :

$X_0 \to X(\#, \#)$; $X(x_1, x_2) \to X(Z_1(x_1), a(x_2))$; $X(x_1, x_2) \to X(Z_2(x_1), \alpha(x_2))$; $X(x_1, x_2) \to D(\beta(x_1, x_2))$; $D(x_1) \to D(b(x_1, x_1))$; $D(x_1) \to b(x_1, x_1)$; $Z_1(x_1) \to Y_1(x_1)$; $Z_1(x_1) \to x_1$; $Z_2(x_1) \to Y_2(x_1)$; $Z_2(x_1) \to x_1$; $Y_1(x_1) \to a(x_1)$; $Y_2(x_1 \to \alpha(x_1)$.

Then $E = \{Z_1, Z_2\}$ and we obtain the rules :

$\bar{X}_0 \to \bar{X}(\bar{\#}, \bar{\#}, \bar{\#}, \bar{\#}, \bar{\#}, \bar{\#})$; $\bar{\#} \to \#$; $\bar{X}(x_1, \ldots, x_6) \to \bar{X}(x_1, \bar{Y}_1(x_1, x_2, x_3), x_3, \bar{a}(x_4, x_5, x_6))$; $\bar{X}(x_1, \ldots, x_6) \to \bar{X}(x_1, x_2, \bar{Y}_2(x_1, x_2, x_3), \bar{\alpha}(x_4, x_5, x_6))$;
$\bar{X}(x_1, \ldots, x_6) \to \bar{D}(\bar{\beta}(x_1, \ldots, x_6), \bar{\beta}(x_1, \ldots, x_6), \bar{\beta}(x_1, \ldots, x_6))$;
$\bar{D}(x_1, x_2, x_3) \to \bar{D}(\bar{b}(x_1, x_2, x_3, x_1, x_2, x_3), \bar{b}(x_1, x_2, x_3, x_1, x_2, x_3), \bar{b}(x_1, x_2, x_3, x_1, x_2, x_3))$; $\bar{D}(x_1, x_2, x_3) \to \bar{b}(x_1, x_2, x_3, x_1, x_2, x_3)$; $\bar{Y}_1(x_1, x_2, x_3) \to \bar{\alpha}(x_1, x_2, x_3)$
$\bar{Y}_2(x_1, x_2, x_3) \to \bar{\alpha}(x_1, x_2, x_3)$; $\bar{a}(x_1, x_2, x_3) \to a(x_1) + a(x_2) + a(x_3)$;
$\bar{\alpha}(x_1, x_2, x_3) \to \alpha(x_1) + \alpha(x_2) + \alpha(x_3)$;
$\bar{\beta}(x_1, \ldots, x_6) \to \beta(x_1, x_4) + \beta(x_1, x_5) + \beta(x_1, x_6) + \beta(x_2, x_4) + \ldots + \beta(x_3, x_6)$;
$\bar{b}(x_1, \ldots, x_6) \to b(x_1, x_4) + b(x_1, x_5) + b(x_1, x_6) + b(x_2, x_4) + \ldots + b(x_3, x_6)$.

Remarks : It is useless to "dilate" some variables as x_2 in X because any subtree like $X(t_1, Z(t_2))$, where $Z \in E$, cannot occur in any derivation, and so is it for

x_1 in D. Other simplifications result of applying immediately in the other rules, the rules that are unique for a non-terminal, like the rule $\bar{Y}_1(x_1, x_2, x_3) \rightarrow \bar{a}(x_1, x_2, x_3)$. Then we obtain :

$X_0 \rightarrow \bar{X}(\#, \#, \#, \#)$; $\bar{X}(x_1,\ldots, x_4) \rightarrow \bar{X}(x_1, \bar{a}(x(x_1, x_2, x_3), x_3, a(x_4))$;
$\bar{X}(x_1,\ldots, x_4) \rightarrow \bar{X}(x_1, x_2, \bar{\alpha}(x_1, x_2, x_3), \alpha(x_4))$; $\bar{X}(x_1,\ldots, x_4) \rightarrow D(\bar{\beta}(x_1, x_2, x_3, x_4))$;
$D(x_1) \rightarrow D(b(x_1, x_1)) + b(x_1, x_1)$; $\bar{a}(x_1, x_2, x_3) \rightarrow a(x_1) + a(x_2) + a(x_3)$;
$\bar{\alpha}(x_1, x_2, x_3) \rightarrow \alpha(x_1) + \alpha(x_2) + \alpha(x_3)$; $\bar{\beta}(x_1,\ldots, x_4) \rightarrow \beta(x_1, x_4) + \beta(x_2, x_4) + \beta(x_3, x_4)$.

Let us define for any w in $\{a, \alpha\}^*$, the set $\phi(w) = \{u = u_1, u_2,\ldots, u_p \mid \exists\, v_0, v_1,\ldots, v_p$ such that $v_0\, u_1\, v_1\, u_2\, v_2,\ldots, u_p\, v_p = w\}$, $F_w = \{\beta(\omega, w) \mid \omega \in \phi(w)\}$, $F = \{F_w \mid w \in \{a, \alpha\}^*\}$. Then all the grammars that are given in this appendix, generate the forest BF.

References

[1] A. ARNOLD, M. DAUCHET, "*Un théorème de duplication pour les forêts algébriques*", J. Comput. System Sci. 13 (1976), pp. 223-244.

[2] A. ARNOLD, M. DAUCHET, "*Forêts algébriques et homomorphismes inverses*", Information and Control. 37 (1978), pp. 182-196.

[3] A. ARNOLD, M. DAUCHET, "*Théorie des magmoïdes*", RAIRO inf. th. 12 (1978), n° 3, pp. 235-257 et 13 (1979), n° 2, pp. 135-154.

[4] A. ARNOLD, M. NIVAT, "*Non deterministic recursive program schemes*", In "Fundamentals of computation theory", (1977), Poznam-Lectures Notes in Computer Sciences n° 56, Springer-Verlag, pp. 12-21.

[5] G. BOUDOL, "*Langages polyadiques algébriques. Théorie des schémas de programme : Sémantique de l'appel par valeur*", Thèse de 3ème cycle, Paris VII (1975).

[6] B. COURCELLE, "*A representation of trees by languages*", Th. Comput. Sci. 6 (1978), pp. 255-279 and 7 (1978) pp. 25-55.

[7] J. ENGELFRIET and E.M. SCHMIDT, "*IO and OI*", J. Comput. System Sci. 15 (1977), pp. 328-353 and 16 pp. 67-99.

[8] M.J. FISCHER, "*Grammars with macro-like productions*", 9th IEE Symp. On switching and automata theory (1968), pp. 131-142.

[9] I. GUESSARIAN, "*Program transformation and algebraic semantics*", Publication interne LITP 78/21 (1978), A paraître dans TCS.

[10] B. LEGUY, "*Réductions, transformations et classification des grammaires algébriques d'arbres*". Thèse de 3ème cycle, Lille (1980).

[11] W.C. ROUNDS, "*Mappings and grammars on trees*", Math. Systems theory. 4 (1968), pp. 257-287.

PROPRIETES DE CLOTURE D'UNE EXTENSION DE TRANSDUCTEURS D'ARBRES DETERMINISTES

Eric LILIN

Résumé

Nous étudions dans cet article des propriétés de clôture d'une extension de transducteurs d'états finis d'arbres déterministes. Nous montrons la fermeture par composition pour certaines sous classes. Nous étudions les inclusions des sous classes introduites.

Abstract

We study closure properties of an extended class of deterministic finite state tree transducers. We shall show the closure under composition for certain sub classes. We compare the different classes introduced.

INTRODUCTION

Les transformations d'arbres réalisées par des transducteurs d'états finis sont étudiés depuis de nombreuses années [8], poussées en cela par de nombreuses motivations (traduction, compilation, schémas de programmes ...) [2], [3].

Afin de pouvoir opérer des transformations commandées par des tests d'égalité de sous-arbres, nous avons été amenés à introduire une généralisation des transducteurs d'états finis, que nous appelons $\bar{\bar{S}}$-transducteurs, en définissant des états munis d'une arité et d'une co-arité (voir notamment [6], [7]). Ceci permet de reconnaitre simultanément et en parallèle un nombre fini de sommets de sous arbres et d'engendrer en sortie un k-uple d'arbres.

Les transducteurs déterministes (à tout moment au cours de la transduction d'un arbre on ne peut appliquer qu'une seule règle de transformation) semblent jouer un rôle important dans l'étude des transformations de programmes. Dans le cas classique (les états sont munis d'une arité supérieure et inférieure égale à 1, et le transducteur ne peut lire le mot vide) les propriétés de clôture par composition ont été étudiées par Engelfriet ([4], [5]).

Nous nous proposons dans ce papier d'étudier les propriétés de fermeture par composition des $\bar{\bar{S}}$-transducteurs déterministes. Nous montrons que dans le cas général les

$\bar{\bar{S}}$-transducteurs ne sont pas fermés par composition. Par contre nous montrons la clôture par composition pour les sous classes suivantes :

- STD dét. complet : sous classe des $\bar{\bar{S}}$-transducteurs descendants dont l'arité supérieure des états est toujours égale à 1 ($\overset{|}{q}$).
- $\bar{S}$TD dét : sous classe des $\bar{\bar{S}}$-transducteurs descendants dont l'arité inférieure des états est toujours égale à 1 ($\overset{/\ldots\backslash}{\underset{|}{q}}$).
- STA dét : sous classe des $\bar{\bar{S}}$-transducteurs ascendants dont l'arité supérieure des états est égale à 1.

La construction du transducteur composé n'est pas simplement réalisée en composant règle par règle, car des problèmes de chevauchement de découpage apparaissent.

Nous comparons dans le chapitre 3 les classes de transformations réalisées par les transducteurs déterministes.

Chapitre I : PRELIMINAIRES

Définition : Un alphabet gradué est la donnée d'un couple (Σ, d) où Σ est un ensemble fini et d une application de Σ dans $\mathbb{N}$. L'application d définit l'arité ou le degré de chaque symbole de Σ. Nous posons $\Sigma_i = d^{-1}(i)$ pour tout i.

Dorénavant nous noterons un alphabet gradué par son support Σ.
Soit $X = \{x_i \ / \ i \in \mathbb{N}\}$ un ensemble de variables.

Définition : L'ensemble $T(\Sigma)^1$ des arbres indexés sur Σ est le plus petit ensemble tel que :

- $\Sigma_o \cup X \subset T(\Sigma)^1$
- $\forall a \in \Sigma_n, \forall t_1,\ldots,t_n \in T(\Sigma)^1$ alors $a(t_1,\ldots,t_n) \in T(\Sigma)^1$.

$T(\Sigma)^p$ désignera l'ensemble des suites ordonnées de p arbres de $T(\Sigma)^1$. $T(\Sigma)^p_q$ désignera l'ensemble des <q, t> où $t \in T(\Sigma)^p$, $q \in \mathbb{N}$ et où toutes les occurrences des variables de t appartiennent à $X_q = \{x_1,\ldots,x_q\}$.

Lorsqu'il n'y aura aucune ambiguité nous désignerons par t les éléments de $T(\Sigma)^p_q$. $\tilde{T}(\Sigma)^p_q$ est l'ensemble des éléments <q, t> de $T(\Sigma)^p_q$ dont la suite des occurrences des variables de t, lues de gauche à droite, est $(x_1,\ldots,x_q)$. On pose $T(\Sigma) = \bigcup_{p,q \in \mathbb{N}} T(\Sigma)^p_q$ et $\tilde{T}(\Sigma) = \bigcup_{p,q \in \mathbb{N}} \tilde{T}(\Sigma)^p_q$. Notons, pour tout $p \in \mathbb{N}$, $[p] = \{1,\ldots,p\}$. Désignons par θ^p_q l'ensemble des applications de [p] dans [q]. $\theta = \bigcup_{p,q \in \mathbb{N}} \theta^p_q$, appelé ensembles des *torsions*, est un sous-ensemble de $T(\Sigma)$. Notons Id_n l'élément <n ; $x_1,\ldots,x_n$> de $\tilde{T}(\Sigma)$. Nous utiliserons continuellement les opérations sur les arbres introduites dans le cadre de la théorie des magmoïdes par Arnold et Dauchet [1]. L'opération binaire de composition sur des n-uples d'arbres est définie comme suit : soit $t \in T(\Sigma)^p_q$ et $u = <r ; u_1,\dots,u_q> \in T(\Sigma)^q_r$. Le composé $t.u \in T(\Sigma)^p_r$ de t par u est obtenu en substituant dans t, u_i à chaque occurrence de x_i pour tout $i \in [q]$. L'opération produit tensoriel noté $t_1 \otimes t_2$, consiste à juxtaposer les séquences finies d'arbres en translatant les indices des variables, figurant dans le deuxième opérande, d'un entier déterminé par le premier opérande.

Définition : Le profondeur d'un arbre $t = <q ; t_1,\dots,t_p>$ de $T(\Sigma)^p_q$, noté prof(t) est définie par :

- $\text{prof}(t) = \sup_i(\text{prof}(t_i))$
- $\text{prof}(x_i) = 0$
- $\text{prof}(a(u_1,\dots,u_n)) = 1 + \sup_j(\text{prof}(u_j)) \ \forall a \in \Sigma_n$

⊥

Intuitivement la profondeur de t est la longueur de sa plus longue branche.

Définition : Un ensemble d'états gradués est la donnée d'un triplet (Q, d_{sup}, d_{inf}) où Q est un ensemble fini, d_{sup} et d_{inf} des applications de Q dans $\mathbb{N}$; d_{sup} et d_{inf} définissent les arités supérieures et inférieures de chaque état de Q.

⊥

Nous noterons ce triplet par Q.

Définition : Un $\bar{\bar{S}}$-transducteur descendant (noté $\bar{\bar{s}}$td) est un 5-uple $<Q, \Sigma, \Delta, R, P>$ où :

- Q est un ensemble d'états gradués ;
- Σ et Δ sont des alphabets gradués finis ;
- P est l'ensemble des états initiaux ; de plus $\forall q \in P$, $d_{sup}(q) = d_{inf}(q) = 1$;
- R est un ensemble fini de règles de la forme : $<q.(a_1 \otimes\dots\otimes a_n)> \to \tilde{u}.\theta.(q_1\otimes\dots\otimes q_m)\theta'$

avec - $q \in Q$, $d_{sup}(q) = n_1$, $d_{inf}(q) = n$;
- $a_i \in \Sigma \cup \{Id_1\}$ pour $i \in [n]$;
- $\tilde{u} \quad \tilde{T}(\Delta)^{n_1}_{n_2}$;
- θ est une bijection de $\theta^{n_2}_{n_2}$, $n_2 = \sum_{i=1}^{m} d_{sup}(q_i)$;
- $q_i \in Q$ pour $i \in [m]$;
- $\theta' \in \theta^{n_3}_{n_4}$, $n_3 = \sum_{i=1}^{m} d_{inf}(q_i)$, $n_4 = \sum_{i=1}^{n} d(a_i)$.

⊥

Définissons la transduction réalisée par le $\bar{\bar{s}}$td T. Pour cela, on définit par induction pour chaque état $q \in Q$, avec $d_{sup}(q) = n_1$, $d_{inf}(q) = n_2$, et pour tout arbre $t \in \tilde{T}(\Sigma)^1_o$, la q-transduction de T notée $T_q(t)$:

1) $\forall a = a_1 \otimes \dots \otimes a_{n_2}$, $a_i \in \Sigma_o$,

$T_q(a) = \{\tilde{v} \in \tilde{T}(\Delta)_o^{n_1} \mid \langle q.a\rangle \to \tilde{v} \in R\}$

2) $\forall\, t = t_1 \otimes \dots \otimes t_{n_2},\ t_i \in \tilde{T}(\Sigma)_o^1$ et $t_i = a_i.t'_i$ avec $a_i \in \Sigma \cup \{Id_1\}$

$T_q(t) = \{\tilde{v}.\theta.(T_{q_1}(\theta'_1.t') \otimes \dots \otimes T_{q_n}(\theta'_n.t')) \mid t' = t'_1 \otimes \dots \otimes t'_{n_2},$

$\langle q.(a_1 \otimes \dots \otimes a_{n_2})\rangle \to \tilde{v}.\theta.(q_1 \otimes \dots \otimes q_n).\langle\theta'_1,\dots,\theta'_n\rangle \in R\}$

$\perp$

Notons que si $A_1,\dots,A_n \subseteq T(\Delta)$ alors $v.(A_1 \otimes\dots\otimes A_n) = \{v.(t_1 \otimes\dots\otimes t_n) \mid t_i \in A_i\}$.

Définition : La transduction réalisée par T est la relation $\hat{T} = \{(t, u) \mid t \in \tilde{T}(\Sigma)_o^1,$ $u \in \tilde{T}(\Delta)_o^1$ et $u \in T_q(t)$ avec $q \in P\}$.

Nous écrirons encore $q.t \overset{*}{\underset{T}{\Rightarrow}} u$.

Définition : Un $\bar{\bar{S}}$-transducteur ascendant (noté $\bar{\bar{s}}$ta) est un 5-uple $\langle Q, \Sigma, \Delta, R, P\rangle$ où Q, Σ, Δ, et P sont les mêmes ensembles que ceux introduits dans la définition d'un $\bar{\bar{s}}$td ; P est l'ensemble des états finaux ; R est un ensemble fini de règles de la forme : $\langle(a_1 \otimes \dots \otimes a_n).\theta.(q_1 \otimes \dots \otimes q_m)\rangle \to q.\tilde{u}.\theta'$ où :

- $a_i \in \Sigma \cup \{Id_1\}$ pour $i \in [n]$;
- θ est une bijection de $\theta_{n_1}^{n_1}$, $n_1 = \sum_{i=1}^{n} d(a_i) = \sum_{i=1}^{m} d_{sup}(q_i)$;
- $q_i \in Q$ pour $i \in [m]$;
- $q \in Q$, $d_{sup}(q) = n$, $d_{inf}(q) = n_2$;
- $\tilde{u} \in \tilde{T}(\Delta)_{n_3}^{n_2}$;
- $\theta \in \theta_{n_4}^{n_3}$, $n_4 = \sum_{i=1}^{m} d_{inf}(q_i)$.

$\perp$

Pour chaque état $q \in Q(d_{inf}(q) = n_1,\ d_{sup}(q) = n_2)$ et pour tout arbre $t \in \tilde{T}(\Sigma)_o^1$ on définit par induction la q-transduction de T, notée $T_q(t)$:

1) $\forall\, a = a_1 \otimes \dots \otimes a_{n_1},\ a_i \in \Sigma_o,$

$T_q(a) = \{\tilde{v} \in \tilde{T}(\Delta)_o^{n_2} \mid \langle a\rangle \to q.\tilde{v} \in R\}$

2) $\forall\, t = t_1 \otimes \dots \otimes t_{n_1}$ avec $t_i = a_i.t'_i$, $a_i \in \Sigma \cup \{Id_1\}$,

$T_q(t) = \{\tilde{v}.\theta'.(u_1 \otimes \dots \otimes u_n) \mid u_i \in T_q(\theta_i^{-1}.(t'_1 \otimes \dots \otimes t'_{n_1}))$

et $\langle(a_1 \otimes \dots \otimes a_{n_1}).\theta.(q_1 \otimes \dots \otimes q_n)\rangle \to q.\tilde{v}.\theta' \in R$ avec

$\theta^{-1} = \langle\theta_1^{-1},\dots,\theta_n^{-1}\rangle$, $u_i \in \tilde{T}(\Delta)_o^{d_{inf}(q_i)}\}$.

Définition : La transduction réalisée par T est la relation $\hat{T} = \{(t, u) \mid t \in \tilde{T}(\Sigma)_o^1,$ $u \in \tilde{T}(\Sigma)_o^1$ et $u \in T_q(t)$ avec $q \in P\}$.

Nous écrirons encore $t \overset{*}{\underset{T}{\Rightarrow}} q.u$.

Définition : Nous dirons que T est un transducteur linéaire si, dans chaque règle de T, la torsion θ' est injective ; T est complet si la torsion θ' est surjective.

Définition : Nous définissons la relation binaire sur les éléments de T(Σ), noté $t \leq u$, par $t \leq u$ si et seulement si il existe v tel que $u = t.v$. Nous dirons que deux arbres t et t' sont compatibles si et seulement si il existe un arbre u tel que $t \leq u$ et $t' \leq u$.

Définition : T est un transducteur déterministe si quel que soient deux parties gauches de règles de T de la forme $q_1.t_1$ et $q_2.t_2$ dans le cas descendant ou $t_1.\vec{q}_1$ et $t_2.\vec{q}_2$ dans le cas ascendant, alors les arbres t_1 et t_2 ne sont pas compatibles.

Cela revient donc bien à ne jamais pouvoir faire de choix dans la dérivation d'un arbre.

Notons que lorsque les règles du transducteur T ne contiennent pas le mot vide en partie gauche et que les états ont tous une arité supérieure et inférieure égale à $1 (d_{sup}(q) = d_{inf}(q) = 1 \; \forall q)$ alors nous retrouverons la définition classique des transducteurs d'arbres [4].

Nous étudierons plus particulièrement les sous-classes suivantes [6] :
- *S-transducteurs* : nous imposons que les états ont une arité supérieure égale à 1 (notés std ou sta)
- *$\bar{S}$-transducteurs* : nous imposons l'arité inférieure des états égale à 1. (notés $\overline{\text{std}}$ ou $\overline{\text{sta}}$).

Notation : Nous noterons par des lettres majuscules la classe des transductions associées à une famille de transducteurs. Par exemple $STD = \{\hat{T} \mid T \text{ est un std}\}$.

Chapitre II : CLOTURES PAR COMPOSITION

Nous étudions dans ce chapitre les propriétés de clôture par composition des diverses classes de transducteurs déterministes. Les classes TD dét. et STD dét. ne sont pas fermées par composition si on n'impose pas que les transducteurs soient complets. En effet les transducteurs de ces classes ne peuvent pas reconnaitre un sous-arbre après l'avoir abandonné. Pour la classe STD dét complet la construction du transducteur composé n'est pas simplement réalisées en composant règle par règle car des problèmes de chevauchement de découpage apparaissent. Comme la classe $\overline{\text{S}}$TD est identique à la classe $\overline{\text{S}}$TD complet, nous montrons que la classe $\overline{\text{S}}$TD dét est fermée par composition.

Dans le cas ascendant les problèmes de complétude n'apparaissent pas car le transducteur reconnait le sous-arbre avant de l'abandonner ; les classes TA dét et STA dét sont fermées par composition. Pour la classe $\bar{S}TA$ un transducteur est dét si et seulement si non seulement toutes les parties gauches des règles sont différentes mais aussi tous les états ont une arité supérieure égale à 1 et dans ces conditions nous retombons sur la classe TA dét. Enfin nous montrons que les classes $\bar{\bar{S}}TA$ et $\bar{\bar{S}}TD$ dét ne sont pas fermées par composition.

Proposition 2.1. : Les classes TD dét et STD dét ne sont pas fermées par composition.

Preuve : Nous allons exhiber 2 td déterministes dont la transduction composée ne peut être réalisée par un std dét, donc à fortiori par un td dét. Il suffira de prendre un transducteur *non complet* abandonnant un arbre non borné en largeur :
Soit le td dét complet $T_1 = \langle Q_1, \Sigma, \Sigma, R_1, P_1\rangle$ tel que $\hat{T}_1 = \{(\alpha.(\bar{a} \otimes t), \alpha.(\bar{a} \otimes t)) \mid t \in \tilde{T}(\{b, \bar{b}\})_o^1\}$. T_1 est défini sur l'alphabet $\Sigma = \Sigma_\Omega \cup \Sigma_2$, $\Sigma_o = \{\bar{a}, \bar{b}\}$, $\Sigma_2 = \{\alpha, b\}$. Soit le td dét $T_2 = \langle Q_2, \Sigma, \Sigma, \Sigma_2, R_2, P_2\rangle$ tel que $\hat{T}_2 = \{(\alpha.(\bar{a} \otimes t), \alpha.(\bar{a} \otimes \bar{a})) \mid t \in \tilde{T}(\Sigma)_o^1\}$. La transduction composée est égale à $\widehat{T_1 \circ T_2} = \{(\alpha.(\bar{a} \otimes t), \alpha.(\bar{a} \otimes \bar{a})) \mid t \in \tilde{T}(\{b, \bar{b}\})_o^1\}$. Or comme t est non borné en largeur, il faudrait pour reconnaitre t introduire des états d'arité arbitrairement grande, ce qui est contraire à la définition du std.

$\perp$

Si nous imposons la propriété supplémentaire de complétude nous obtenons :

Proposition 2.2. : La classe des transducteurs déterministes complets est fermée par composition.

Ce résultat est du à Engelfriet [4].

Proposition 2.3. : La classe des S-transducteurs déterministes complets est fermée par composition.

Nous allons donner une idée de la construction du std déterministe composé. Soient $T_1 = \langle Q_1, \Sigma_1, \Delta_1, R_1, P_1\rangle$ et $T_2 = \langle Q_2, \Sigma_2, \Delta_2, R_2, P_2\rangle$ deux std déterministes complets. Construisons le std déterministe complet composé $T = \langle Q, \Sigma_1, \Delta_2, R, P_1\rangle$ tel que $\hat{T} = \widehat{T_1 \circ T_2}$. Pour chaque règle de T_1 : $\langle q.\vec{a}\rangle \to \tilde{u}.\vec{q}.\theta$ où q est un état initial de T_1, $\tilde{u}$ un arbre de $\tilde{T}(\Delta_1)_m^1$, θ une torsion surjective, $\vec{a}$ et $\vec{q}$ respectivement un n-uple d'éléments de $\Sigma_1 \cup \{Id_1\}$ et un m-uple d'états de T_1. Nous allons, en utilisant des règles de T_2, dériver le plus loin possible l'arbre $\tilde{u}$: soit donc (r) la plus longue suite de règles de T_2 (cette suite est unique) telle que $q'.\tilde{u} \overset{*}{\underset{(r)}{\Rightarrow}} \tilde{v}.\vec{q}'.\theta'.\vec{u}'$, où q' est un état initial de T_2, $\tilde{v}$ un arbre de $\tilde{T}(\Delta_2)_{m'}^1$, θ' une torsion surjective, $\vec{q}'$ un m'-uple d'états de T_2 et $\vec{u}'$ un p-uple de sous arbres de $\tilde{u}$ non encore dérivés de T_2.

Nous aurons alors la règle de T : $<q.\vec{a}> \to \tilde{v}.(\tilde{q}_1 \otimes \ldots \otimes \tilde{q}_{m'}).\tilde{\theta}$ sachant que $(\tilde{q}_1 \otimes \ldots \otimes \tilde{q}_{m'}).\tilde{\theta} = \vec{q}'.\theta'.\vec{u}'.\vec{q}.\theta$. $\tilde{\theta}$ est une torsion surjective. Les états $\tilde{q}_1, \ldots, \tilde{q}_{m'}$ sont des états de T, de la forme :

(1) $\tilde{q}_i = q'_\ell.t.(q_1 \otimes \ldots \otimes q_s)$ où $q'_\ell \in Q_2$ avec $d(q'_\ell) = m$, $t \in \tilde{T}(\Delta_1)^m_s$ et $q_j \in Q_1$ pour $j = 1, \ldots s$. L'arité de l'état $\tilde{q}_i$ est égal à $\sum_{j=1}^{s} d(q_j)$. Comme nous avons dérivé l'arbre $\tilde{u}$ le plus loin possible, pour chaque nouvel état $\tilde{q}_i$ il existe un j tel que $prof(t_j) = 0$ (sinon $\tilde{q}_i$ est un état puits) où t_j est la $j^{ème}$ composante du m-uple t. Soit donc q_j l'unique état de T_1 "accroché" à l'arbre t_j ; pour pouvoir continuer, il suffit d'appliquer une règle de T_1 ayant en partie gauche l'état q_j, puis de dériver le plus loin possible l'arbre $t.u_j$ à partir de l'état q'_ℓ. Nous introduisons éventuellement de nouveaux états de la forme (1). Les éléments constitutifs de $\tilde{q}_i$ étant tous finis, le nombre d'états et de règles de T est fini. Il est clair que T est déterministe et complet.

⊥

Nous avons montré dans [6] que la classe STA n'était pas fermée par composition, et ceci à cause de la non linéarité et du non déterminisme des transducteurs : le premier transducteur recopie en nombre non borné un sous-arbre, tandis que le second transforme différemment ces sous-arbres.

Dans le cas où nous conservons l'une seulement des 2 propriétés, alors nous pouvons montrer la fermeture par composition :

1) La classe des S-transducteurs ascendants *linéaires* est fermée par composition [6].

2) *Proposition 2.4.* : La classe TA dét des transducteurs ascendants déterministes est fermée par composition.

Ce résultat est du à Engelfriet [4].

Proposition 2.5. : La classe des S-transducteurs ascendants déterministes est fermée par composition.

Preuve : Soient $T_1 = <Q_1, \Sigma_1, \Delta_1, R_1, P_1>$ et $T_2 = <Q_2, \Sigma_2, \Delta_2, R_2, P_2>$ deux sta déterministes. Montrons qu'il existe un sta déterministe $T = <Q, \Sigma_1, \Delta_2, R, P>$ tel que $\hat{T} = \widehat{T_1 \circ T_2}$. Nous construisons T de la façon suivante :

- $Q = \{\tilde{q} \mid \tilde{q} = q.(q'_1 \otimes \ldots \otimes q'_m)$ avec $m = d(q)$, $q \in Q_1$, $q'_i \in Q_2\}$
- $P = \{\tilde{q} \quad \tilde{q} = q.(q'_1 \otimes \ldots \otimes q'_m)$ avec $q \in P_1$, $q'_i \in P_2\}$
- construction de l'ensemble des règles R :

Pour toute règle de T_1 : $<a.\vec{q}> \to q.\tilde{u}.\theta$ où $a \in \Sigma \cup \{Id_1\}$, $\vec{q}$ est un n-uple d'états de T_1, $q \in Q_1$, $\tilde{u} \in \tilde{T}(\Delta_1)$ et θ est une torsion, considérons toutes les suites possibles d'états de T $\tilde{q}_1, \ldots, \tilde{q}_n$ telles que :

a) $\tilde{q}_i = q_i.\vec{q}'_i$ où $\vec{q}'_i$ est un n_i-uple d'états de T_2, pour $i = 1,\ldots,n$.

b) il existe une suite (r') de règles de T_2 telle que $\tilde{u}.\theta.(\vec{q}_1 \otimes \ldots \otimes \vec{q}_n) \underset{(r)}{\Rightarrow} (q'_1 \otimes \ldots \otimes q'_m).\tilde{v}.\theta'$.

Sous ces conditions la règle $<a.(\tilde{q}_1 \otimes \ldots \otimes \tilde{q}_n)> \rightarrow \tilde{q}.\tilde{v}.\theta'$ avec $\tilde{q} = q.(q'_1 \otimes \ldots \otimes q'_m)$ est une règle de T. Comme Q_1 et Q_2 sont finis, le nombre de règles est fini. Nous laissons au lecteur le soin de vérifier que $T = \widehat{T_1 \circ T_2}$.

⊥

Proposition 2.6. : La classe des $\bar{S}$-transducteurs déterministes descendants est fermée par composition.

Donnons une idée de la construction du transducteur composé. Soient $T_1 = <Q_1, \Sigma_1, \Delta_1, R_1, P_1>$ et $T_2 = <Q_2, \Sigma_2, \Delta_2, R_2, P_2>$ deux $\bar{s}$td déterministes que nous supposerons complets (on peut toujours construire un $\bar{s}$td complet équivalent à un $\bar{s}$td quelconque). Considérons chaque règle de T_1 : $<q_o.a> \rightarrow \tilde{u}.\theta_1.\vec{q}.\theta'_1$ où $q_o \in Q_1$, $a \in \Sigma_1 \cup \{Id_1\}$, $\tilde{u} \in \tilde{T}(\Delta_1)$, $\vec{q}$ est un m-uple d'états de T_1, θ_1 et θ'_1 des torsions. L'arbre $\tilde{u}$ se dérive de manière unique par T_2 : $q'_o.\tilde{u} \underset{T_2}{\Rightarrow} \tilde{v}.\theta_2.\vec{q}'.\theta'_2$ où $q'_o \in Q_2$, $\tilde{v} \in \tilde{T}(\Delta_2)$, $\vec{q}'$ est un m'-uple d'états de T_2, θ_2 et θ'_2 des torsions. Sous ces conditions la règle : $<q_o.a> \rightarrow \tilde{v}.\theta_2.(\tilde{q}_1 \otimes \ldots \otimes \tilde{q}_m).\theta'_1$ est une règle de T, sachant que $\tilde{q}_1 \otimes \ldots \otimes \tilde{q}_m = \vec{q}'.\theta'_2.\theta_1.\vec{q}$. Les états de T sont donc de la forme $\tilde{q} = (q'_1 \otimes \ldots \otimes q'_p).\theta.q$ où $q'_i \in Q_2$, $q \in Q_1$ et θ est une torsion. Notons que $d_{sup}(\tilde{q}) = \sum_i d_{sup}(q'_i)$. Pour ces nouveaux états on considère les règles de T_1 ayant en partie gauche l'état q, et on applique à nouveau le procédé décrit plus haut.

⊥

Proposition 2.7. : La classe $\bar{\bar{S}}$TD dét n'est pas fermée par composition.

Preuve : On peut réaliser par des $\bar{\bar{s}}$td dét les transformations

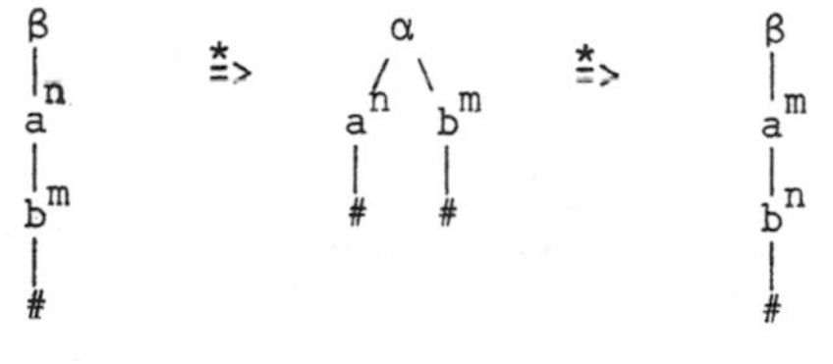

Il est clair qu'un seul $\bar{\bar{s}}$td ne peut réaliser cette transformation.

⊥

Chapitre III : COMPARAISON DES CLASSES DE TRANSDUCTEURS DETERMINISTES

Nous nous proposons dans ce chapitre d'étudier les propriétés d'inclusion ou de non inclusion des différentes classes de transducteurs introduites. Nous retrouvons les mêmes inclusions ou non inclusions établies pour les différentes classes de transducteurs non linéaires et non déterministes [6] sauf que la classe $\bar{S}$TD dét. est incluse strictement dans la classe STA dét.

Proposition 3.1. : La classe TA dét complet n'est pas incluse dans la classe $\bar{\bar{S}}$TD dét.

Preuve : Soit la relation $R = \{(a^n.\bar{a}, a^n.\bar{a}), (a^n.\bar{b}, b^n.\bar{b}) \mid n \geq 0\}$. Il est clair qu'il existe un ta dét. complet tel que $\hat{T} = R$. Or le $\bar{\bar{s}}$td T' tel que $\hat{T}' = R$ ne peut être que non déterministe : en effet lorsque T' lit le symbole a, il ne sait pas encore s'il doit engendrer un a ou un b.

Proposition 3.2. : La classe TD dét. complet n'est pas incluse dans la classe TA dét.

Preuve : Soit la relation $R = \{(a.b^m.\#, a^{m+1}.\#), (c.b^m.\#, c^{m+1}.\#) \mid m \geq 0\}$. Il est clair qu'il existe un td dét. tel que $\hat{T} = R$. Or le ta T' tel que $\hat{T}' = R$ ne peut être que non déterministe, en effet lorsque T' lit un symbole b, il ne peut pas encore savoir s'il doit transformer b en a ou c.

Proposition 3.3. : La classe $\bar{S}$TD dét est incluse dans la classe STA dét.

Preuve : Soit $T = \langle Q, \Sigma, \Delta, R, P\rangle$ un $\bar{s}$td dét , construisons un sta dét $T' = \langle Q', \Sigma, \Delta, R', P'\rangle$. Nous supposerons T complet. Les états de Q' forment une partition des états de Q.
$P' = \{\{q_i\} \mid q_i \in P\}$. On construit les règles de R' de la façon suivante ;
Pour tout $a \in \Sigma$ on considère les p_a règles de T ayant en partie gauche la lettre a :
$\langle q_i.a\rangle \to \tilde{u}_i\ \theta_i.\vec{q}_i.\theta'_i \quad i = 1, \ldots, p_a$ avec

- $d(a) = n$
- $\vec{q}_i$ un m_i-uple d'états de T
- θ_i une bijection
- θ'_i une surjection de $[m_i] \to [n]$

on sait que

$\theta'^{-1}_i(j) = \{x_{j_1}, \ldots, x_{j_k}\}$ pour $j \in [n]$. On associe à $\theta'^{-1}_i(j)$ l'ensemble $\{q_{j_1},\ldots,q_{j_k}\} = Q^{-1}_i(j)$.
L'ensemble des règles de la forme : $\langle a.(\bar{q}_1 \otimes \ldots \otimes \bar{q}_n)\rangle \to \bar{q}.(\tilde{u}_1 \otimes \ldots \otimes \tilde{u}_{p_a}).\bar{\theta}$ appartiennent à R', où $\bar{q}_j = \{q_{\alpha^j_1},\ldots, q_{\alpha^j_{\ell_j}}\} \supseteq Q^{-1}_i(j)\ \forall\, i \in [P_a]$ pour $j = 1,\ldots,n$;

notons que $d_{inf}(\bar{q}_j) = d_{sup}(q_{\alpha^j_1}) + \ldots + d_{sup}(q_{\alpha^j_{\ell_j}})$; $\bar{q}$ est l'ensemble $\{q_1,\ldots,q_{p_a}\}$ et $d_{inf}(\bar{q}) = \sum_{i=1}^{p_a} d_{sup}(q_i)$; $\bar{\theta} = \langle\bar{\theta}_1,\ldots, \bar{\theta}_{p_a}\rangle$ est défini par :

$$\bar{\theta}_i(k) = d_{inf}(\bar{q}_1) + \dots + d_{inf}(\bar{q}_{j-1}) + d_{sup}(q_{\alpha_1^j}) + \dots + (q_{\alpha^j_{\ell''-1}}) + \ell'$$

sachant que :

- $\theta_i(k) = d_{sup}(q_{i_1}) + \dots + d_{sup}(q_{i_{\ell-1}}) + \ell'$ avec $\ell' \leq d_{sup}(q_{i_\ell})$
- $\theta'_i(\ell) = j$
- $q_{\alpha_{\ell''}} = q_{i_\ell}$

Nous laissons au lecteur le soin de vérifier que $\hat{T} = \hat{T}'$.

$\perp$

En résumé nous obtenons le diagramme des inclusions de classe de transducteurs déterministes suivant :

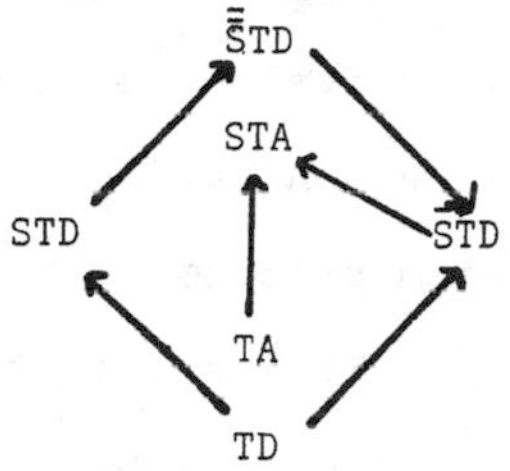

(A → B signifie $A \subsetneq B$)

REFERENCES

[1] ARNOLD A. et DAUCHET M., "*Théorie des magmoïdes*" RAIRO 12 p 235-257 et RAIRO 13 p 135-154.

[2] DAUCHET M., "*Transductions de forêts - Bimorphismes de magmoïdes*", Thèse d'Etat, Université de Lille (1977).

[3] DONER J., "*Tree acceptors and some of their applications*", J. Comput System Sci. 4 (1970) p 406-451.

[4] ENGELFRIET J., "*Bottom up and top-down tree transformations - a comparison*", Math System Theory 9 (1975) p 198-231.

[5] ENGELFRIET J., "*Top down tree transducers with regular look-ahead*" Math System Theory 10 (1977) p 289-303.

[6] LILIN E., "*Une généralisation des transducteurs d'états finis d'arbres : les S-transducteurs*" Thèse de 3ème cycle, Université de Lille (1978).

[7] LILIN E., "*S-transducteurs de forêts*" (1978), communication au 3e CLAAP.

[8] ROUNDS W.C., "*Tree transducers and transformation*", Ph D. Dissertation Stanford University (1968).

CERTAIN ALGORITHMS FOR SUBGRAPH ISOMORPHISM PROBLEMS

Andrzej Lingas

Uniwersytet Warszawski, Instytut Matematyki

and

Massachusetts Institute of Technology,

Laboratory for Computer Science +)

Introduction

The subgraph isomorphism problem (SI for short) is to determine whether an input graph G is isomorphic to a subgraph of another input graph H . For instance, if G is an n-vertex circuit and H is an n-vertex planar graph then SI is equivalent to the NP-complete problem of determining whether a planar graph has a Hamiltonian circuit [3] . Thus, the subgraph isomorphism problem is NP-complete even if G and H range only over planar graphs. It seems that this distinguishes SI from the graph isomorphism problem (GI for short) which is determining whether two input graphs are isomorphic each to other. Of course, there are no known polynomial time algorithms for GI and the possibility that GI is NP-complete is not excluded. However, GI restricted to planar graphs is solvable even in time linear in the number of vertices of input graphs [6] . Moreover, polynomial time algorithms for GI restricted to graphs of a fixed genus has been presented recently [2,12] . Perhaps, the higher intractability of SI in comparison to GI causes the gap between the number of articles devoted to SI and GI , respectively [13] . Of course, there are several articles on such famous instances of SI as the clique problem [1] or the Hamiltonian circuit problem [1], however, here I mean articles presenting general approach to SI, using such general properties of graphs as planarity . Note that from the practical viewpoint, SI is of no less importance than GI. For instance, in organic chemistry SI corresponds to the problem of determining whether a chemical structure is a piece of another, whereas

+) This research was supported in part by NSF grant MCS 7805849.

GI corresponds to the problem of determining whether two chemical structures are identical . Moreover, SI is closer related to the pattern recognition problems [4] than GI . This paper consists of two independent sections .

In Section 1, two refinements of a unique algorithm of Ullmann [15] for the (general) subgraph isomorphism problem are given . One of the refinements relies on an algorithm for maximum matchings in bipartite graphs and develops the main idea of Ullmann's algorithm .

In Section 2, an algorithm for determining the number of subgraphs of input graph that are so called monotonously or cyclically isomorphic to a fixed pattern graph is presented. The algorithm is described in terms of certain pebbling games. For n-vertex input graph it runs in time $O(\binom{n}{m}n^2)$ provided that the pattern graph has so called m-pebbling. The constant m can be small for the pattern graphs which are members of classes of graphs with good separator theorem [9,10] . Hence, it is concluded that if a graph G is a member of such class of graphs then there is a graph G^* satisfying two following conditions :

(a) G^* is isomorphic to G,

(b) the algorithm with G^* as the pattern graph is much faster than exhaustive search .

For instance, if G is a planar graph on k vertices then we can test whether the graph G^* can be monotonously embedded in another graph on n vertices in $O(\binom{n}{c\sqrt{k}})$ steps whereas exhaustive search requires $\Omega(\binom{n}{k})$ steps .

Preliminary Notations and Notions

We shall adhere to certain standard graph and algorithm notations and notions from [5,1], respectively . We choose the following notations for economy :

For a set A, #A denotes the cardinality of the set . Next, for a map f of B into C, where $A \subset B$, f(A) is the image of A, i.e. $\{ f(a) \mid a \in A \}$.

N denotes the set of all natural numbers . The time complexity of algorithm is understood in terms of the time complexity of random access machine (RAM , see [1]), in which multiplication is not allowed and additions are counted as single steps .

1. The Refinements of Ullmann's Algorithm for Subgraph Isomorphism.

In [15] Ullmann presented a general, tree search enumeration algorithm for the subgraph isomorphism problem. The algorithm can be immediately generalized for the subdigraph isomorphism problem. Here two refinements of Ullmann's algorithm will be shown. The first, rather trivial refinement follows from well known methods of vertex classification used in the graph isomorphism problem [13,14] . The second one improves the main idea of Ullmann's algorithm by using an algorithm for maximum matchings in bipartite graphs [8]. In order to present the refinements Ullmann's algorithm will be outlined first .

Ullmann's Algorithm

Let G1=(V1,E1), G2=(V2 ,E2) be two input undirected graphs. In Ullmann's algorithm, which tests whether G1 is a subgraph of G2, the graphs G1, G2 are represented by the adjacency matrices $A=(a_{ij})$, $B=(b_{kl})$, respectively. Let $M'=(m'_{ij})$ be a $(\#V1) \times (\#V2)$ matrix of 0's and 1's, in which each row contains exactly one 1 and no column contains more than one 1. Clearly, M' specifies an one-to-one map of $\{1,\ldots,\#V1\}$ into $\{1,\ldots,\#V2\}$. Next, let $C=(c_{ij})=M'(M'B)^T$, where T denotes transposition. If the following condition

(1.1) $\forall\ 1 \leq i \leq \#V1 \ \forall\ 1 \leq j \leq \#V2 \qquad ((a_{ij} = 1) \rightarrow (c_{ij} = 1))$

holds then M' specifies an isomorphism between G1 and a subgraph of G2, in which the i-th vertex of G1 corresponds to the j-th vertex of G2 provided that $m'_{ij}=1$.

At the start of Ullmann's enumeration algorithm, a $(\#V1) \times (\#V2)$ matrix $M^*=(m^*_{ij})$ is defined as follows :

$$(\forall i,j) \quad m^*_{ij} = \underline{\text{if}}\ \deg(i) \leq \deg(j)\ \underline{\text{then}}\ 1\ \underline{\text{else}}\ 0$$

Ullmann's algorithm enumerates matrices $M'=(m'_{ij})$ satisfying the following condition :

$$(\forall i,j) \qquad m'_{ij} = 1 \rightarrow m^*_{ij} = 1$$

Such matrices M' are tested for isomorphism by verifying the condition 1.1 . In the search tree of the algorithm, all leafs which are at the depth #V1 correspond to distinct matrices M'. With each node of the tree which is at the depth d < #V1 , a distinct (#V1) $\times$ (#V2) matrix M, having the #V1-d last rows in common with the matrix M*, is associated. The first d rows of M is a result of changing to 0 all but one of the 1's in each of the corresponding d rows of M*. Thus the root of the search tree corresponds to M* . A matrix M at the depth d < #V1 may give rise to many matrices at the depth d+1 which are generated by systematically changing certain 1's to 0 in the d+1st row of M,i.e.M*. Such matrices correspond to sons of the node of M in the tree .

The main idea of Ullmann's algorithm consists in eliminating some of the descendants of M by changing some of the 1's to 0 in M. The elimination is based on the following observation :

If, in the search tree the matrix M is an "ancestor" of a matrix M' which specifies an isomorphism between G1 and a subgraph of G2 with the i-th vertex of G1 corresponding to the j-th vertex of G2, i.e. $m_{ij}=1$, then the following condition holds :

$$(1.2) \qquad \forall\ 1 \leq x \leq \#V1 \qquad (a_{ix}=1) \rightarrow (\exists\ 1 \leq y \leq \#V2 \quad m_{xy} \cdot b_{yj}=1)$$

The above condition results from the following fact. If the i-th vertex of G1 is adjacent to the $x = i_1,\ldots,i_m$ - th vertex of G1,i.e. $a_{ix}=1$, then for each $x = i_1,\ldots,i_m$ there exists y, $1 \leq y \leq \#V2$, such that the x-th vertex of G1 corresponds to the y-th vertex of G2 . The elimination of 1's consists in changing each $m_{ij}=1$ which does not satisfies the condition (1.2) to $m_{ij}=0$. Such changes of M may cause further and further changes of M and therefore the elimination of 1's ends either when no further change is possible or when M contains a row of zeros. In the latter case, the node of M is blind,i.e, has no sons in the tree .

Refinement 1 .

Let us consider the matrices of the shortest paths for graphs G1, G2. They will be denoted by (c_{ij}), (d_{ij}), respectively . That is, c_{ij} is the length of a shortest path joining the i-th vertex with the j-th vertex in G1, and d_{ij} is analogously defined for G2 . Next, let us introduce a (#V1) × (#V1-1) matrix (e_{iu}) with entry e_{iu} equal to $\# \{j \mid c_{ij} \leq u\}$ and a (#V2) × (#V-1) matrix (f_{iu}) with entry f_{iu} equal to $\# \{j \mid d_{ij} \leq u\}$.

Using the vertex classification determined by the matrices (e_{iu}) , (f_{iu}), the initial matrix M* in Ullmann's algorithm may be refined as follows :

$$(\forall\, i,j) \quad m^*_{ij} = \underline{\text{if}}\ (\forall\ 1 \leq u \leq \#V1-1 \quad e_{iu} \leq f_{ju})\ \underline{\text{then}}\ 1\ \underline{\text{else}}\ 0 .$$

Simply, if there is an isomorphism between G1 and a subgraph of G2 in which the i-th vertex of G1 corresponds to the j-th vertex of G2, and, on the other hand, there are e_{iu} vertices of G2 which are placed at the distance not greater than u from the i-th vertex in G1, then in G2 as a "overgraph" of G1 there are at least e_{iu} vertices placed at the distance not greater than u from the j-th vertex of G2, i.e. $e_{iu} \leq f_{ju}$. Of course, if G1 and G2 are to be tested for isomorphism then we should rather substitute = for ≤ in the definition of matrices (e_{iu}),(f_{ju}) obtaining vertex classification analogous to that presented for digraphs in [14] and then change $e_{iu} \leq f_{iu}$ to $e_{iu} = f_{iu}$ in the new definition of the matrix M* .

The evaluation of the refined matrix M* takes $O((\#V1)^3)$ steps. To see this, first let us notice that the matrix (c_{ij}) can be computed in $O((\#V1)^3)$ steps, see p.200 in [1] . Given (c_{ij}) and the #V1 × #V1 matrix Z of zeros, the matrix (e_{iu}) can be evaluated in $O((\#V1)^2)$ steps as follows :

(e_{iu}) := Z ;
<u>for</u> i = 1 <u>step</u> 1 <u>until</u> #V1 <u>do</u>

```
begin
  for j = 1 step 1 until #V1 do
    begin
      u := c_ij ;
      e_iu := e_iu + 1
    end
  for u = 2 step 1 until #V1 - 1 do
  e_iu := e_i,(u-1) + e_iu
end
end
```

Likewise, the matrix (f_{iu}) can be computed in $O((\#V2)^3)$ steps .

Next, let us suppose that we are at the depth $d < \#V1$ in the search tree of Ullmann's algorithm and we select the succeding 1 from the ℓ-th column of M in order to define a new matrix at the depth d + 1 . Then we may also require the following condition to be fulfilled :

$$(1.3) \quad \forall\ 1 \le i \le d\ \forall\ 1 \le j \le \#V2 \qquad (m_{ij} = 1 \rightarrow c_{d+1,i} \ge d_{\ell j})$$

The condition results from the following observation . If the d+1 st vertex of G1 corresponds to the ℓ-th vertex of G2 in an isomorphism which is an extension of the partial map specified in the first d rows of M' then for each i=1,...,d the distance between the i-th and the d+1 st vertex in G1 is not less than the distance between corresponding vertices in G2. Of course, we can verify (1.3) in O(d) steps .

Refinement 2.

Let us assume again that we are at the depth $d < \#V1$ in the search tree of Ullmann's algorithm and apply the elimination of 1's in the matrix M by checking the condition 1.2. This condition is equivalent to the conjunction of two following conditions :

$$(1.4)\ \forall\ 1 \le x \le d \quad ((a_{ix}=1) \rightarrow (\exists\ 1 \le y \le \#V2 \quad m_{xy} \cdot b_{yj}=1))$$

$$(1.5)\ \forall\ d < x \le \#V1 \quad ((a_{ix}=1) \rightarrow (\exists\ 1 \le y \le \#V2 \quad m_{xy} \cdot b_{yj} = 1))$$

In (1.4), for each x=1,...,d there is one and only one $y \in \{1,...,\#V2\}$ satisfying $m_{xy}=1$ since each of the d first rows of M contains exactly one 1 in a distinct column. Thus the condition 1.4 is equivalent to :

$$(1.6)\ \forall\ 1 \le x \le d \quad ((a_{ix}=1) \rightarrow m_{xf(x)} \cdot b_{f(x)j}=1)$$

where f is one-to-one map of $\{1,\ldots,d\}$ into $\{1,\ldots,\#V2\}$ specified by M, i.e., for each $x = 1,\ldots,d$, $f(x)=y$ if and only if $m_{xy}=1$. Note that if in the search tree the matrix M is an ancestor of any matrix M' defining an isomorphism between G1 and a subgraph of G2 , then the isomorphism specified by M', say g, is an extension of the one-to-one map f to one-to-one map g of $\{1,\ldots,\#V1\}$ into $\{1,\ldots,\#V2\}$ for which the following condition holds :

$$(1.7) \quad \forall\, 1 \le x \le \#V1 \quad ((a_{ix}=1) \rightarrow m_{xg(x)} \cdot b_{g(x)j} = 1)$$

Therefore we may require the following condition to be satisfied :

$$(1.8) \quad \exists\, g:\{1,\ldots,\#V1\} \overset{1-1}{\rightarrow} \{1,\ldots,\#V2\} \quad (\forall\, 1 \le x \le d \quad g(x) = f(x))$$
$$\& \quad (\forall\, 1 \le x \le \#V1 \quad ((a_{ix} = 1) \rightarrow m_{xg(x)} \cdot b_{g(x)j} = 1))$$

Note that, according to the definition of the map f, the condition 1.8 is equivalent to the conjunction of the condition 1.4 with the following condition :

$$(1.9) \quad \exists\, h:\{d+1,\ldots,\#V1\} \overset{1-1}{\rightarrow} \{1,\ldots,\#V2\}) \quad (\forall\, 1 \le x \le d \;\forall\, d<z=\#V1 \;\forall\, 1 \le y \le \#V2$$
$$(m_{xy}=1 \rightarrow h(z) \ne y)) \;\&\; (\forall\, d<x=\#V1 \quad ((a_{ix}=1) \rightarrow m_{xh(x)} \cdot b_{h(x)j}=1))$$

Therefore we may replace the test of (1.2) by a test of the conjunction of (1.4) with (1.9) in the procedure of elimination of 1's . Of course, this conjunction is much stronger than (1.2) (if #V1-d is large enough) and hence such refinement can cause more frequent changes of 1's to 0 in M, which can result in more effective pruning the search tree . The test of (1.4) can be easily performed in O(d) steps . To verify (1.9) in polynomial time we shall reduce this task to finding a maximum matching in bipartite graph F_{ij} .

Let $D = \{y \mid \exists\, 1 \le x \le d \;\; m_{xy} = 1\}$. The graph F_{ij} consists of vertices v_x where $x \in \{d+1,\ldots,\#V1\}$, vertices w_y where $y \in \{1,\ldots,\#V2\}$ & $y \notin D$, and such edges (v_x,w_y) that the implication $a_{ix}=1 \rightarrow m_{xy} \cdot b_{yj}=1$ holds . Clearly, F_{ij} can be constructed in $O(\#V1 \cdot \#V2)$ steps . It is also obvious that the one-to-one map h from (1.9) exists if and only if, there is a matching composed of #V1-1 edges in F_{ij} . Now it suffices to start the maximum matching algorithm of Hopcroft and Karp [8] , and then to check whether the resulting maximum matching of F contains #V1-d edges . If not , then $m_{ij}=1$ is changed to $m_{ij}=0$. According to [8], the matching algorithm applied to F_{ij} and hence the whole test of (1.4) and (1.9) can be performed in $O(\#V1 \cdot \#V2 \cdot (\#V1-d)^{0.5}))$, i.e. $O((\#V1)^{1.5}\, \#V2)$, steps .

2. The algorithm for the fixed subgraph monotone isomorphism problem.

Here by a graph $F = (V_F, E_F)$ we shall mean an undirected graph [5] with the set of vertices V_F and the set of edges E_F. Moreover we assume that V_F is a subset of natural numbers.
An isomorphism f between the graph F and another graph K will be called monotone if for each $i,j \in V_F$, $i < j$ implies $f(i) < f(j)$.

In this section, the algorithm determining the number of distinct subgraphs of input graph that are monotonously (or more general, cyclically) isomorphic to a fixed pattern graph is presented. The algorithm is described in terms of a pebbling game played on the vertices of the input graph according to a fixed strategy of another pebble game played on the vertices of the pattern graph. It is observed that for any graph G which is a member of class of graphs with good separator theorem [9,10] there is an isomorphic graph G^* such that the algorithm finds the number of monotonous (or cyclic) embeddings of G^* in the input graph much faster than exhaustive search. The idea of the algorithm is simple. Hence, in the case of "regular" pattern graphs further speed up is possible. The algorithm can be easily generalized for the case of isomorphism without any constraints. Simply, it suffices to perform the algorithm for each pattern graph on V_G which is isomorphic to the fixed graph G. However, this generalization seems to be of little value since even for a simple, connected graph G there are always pattern graph isomorphic to G for which the algorithm is slow.

To define the mentioned pebbling games we need the following definitions :

For $\ell, q \in N \cup \{0\}$, let $(\ell,q) = \{ m \in N \mid \ell < m < q \}$. Furthermore let $[\ell,q] = (\ell,q) \cup \{\ell,q\}$. For $r \in N, \ell, q \leq r$, $(\ell,q)_r$ denotes the open segment of natural numbers lying between 1 and q in the cyclic order $1,2,\ldots,r,1,2,\ldots$, i.e, $(\ell,q)_r$ = <u>if</u> $\ell < q$ <u>then</u> (ℓ,q) <u>else</u> $(\ell,r) \cup \{r\} \cup (0,q)$. Fig.2.1 presents the segment $(4,2)_5$.

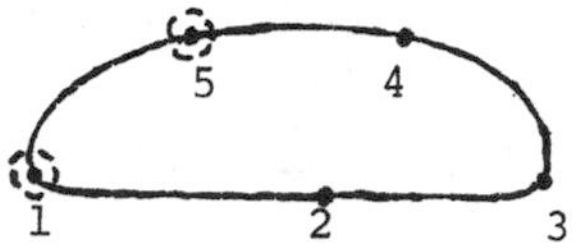

Fig.2.1

Further it will be assumed that k is the number of vertices of the pattern graph G.

Given m pebbles, we place pebbles on the vertices of the graph G by applying the following rules :

(i) A pebble may be placed on a vertex i of G at step t if either t is the first step of the game or there are such two vertices ℓ, q of G ($\ell=q$ is allowed) that $i \in (\ell,q)_k$, ℓ and q hold pebbles and no vertex from $(\ell,q)_k$ has been pebbled in the previous steps.

(ii) A pebble may be removed from a vertex at any step .

Note that according to the rule (i) a pebble may be placed only once on each vertex .

A configuration of the pattern graph G specifies the vertices which hold pebbles in G . Of course, the number of pebbles placed on G is not greater than m .

The objective of the game is to start with no pebbles on G and to find such sequence of steps that for each edge $\{i,j\}$ of G there is a step of the sequence which results in a configuration of G holding pebbles on i and j simultaneously . Such sequence will be called an m-pebbling of the pattern graph . Note that an m-pebbling of G consists of 2k steps at most .

Further it will be assumed that n is the number of vertices of the input graph H and that $k \leq n$.

The rules of the pebbling game played on H in steps, according to an m-pebbling P of the pattern graph are the following :

(a) The collection of pebbles consists of k pebbles p_i where $i \in V_G$.

(b) A pebble p_i may be placed on a vertex i* of H at step t if a pebble is placed on the vertex i of G just at the t-th step of P and, either t is the first step of the game, or there are pebbles p_ℓ, p_q on vertices ℓ^*, q^*, respectively (ℓ may be equal to q) such that :

(b1) $i \in (\ell,q)_k$ and no vertex of G from $(\ell,q)_k$ has been pebbled in the previous steps of P,

(b2) $i^* \in (\ell^*,q^*)_n$ and the t-1 st step results in such configuration of the input graph H that no vertex of H from $(\ell^*,q^*)_n$ holds a pebble,

and moreover, in the case of $t > 1$ the following condition must be satisfied :

(b3) for each pebble held by a vertex j* of H , $\{i^*,j^*\}$ is an edge of H whenever $\{i,j\}$ is an edge of G .

(c) A pebble may be removed from a vertex of H at step t if a pebble is removed from the vertex i of G at the t-th step of the pebbling P .

Note that during the game, at most m pebbles may be placed on the input graph like in the m-pebbling P of the pattern graph according to which the game is played .

Let s be the number of steps of the m-pebbling P . The game is won if it takes s steps . To show that the existence of winning strategy for the game is equivalent to the existence of so called cyclic isomorphism between G and a subgraph of H we need the following definitions:

For a finite subset B of N, $<_B$ denotes the relation of successor in the cyclic order induced by B and $<$,i.e., for $\ell, q \in N$, $\ell <_B q$ if and only if, either $\ell = \max(B)$ & $q = \min(B)$ or $\ell < q$ & $(\ell, q) \cap B = \emptyset$.

An isomorphism f between G and a subgraph of H is <u>cyclic</u> if for each $i,j \in V_G$, $i <_{V_G} j$ implies $f(i) <_{f(V_G)} f(j)$. In other words, f is cyclic if it preserves the cyclic order $<_{V_G}$.

The following lemma establishes a connection between the pebbling games and the subgraph cyclic isomorphism problem .

<u>Lemma 2.1.</u> The pebbling game played on the input graph H according to an m-pebbling P of the pattern graph G is won if and only if there is a cyclic isomorphism between G and a subgraph of H .

<u>Proof</u>. First suppose that there is a winning strategy Q . It follows from the rules (i), (b) that the number of pebbling steps of Q, i.e. such steps in which a pebble is placed on a vertex, is equal to k $^{+}$.

Hence, by virtue of the same rules each pebble p_i, $1 \le i \le k$, is placed on H (only) once. Therefore, we may define a map f of V_G into V_H by setting f(i) equal to the vertex i* of H on which the pebble is placed at a step of the winning strategy Q .

Let B_t be the set of such vertices i of G that the pebble p_i has been used during the first t steps of Q. Next, let G_t be the subgraph of G induced by the set B_t . We shall show by induction on t that f restricted to B_t is a cyclic isomorphism between G_t and a subgraph of H, which, for t=s, will imply that f is a cyclic isomorphism between G and a subgraph of H .

The case t=1 is trivial since then G_t consists exactly of one vertex. So, we may consider the case $1 < t \le s$ and inductively assume that f restricted to B_t is a cyclic isomorphism between G_{t-1} and a subgraph of H . Next, we may assume without loss of generality that at the t-th step of Q, a pebble p_i is placed on a vertex i* since otherwise $B_t = B_{t-1}$ and consequently $G_t = G_{t-1}$.

Note that f(i) = i* . Of course, at the t-th step of Q only the non-

---

+) We may assume without loss of generality that G has no isolated vertex .

trivial part of the rule (b) may be applied . By this part and the definition of G_t and f it immediately follows that there are vertices ℓ, q of G_t satisfying two following conditions :

(B1) i ε $(\ell,q)_k$ and no vertex of G_t from $(\ell,q)_k$ has been pebbled in the previous steps of P,

(B2) f(i) ε $(f(\ell),f(q))_n$ and the t-1 st step of Q results in a configuration of the input graph H with no pebble on any vertex from $(f(\ell),f(q))_n$.

The condition B1 may be restated as follows :

(2.1) $\quad \ell <_{B_t} i \ \& \ i <_{B_t} q$

We shall prove that the following, corresponding condition also holds:

(2.2) $\quad f(\ell) <_{f(B_t)} f(i) \ \& \ f(i) <_{f(B_t)} f(q)$

Suppose otherwise. First consider the case in which $\neg(f(\ell) <_{f(B_t)} f(i))$. Then, there is such r εB_{t-1} that $f(\ell) <_{f(B_t)} f(r)$. Thus $f(\ell) <_{f(B_{t-1})} f(r)$ since ℓ, r ε B_{t-1} . On the other hand, the condition 2.1 implies $\ell <_{B_{t-1}} q$ and f is cyclic on B_{t-1} . Hence, $f(\ell) <_{f(B_{t-1})} f(q)$. Thus f(r) equals f(q) . However f is one-to-one and which yields r=q . This contradicts the membership of f(r) in $(f(\ell),f(q))_n$. In the case $\neg(f(i) <_{f(B_t)} f(q))$ we get a contradiction analogously. Thus the condition 2.2 holds . Now it is easily seen that f is cyclic on B_t, i.e., for each j_1, j_2 from B_t , if $j_1 <_{B_t} j_2$ then $f(j_1) <_{f(B_t)} f(j_2)$. Namely, if j_1, j_2 are in B_{t-1} and $j_1 \neq \ell$ then, according to our induction assumption, $j_1 <_{B_t} j_2$, i.e. $j_1 <_{B_{t-1}} j_2$, implies $f(j_1) <_{f(B_{t-1})} f(j_2)$ which yields $f(j_1) <_{f(B_t)} f(j_2)$ since $f(j_1) \neq f(\ell)$ and f is one-to-one. Next, if $j_1 = \ell$ or $j_1 = i$ then the conditions 2.1, 2.2 guarantee this implication too. See Fig. 2.2.

Fig.2.2.

Hence, to show that f restricted to B_t is a cyclic isomorphism between G_t and a subgraph of H it remains to prove that f preserves the adjacency of vertices on B_t . By virtue of the inductive hypothesis it suffices to show that for any $j \in B_{t-1}$, { f(i), f(j)} is an edge of H whenever { i,j } is an edge of G.
If the vertex j holds a pebble in the configuration of G resulting from the t-1 step of the pebbling P then this is guaranteed by the subrule (b3) of the rule (b) . So we may assume that there is no pebble on the vertex j in this configuration of G . It follows from the membership of j in B_{t-1} that a pebble has been placed on the vertex j of G and removed from it before the t-th step of P. Hence, according to the rule (i) of the pattern pebble game, the vertex j does not hold any pebble throughout the steps t, t+1,...,s of P .However, the definition of m-pebbling of the pattern graph says that for each edge $\{j_1, j_2\}$ of G there is a step of the pebbling resulting in a configuration with pebbles on j_1 and j_2 simultaneously. Hence, {i,j} cannot be an edge of G in this case .

Conversely, assume that there is a cyclic isomorphism f between between the pattern graph G and a subgraph of H . It is easy to see that the following strategy played according to the m-pebbling P is won :

<u>for</u> t = 1 <u>step</u> 1 <u>until</u> s <u>do</u>
<u>if</u> at the t-th step of P, a pebble is placed on (removed from) the vertex i of G <u>then</u> place (remove, respectively) the pebble p_i on (from) the vertex f(i) of H □

Of course, a monotone isomorphism is an instance of cyclic isomorphism . We can also obtain an analogue of Lemma 2.1 with monotone isomorphism substituted for cyclic isomorphism by setting certain constraints in the pebbling games played on the pattern and input graph, respectively. The constraints consist in substituting the ordinary open segments (ℓ,q), (ℓ^*,q^*) for $(\ell,q)_k$, $(\ell^*,q^*)_n$,respectively, in the rule (i) of the pattern pebbling game and in the subrules (b1), (b2) of the rule (b) of the input pebbling game, respectively .
In order to initiate the modified games suitably we add the following further constraints :

(1) in the first and the second step of the pattern pebbling game, a pebble is placed on the vertex 1 and k of G, respectively,

(2) in the second step of the input pebbling game, the pebble p_k may be placed only on a vertex which is greater than the

vertex holding the pebble p_1.

Let us call the resulting, restricted variants of the pebbling games and the corresponding variant of m-pebbling monotone . The proof of the following monotone analogue of Lemma 2.1 is left to the reader .

Lemma 2.2. The monotone pebbling game played on the input graph H according to a monotone m-pebbling of the pattern graph G is won if and only if there is a monotone isomorphism between G and a subgraph of H .

Of course, we cannot obtain an analogue of Lemma 2.1 with isomorphism substituted for cyclic isomorphism . Let I(G) be the set of all graphs on vertices 1,...,k that are isomorphic to the pattern graph G. Clearly, there are $k!/\#\Gamma(G)$ graphs in I(G), where $\Gamma(G)$ is the set (group) of all automorphisms of the pattern graph G, see [5] . Note that there is an isomorphism between G and a subgraph of H if and only if there is a graph $G^* \in I(G)$ and a monotone isomorphism between the graph G* and a subgraph of H . Therefore, if $m \in N$ and for each graph from I(G) there is a monotone m-pebbling of the graph then G can be embedded in H if and only if there is $G^* \in I(G)$ and a monotone m-pebbling P* of G* such that the monotone pebbling game played on H according to P* is won . However this rather does not lead to a fast algorithm for the general fixed subgraph isomorphism problem. Simply, our algorithm for the cyclic or monotone subgraph isomorphism problem works in $O(\binom{n}{m}n^2)$ steps and the constant m can be large for some members of I(G) even if G is a trivial connected graph. For instance, the k-vertex line from Fig.2.3 has monotone 3-pebbling but it is possible to construct such k-vertex lines which has no monotone m-pebbling for $m < O(\sqrt{k})$. The constant m is greatest,i.e. equal to k, when G is k-clique [1] . Therefore, our algorithm applied to determining whether the input graph has k-clique is not better than exhaustive search (note that the application is quite straightforward since I(G) has only one element) .

The constant m can be small with respect k even for non-trivial pattern graphs provided that they have a good separator theorem [9,10] and are suitably numbered . Following Lipton and Tarjan [10] it will be convenient to assume the following form of an f(k) separator theorem for a class of graphs S .

There is a constant α_S , where $0 < \alpha_S < 1$, with the following property : if F is a graph on k vertices in S , then the vertices of F can be partitioned into three sets A, B, C such that no edge joins a vertex in A with a vertex in B, neither A nor B contains more than $\alpha_S k$ vertices , C contains no more than f(k) vertices . Moreover, the two subgraphs of F induced by A and B are also in S.

Some of the recently established separator theorems are enumerated in Table 2.1 . For details and references see [9,10] .

Class of graphs	f(k)
forests	1
outerplanar graphs	O(1)
X-trees ($k=2^j$)	$O(\log_2 k)$
graphs of genus i	$O(i\sqrt{k})$
i-dimensional mesches for i > 2	$O(k^{1-1/i})$

Tab. 2.1.

Using such separator theorem one can prove the following lemma.

Lemma 2.3. Suppose S is a clas of graphs for which the f(k) separator theorem holds . Let G be a k-vertex graph from S where $k \geq 3$. There is a graph G* isomorphic to G that has a monotone A(k-2)+2 - pebbling, where A(1) is a positive constant and A(j) is defined for any positive integer j > 1 by : $A(j) = \max_{1 \leq \ell \leq \alpha_S \cdot j} A(\ell) + f(j)$.

Proof. The idea of the prof is as follows :
If we have a set of vertices which separates two parts of G then we can place pebbles on all vertices of the separating set, next pebble the first part, further remove all pebbles from this part and pebble the second part . The assumed separator theorem guarantees finding a suitable separating set . To get a monotone pebbling G can be renumbered . The idea is recursively implemented by a procedure called Div . This procedure has two parameters : an input graph F and a set S of natural numbers which is used to renumber the vertices of F. We assume #S = #F. Div is defined as follows :

```
Div (F,S)
begin
  if  F  has one vertex  then  pebble this vertex,and
      renumber it using the only element of S
      else
```

```
begin
   find the sets A, B, C for F according to
   the assumed separator theorem ;
   place pebbles on all vertices from C and,
   in the same order, assign the least #C
   numbers of S to vertices from C, respectively;
   delete the least #C numbers from S;
   Div ( "The subgraph of F induced by A",S );
   remove all pebbles from A ;
   Div ( "The subgraph of F induced by B ", S )
  end
end
```

The reader should remember that in the monotone variant of the pebble game on a pattern graph on vertices 1,...,k , the vertices 1, k hold pebbles in the configuration resulting from the second step of the game. Therefore, to get the desired graph G* and its monotone A(k-2)+2 pebbling it suffices to perform the following block :

```
place two pebbles on the vertices 1, k ;
assign 1, k to the vertex 1, k of G, respectively ;
Div ( "The subgraph of G induced by V_G-{ 1,k} ",(1,k))
```

Note that during the execution of the block the set of vertices of G which have not been pebbled is always of the form (j,k), where j is a vertex of G* pebbled in the last (pebbling) step . This follows from the method of renumbering the set C in Div . On the other hand, it is easily seen that $A(\ell)$ is not less than the number of pebbles needed to perform the pebbling defined by Div (F,S) . Thus the block really defines a monotone A(k-2)+2 pebbling of the renumbered graph G, i.e. G* .□

Upper bounds on the number of pebbles needed by monotone pebblings for many graphs can be derived as consequences Lemma 2.3 and Table 2.1 . It suffices only to solve simple recurrence equations . The upper bounds are enumerated in Table 2.2 (below) .

Class of graphs	the number of pebbles
forests	$O(\log_2 k)$
outer planar graphs	$O(\log_2 k)$
X-trees ($k=2^j$)	$O((\log_2 k)^2)$
graphs of genus i	$O(i\sqrt{k})$
i-dimensional mesches for $i > 2$	$O(k^{1-1/i})$

Finally,we shall discuss the algorithm for determining the number of subgraphs of the input graph H that are cyclically isomorphic to the fixed, pattern graph G . The algorithm can be easily adjusted to the case of monotone isomorphism .

Let us recall that k, n denote the number of vertices of G and H ,respectively . Our algorithm can be performed in $O(\binom{n}{m}n^2)$ steps, according to a fixed m-pebbling P of G . It simulates all possible pebbling games played on H according to P, by applying the method of breadth-first seach . The potential gain of time achieved by our algorithm (in comparison with exhaustive search) follows from the fact that there are at most $\binom{n}{m}m!$ different configurations resulting from a given step t of all possible pebbling games played on H according to P . At the t-th step of the simulation all such distinct configurations are stored in a list L of records . Each record of L consists of two parts . We shall call them the head and the tail of the record . In the head, one of the mentioned configurations is encoded in the form of a lexicographically ordered sequence of distinct pairs (i,i*) +) . The presense of a pair (i,i*) means that the pebble p_i is placed on the vertex i* of H in this configuration . Of course, there are no more than m pairs in the head . The tail encodes the number of distinct strategies of the pebbling game played on H according to P, which lead to the configuration encoded in the head, at their t-th step . At the next step, the list L is updated according to the rules of the pebbling game on H . Thus, if at the t+1 st step of P a pebble is removed from a vertex i of G, then all the pairs of the form (i,i*),where $i^* \in [1,n]$, are deleted from the heads in accordance with the rule (c). Next, suppose that a pebble is placed on a vertex i of G at the t+1 st step of P . By the rule (i), there are such vertices ℓ, q of G that $i \in (\ell,q)_k$, ℓ and q hold a pebble and no vertex of G from $(\ell,q)_k$ has been pebbled in the previous steps of P. It is easily seen that the vertices ℓ, q are determined uniquely. On the other hand, by the rules (b), (c), if a vertex j of G holds a pebble in the configuration resulting from the t-th step of P, then a vertex of H must hold the pebble p_j in the corresponding configuration of any pebbling game played on H according to P . Hence, each head in the list L contains two pairs of the form (ℓ,ℓ^*) , $\ell^* \in [1,n]$, and (q,q^*), $q^* \in [1,n]$, respectively . The only possibility to continue the game on H is to choose i* from $(\ell^*,q^*)_n$ satisfying the rules (b2-3) and place the pebble p_i

---

+) It follows from the rules of the pebbling games that at each step of the simulation, the first coordinates of the pairs are uniquely determined and hence only the second ones need to be stored really.

on the vertex i* of H . Therefore, for each record of L and each i* from (ℓ^*, q^*) satisfying the rule (b3), where the pairs (ℓ, ℓ^*), (q, q^*) occur in its head, we add the record with the pair (i,i*) inserted in its head to the top of an auxiliary list K . Next we pop the original record off the list and consider the current top record of L unless L is empty . In such way, all possible configurations arising from the t+1 st step of the game on H are collected in the list K . Of course, if K is empty then zero is printed and the algorithm halts . Note that K contains $O(\binom{n}{m}n)$ records and it can be generated in $O(\binom{n}{m}n^2)$ steps since $\#(\ell^*, q^*) \leq n$ and the rule (b3) can be tested in O(n) steps . However the list K is not convenient since some configurations on H, i.e. heads, can repeat many times in it . Therefore, to take the advantage of the breadth-first search method, K is sorted according to the heads. Since the heads are of the length bounded by the constant m and the list K may really hold only the heads with pointers to the corresponding tails, we may assume that K is sorted in $O((\binom{n}{m}n+n)m)=O(\binom{n}{m}n)$ steps, see Theorem 3.1 in [1] . Next, we pop step by step the list K adding to the top of the list L all such records (A,B) that the head A occurs in K and the tail encodes the sum of the values represented by the tails C for which (A,C) is a record of K . Thus, the whole updating of the list L can be performed in $O(\binom{n}{m}n^2)$ steps . Let s be the number of steps of the m-pebbling P . After s steps of such simulation, the sum of the values encoded by the tails in the list L is outputed . It is left to the reader to show, using Lemma 2.1, that this sum denotes the number of the subgraphs of H that are cyclically isomorphic to to the pattern graph G . Since s is fixed, the whole algorithm takes also $O(\binom{n}{m}n^2)$ steps .

A monotone m-pebbling of a graph is an instance of m-pebbling of the graph . Combining this with Table 2.2 and with the running time of the algorithm we obtain the following corollary :

<u>Corollary 2.1.</u> If a k-vertex graph G is in the class enumerated in the i-th row of Table 2.2 then there is a graph G* isomorphic to G such that the number of subgraphs of n-vertex input graph which are monotonously (or cyclically) isomorphic to G* can be determined in $O(\binom{n}{i(k)}n^2)$ steps, where i(k) stands for the upper bound occurring in in the i-th row of Table 2.2 .

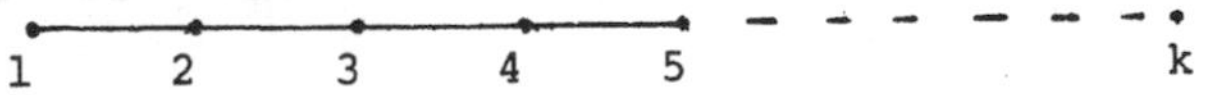

Fig.2.3.

References

[1] Aho A.V.,Hopcroft J.E.,Ullman J.D., The Design and Analysis of Computer Algorithms, Addison-Wesley, 1976 .

[2] Filotti I.S.,Mayer J.N., Apolynomial-time algorithm for determining the isomorphism of graphs of fixed genus, ACM symposium on Theory of Computing, 1980 .

[3] Garey M.R.,Johnson D.S. and Tarjan R.E., The planar Hamiltonian circuit problem is NP-complete, SIAM J.Compt. vol 5, no 4, 1976 .

[4] Gonzalez R.C., Tou J.T., Pattern Recognition Principles, Addison-Wesley, Massachusetts, 1974 .

[5] Harary F., Graph Theory, Addison-Wesley, 1969 .

[6] Hopcroft J.E., Wong J., Linear time Algorithm for Isomorphism of Planar Graphs, 6th ACM symp. on Theory of Computing, 1974.

[7] Hopcroft J.E.,Tarjan R.,Isomorphism of planar graphs, 4th An. Symp. on the Theory of Computing,1972 .

[8] Hopcroft J.E.,Karp R.M., An $n^{5/2}$ algorithm for maximum matchings in bipartite graphs,SIAM J.Compt. vol 2, 1973 .

[9] Leiserson C.E., Area-Efficient Graph Layouts, 21 st FOCS symposium, 1980 .

[10] Lipton R.J., Tarjan R.E., A separator theorem for planar graphs, Stanford Univ. 1977, CS-77-627 .

[11] Lipton R.J., Tarjan R.E., Applications of a planar separator theorem, Stanford Univ. 1977, CS-77-628 .

[12] Miller G., Isomorphism testing for graphs of bounded genus, ACM symposium on Theory of Computing, 1980 .

[13] Read R.C., Corneil D.G., The graph isomorphism disease, J. of Graph Theory, no 4, 1977 .

[14] Schmidt D.C., Druffel L.E., A fast backtracking algorithm to test directed graphs for isomorphism using distance matrices, J.ACM, 1976, 433-445 .

[15] Ullmann J.R., An algorithm for subgraph isomorphism, J.ACM, vol 23, no 1, 1976 .

A # P-COMPLETE PROBLEM OVER ARITHMETICAL TREES

Giancarlo Mauri°^ - Nicoletta Sabadini°

°) Istituto di Cibernetica - Università di Milano

^) Istituto di Matematica, Informatica e Sistemistica - Università di Udine

1. INTRODUCTION

In recent years, the topic of computational complexity, formerly concerned only with the complexity of decision problems on Turing Machines (deterministic and non deterministic), has been widened to consider also the complexity of enumeration problems and different models of computation, such as Random Access Machines (RAM's)with arithmetical and boolean operations.

The first generalization, from decision problems to enumeration problems, has been carried out by Simon [8] and Valiant [10,11], who introduced the concept of Counting Turing Machine as a formal model for describing enumeration problems and defined the class #P.

On the other hand, Hartmanis and Simon [4], Simon [9], Schönhage [7], Bertoni, Mauri and Sabadini [2], studied the power of RAM's with different sets of operations as far as the solution of decision problems is concerned; in particular, in [7] and [2] the complexity on RAM's of problems in the classes NP and P-SPACE, respectively, is analyzed.

Furthermore, in [2] and [8] RAM's are applied to the solution of enumeration problems, and it is shown that they are a very natural and powerful tool to solve them.

In solving enumeration problems, polynomials play a central role; in fact, for some given enumeration problems, there exists a multivariate polynomial that describes their solutions, in the sense that the coefficients of the polynomial represent the solutions for the various instances of the problem. Hence, a particular instance of the problem is reduced to the computation of a particular coefficient of the polynomial.

In this paper, it is proved that the problem of computing a coefficient in a multivariate polynomial, represented as an arithmetical tree, i.e. a tree whose internal nodes and whose leaves are labeled respectively by arithmetical operations and by variables or constant values, is #P-complete (Sect.3).

Part of the proof uses a result implicitly given in Valiant [12]. However, in [12] the emphasis is on a different kind of aspects, i.e. reducibility in some sense "mathematical" between polynomials. For instance, with respect to the notions in [12], deducing a coefficient from a small formula is difficult even in the univariate case, while in our case the analogous problem for univariate polynomials is in P.

Finally, as an application of the previous result, we are able to prove (Sect.3) that every #P-problem can be solved on RAM's with a polynomial number of operations of sum and product and only two divisions and one subtraction.

2. PRELIMINARY DEFINITIONS

We consider the alphabet $A = \{0,1,+,*\} \cup \{x_0,x_1,\dots,x_n,\dots\}$, where:

- 0,1 are interpreted as constant symbols (arity 0);
- $x_0,x_1,\dots,x_n,\dots$ as variable symbols;
- + and $*$ as operation symbols of arity 2.

Let now $\mathcal{F}$ be the free algebra generated by A.

Def.2.1 - An <u>arithmetical tree</u> is any element $t \in \mathcal{F}$; the <u>dimension</u> $|t|$ of t is the number of vertices of t.

Example:

t = 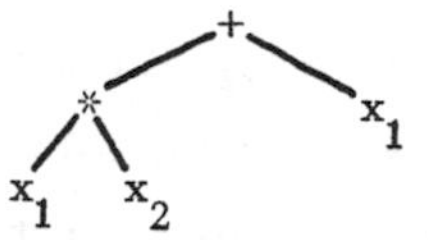

, $|t| = 5$

A model of the above structure is the semiring $< N,+,*,0,1 >$ of integer positive numbers with the usual operations of sum and product. If we interprete the symbols $x_0,\dots,x_n,\dots$ as variables over N, then every tree can be interpreted as a multivariate polynomial with non negative coefficients, that is as the generating polynomial of some enumeration problem [6]. So, the tree can be looked at as a "program" for computing the multivariate polynomial. However, we remark that this concept is not (polynomially) equivalent to the usual definition of "straight line program" [1]; for instance, the straight line program of length n:

$$< x_o = x\ ,\ x_1 = x_o * x_o\ , \ldots ,\ x_n = x_{n-1} * x_{n-1} >$$

defines the polynomial x^{2^n} which is describable only by means of trees with an exponential number of vertices.

Def.2.2 - A <u>Counting Turing Machine</u> (CTM) [10] is a non deterministic Turing Machine with an auxiliary device that (magically) prints on a special tape the number of accepting computations induced by the input.

Therefore, a CTM defines a (partial) function f: $\Sigma^* \rightarrow N$, where Σ is the input alphabet, called <u>counting function</u>, where f(x) is the number of accepting computations induced by x.

Def.2.3 - $\#P$ is the set of counting functions defined by CTM's with polynomial time complexity (worst case). The <u>time complexity</u> of a CTM $\mathcal{T}$ on input x is the length of the longest computation accepting x.

The notions of polynomial reducibility and complete problems can be introduced in $\#P$ as usual. In particular, Valiant [10] proved the following fact:

Th. 2.1 - Let (A_{ij}) be a 0,1 square matrix of dimension n, and let the permanent $Per(A_{ij})$ be defined by $Per(A_{ij}) = \sum_{P} \prod_{i=1}^{n} A_{P(i),i}$, where P denote a permutation of the n-uple (1,...,n). Then the problem "Calculus of the permanent of A" is a $\#P$-complete problem.

For a formal notion of Random Access Machine (RAM), we will refer to [1]. Informally, we define RAM models which have a countable number of storage locations with addresses 0,1,2,...,n,..., each of which can store a natural number, denoted by $<n>$, and an extra accumulator with current content z. The finite control is given by a deterministic program consisting of a sequence of labelled instructions of the usual types: goto, i-load, store,..... We are interested in RAM's which have the following extra instructions: add, sub, mult, div, with the obvious semantics:

$$\text{add } n \;\; ; + ; \;\; z \leftarrow z + <n>$$

$$\text{sub } n \;\; ; \dot{-} ; \;\; z \leftarrow \text{Max}\,(0, z - <n>)$$

$$\text{mult } n \;\; ; * ; \;\; z \leftarrow z * <n>$$

$$\text{div } n \;\; ; \div ; \;\; z \leftarrow \lfloor z : <n> \rfloor \text{ , HALT if } <n> = 0$$

We will denote such a class as RAM($+, \dot{-}, *, \div$).

3. A #P-COMPLETE PROBLEM

Generally, the solution of an enumeration problem by means of a generating function, requires to determine the coefficients in a suitable expansion of such a function.

In our case, since we consider multivariate polynomials, we can define the problem:

PROBLEM: CALCULUS OF A COEFFICIENT

INSTANCE: (t,m) where t is an arithmetical tree, m is a monomial over the variables $x_1,\dots,x_n,\dots$

QUESTION: which is the coefficient of m in the polynomial defined by t?

For a deeper analysis, we will associate to every arithmetical tree t both the dimension $|t|$ and the number Var(t), i.e. the number of variables which appear in t. We will define the following class of arithmetical trees:

Def.3.1 - For all $1 > \epsilon > 0$, let $\mathscr{C}_\epsilon \equiv \{ t \mid t \in \mathscr{F}, \mathrm{Var}(t) \leq |t|^\epsilon \}$ and $\mathscr{C}_0 = \{ t \mid t \in \mathscr{F}, \mathrm{Var}(t) = 1 \}$, i.e. the set of trees which can be interpreted as univariate polynomials .

For every $\epsilon \geq 0$, we can consider the following problem:

PROBLEM: CALCULUS OF A COEFFICIENT IN $\mathscr{C}_\epsilon$

INSTANCE: (t,m), where $t \in \mathscr{C}_\epsilon$, m is a monomial in the variables $x_1,\dots,x_n,\dots$

QUESTION: which is the coefficient of m in the polynomial defined by t?

The following theorems hold:

Th.3.1 - For every $\epsilon > 0$, CALCULUS OF THE COEFFICIENT IN $\mathscr{C}_\epsilon$ is a #P-complete problem.

Proof - a) The more general problem CALCULUS OF THE COEFFICIENT is in #P.

We can construct a CTM which solves the problem in the following way:

1) By the following top-down nondeterministic tree-transductor we obtain, nondeterministically, all the monomials of the polynomial in time $|t|$, where t is the tree.

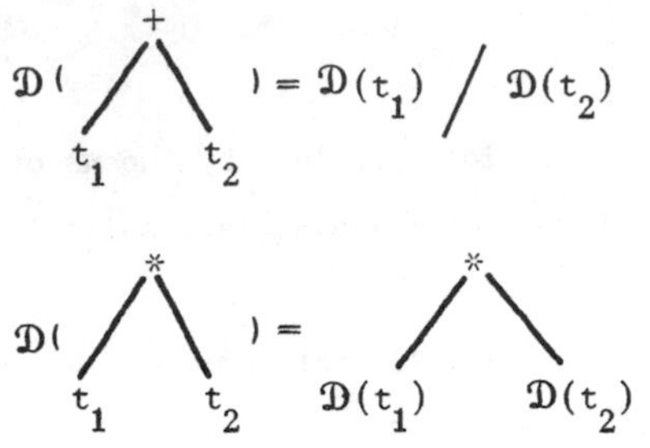

$\mathcal{D}(0) = 0 \qquad \mathcal{D}(1) = 1 \qquad \mathcal{D}(x_k) = x_k$

2) Deterministically, the CTM compares the result m' of a computation with the monomial m; if $m = m'$, the CTM accepts the computation, otherwise it doesn't.

b) The problem is #P-hard.

The proof technique is by inclusion [3] , with a simple padding argument. The case CALCULUS OF THE COEFFICIENT is implicit in a result of Valiant [12] .

Let A be a 0,1 square matrix (A_{ik}) of dimension n; we define $|A| = n^2$

We consider the associate problem $<t_A, x_1 \dots x_n>$, where:

$$t_A = \prod_i \sum_k A_{ik} x_k + x_1^{n^{\lceil \frac{1}{\epsilon} \rceil}}$$

We have:

1) The translation $A \longmapsto <t_A, x_1 \dots x_n>$ requires polynomial time with respect to the dimension of A. More precisely, it requires $O(|A|^{\lceil \frac{2}{\epsilon} \rceil})$ time.

2) Coeff of $x_1 \dots x_n$ in t_A = Per A

3) $|t_A| \geq n^{\lceil \frac{1}{\epsilon} \rceil}$, $Var(t_A) = n \Longrightarrow Var(t_A) = n \leq (n^{\lceil \frac{1}{\epsilon} \rceil})^{\epsilon} \leq |t_A|^{\epsilon}$

i.e. $t_A \in \mathcal{C}_\epsilon$

1), 2), 3), prove that the #P-complete problem CALCULUS OF THE PERMANENT is polynomially reducible to CALCULUS OF THE COEFFICIENT IN $\mathcal{C}_\epsilon$

Th.3.2 - The problem CALCULUS OF THE COEFFICIENT IN $\mathcal{C}_0$ is solvable in polynomial time by a deterministic Turing Machine.

Proof - An $O(n^3 \cdot \log^2 n \cdot \log\log n)$ algorithm is easily obtained.

First , let t be an arithmetical tree interpreted as a univariate polynomial, let c' be an arbitrary coefficient of every arithmetical subtree t' of t. The following bounds hold:

a) degree of $t' \leq |t|$

b) $c' \leq 2^{|t|}$

To obtain a normal form $\sum_K C_k x^k = t$ (hence, a fortiori, the coefficient of x^k in t), we must perform $\leq |t|$ operations between polynomials.

Every operation can be performed, for instance, by Fast Fourier Transform techniques in $\leq O(|t| \log |t|)$ operations + or * between integer numbers and every one of these operations require $\leq O(|t| \log |t| \log \log |t|)$ time, for instance by Schönage-Strassen algorithm. (Total time $|t|^3 \log^2 |t| \log\log |t|$).

Remark 1 - In an analogous way, it is possible to find a polynomial algorithm to determine the coefficient of m in t if Var(t)$\leq$ k, for a fixed k.

Remark 2 - The difference between Th.2.2 and the assertion of Valiant [12] i.e. that deducing a coefficient from a small formula is difficult even in the univariate case, is a consequence of the different notion of reduction and size in [12] . For instance, the formula $t = \prod_i \sum_K A_{ik} x^{2^k}$, used by Valiant, for which holds the fact that coefficient of $x^{2^{n+1}-1}$ in t = Per A, in our case has a great dimension with respect to n (in fact $|t| \geq 2^n$).

4. AN APPLICATION

In this section, we prove that every #P-problem can be computed in polynomial time with a RAM(+,*,$\dot{-}$,÷) [7] , which uses a polynomial number of operations of sum and product, only two divisions and one substraction.

So, we generalize the result of Schönage over NP-problems [7]. An analogous result is in [8] .

Given a 0,1 square matrix A of size n, let $P_A(x_1,\dots,x_n)$ be the following multivariate polynomial:

$$P_A(x_1,\dots,x_n) = \prod_i \sum_K A_{ik} x_k$$

An alternative representation of Per A is:

$$\text{Per } A = \text{coeff of } x_1 \dots x_n \text{ in } P_A(x_1,\dots,x_n)$$

We need the following lemmas:

LEMMA 4.1 - Coeff of $x_1 \dots x_n$ in $P(x_1,\dots,x_n)$ = coeff of $x^{2^{n+1}-1}$ in $P(x,x^2,x^4,\dots,x^{2^{n-1}})$ (Implicit in [12]).

Lemma 4.2 - Given $P(x) = \sum_{k=0}^{n} p_k x^k$, where p_k integer, $p_k \geq 0$, $p_n > 0$, then:

$$p_k = (P(P(1)) \div P^k(1)) - (P(P(1)) \div P^{k+1}(1)) * P(1)$$

Proof - In fact, by induction on k, we first prove that $\sum_{j=0}^{k-1} p_j P^j(1) < P^k(1)$:

a) $\sum_{j=0}^{0} p_j P^j(1) = p_o < p_o + \ldots + p_n = P(1)$

b) $\sum_{j=0}^{k} p_j P^j(1) = \sum_{j=0}^{k-1} p_j P^j(1) + p_k P^k(1) < P^k(1) + p_k P^k(1) < (p_o + \ldots + p_n) P^k(1) = P^{k+1}(1)$

From this, we have:

$$P(P(1)) \div P^k(1) = p_k + p_{k+1} P(1) + \ldots + p_n P^{n-k}(1)$$

$$P(P(1)) \div P^{k+1}(1) = p_{k+1} + \ldots + p_n P^{n-k-1}(1)$$

and the thesis easily follows.

Th.4.3 - Let A be a 0,1 square matrix of size n. Per A can be evaluated in $O(n^2)$ arithmetical operations of sum and product, only two divisions and one subtraction.

Proof - We have Per A = coeff of $x^{2^{n+1}-1}$ in $\prod_i \sum_k A_{ik} x^{2^{k-1}}$ (Lemma 4.1)

We can compute :

a) $\bar{x} = P_A(1,1,\ldots,1)$ in $O(n^2)$ operations of + and *

b) $\bar{x}, \bar{x}^2, \ldots, \bar{x}^{2^{n+1}}$ and $\bar{x}^{2^{n+1}-1}$ in $O(n)$ operations of *

c) $P_A(\bar{x},\ldots,\bar{x}^{2^{n-1}})$ in $O(n^2)$ operations of + and *

d) $\text{Per } A = (P_A(\bar{x},\ldots,\bar{x}^{2^{n-1}}) \div \bar{x}^{2^{n+1}-1}) \dot{-} (P_A(\bar{x},\ldots,\bar{x}^{2^{n-1}}) \div \bar{x}^{2^{n+1}})\bar{x}$

in two divisions and one subtraction.

It is easy to program the above algorithm on a $\mathscr{R} \in \text{RAM}(+,*,\dot{-},\div)$ which uses a polynomial number of operations of sum and product, and only two divisions and one subtraction.

By remembering that CALCULUS OF THE PERMANENT is a #P-Complete problem, that every polynomial reduction can be simulated in polynomial time on a RAM without extra instructions [1] , we can conclude:

Th.4.4 - Every # P-Problem can be solved on a $\text{RAM}(+,*,\dot{-},\div)$ in polynomial time, by using moreover only two divisions and one subtraction.

REFERENCES

[1] Aho,A.V., Ullman,J.E., Hopcroft,J.D., The Design and Analysis of Computer Algorithms, Addison Wesley, Reading, Mass., 1974

[2] Bertoni,A., Mauri,G., Sabadini,N., A characterization of the class of functions computable in polynomial time on RAM's, Proc. ACM STOC 81, to appear, 1981

[3] Garey,M.R., Johnson,D.S., Computers and Intractability, W.H. Freeman and Co., San Francisco, 1979

[4] Hartmanis,J., Simon,J., On the power of multiplication in Random Access Machines, IEEE Conf. Rec. 15th Symp. on Switching Automata Theory, 1974, 13-23

[5] Percus,J.K., Combinatorial Methods, Springer, Berlin, 1971

[6] Riordan,J., An introduction to combinatorial analysis, Wiley, New York, 1958

[7] Schönhage,A., On the power of Random Access Machines, Proc. 6th ICALP, Lect. Not. in Comp. Sci. 71, Springer, Berlin, 1979, 520-529

[8] Simon,J., On the difference between the one and the many, in Lect. Not. in Comp. Sci. 52, Springer, Berlin, 1977, 480-491

[9] Simon,J., Division is good, Comp. Sci. Dept., Pennsylvania State University, 1979

[10] Valiant,L.G., The complexity of computing the permanent, Theoretical Computer Science 8, 1979, 189-202

[11] Valiant,L.G., The complexity of enumeration and reliability problems, Res. Rep. CSR-15 77, Dept. of Comp. Sci., Univ. of Edinburgh, 1977

[12] Valiant,L.G., Completeness classes in algebra, 11th ACM STOC, 1979, 249-261

ACKNOWLEDGEMENTS

We wish to thank Prof. Flajolet and Prof. Bertoni for their useful suggestions. This research has been supported by the italian Ministero della Pubblica Istruzione and by the CP Project of Università di Milano and Honeywell Information Systems Italia.

Trees in Kripke models and in an intuitionistic refutation system.

P. Miglioli, U. Moscato, M. Ornaghi

Istituto di Cibernetica dell'Università di Milano

1. Introduction.

An important field of Mathematical Logic where the tree-representation is very perspicous is the semantics of Intuitionism, both for propositional and predicate calculus, in terms of Kripke models. In this area, very well known completeness results are available together with the possibility of using, at least in principle, formal calculi in order to construct counter-models for intuitionistically unprovable formulas [see e.g. 6,11,17]. However, the emphasis has been mainly devoted to the possibility itself of making intuitionistic logic complete with respect to an appropriate semantics rather than to the possibility of efficiently building up proofs and counter-models.

On the contrary, the aim of this paper is to introduce an adequate refutation calculus in order to efficiently carry out the above tasks: this calculus is an appropriate extension of the one expounded in [6] , which in turn is proposed as an improvement of Beth's [1] with the introduction of Smullyan's notation [18] . As far as the completeness of just our "efficient" calculus is involved, the completeness proof requires an appropriate construction in terms of trees, both for the propositional and the predicative case. We will treat in detail only the first case and will show how our construction provides a flexible tool to concretely analyze the "tree-geometry" of Kripke models.

We believe that this kind of investigations is of potential interest for Computer Science. For, classical refutation calculi have been introduced in Artificial Intelligence to mechanically prove theorems and, more generally, as a "technique of computation" [see e.g. 2,3,9]. In this line, the problem arises of investigating the relevance for Artificial Intelligence of intuitionistic refutation calculi, when a "reasonably efficient" basis is provided: we believe that our calculus is a good starting point for such an investigation.

For, intuitionistic logic is a constructive logic which has been shown by variuos authors [4,5,7,8,10,12,13,14] to be interesting for program construction. From this point of view, the possibility of using direct (Gentzen-like) intuitionistic proof-systems as formal frames where one can define correct and general algorithms solving tasks specified by first order formulas is rather well understood.

To better explain, let π be an intuitionistic proof of a closed formula $\forall x \exists z\, A(x,z)$. If π is carried out in a suitable calculus (such as, e.g., the Prawitz's one [16]) then, by means of well known techniques (e.g. Prawitz's normalization) one can do the following:

1) for every closed term t, from π one immediately obtains, by particuarization, a proof π' of $\exists z\, H(t,z)$;

2) from π' one can construct, in a quite automatical way, a proof π'' of a formula $H(t,t')$, t' being a suitable closed term.

In other words, the starting proof π can be seen as an uniform and general algorithm allowing to solve the problem "$\forall x \exists z\, H(x,z)$", intended as: "for every input value $\bar{x}$ (in an appropriate frame) find an output value $\bar{z}$ such that $H(\bar{x},\bar{z})$ holds (in the considered frame)".

Another kind of problems which can be profitably solved in an intuitionistic frame are problems of the form "$\forall x(H(x) \vee \neg H(x))$", intended as: "for every input value $\bar{x}$, decide whether $H(\bar{x})$ holds or $\neg H(\bar{x})$ holds". Here an intuitionistic proof of $\forall x(H(x) \vee \neg H(x))$ allows to solve the problem.

Thus, direct intuitionistic proof-systems correspond to a way of using first order logic as a tool to analyze and specify algorithms. Now, to be able to make a reasonable comparison between this point of view and the above quoted attitude of Artificial Intelligence [2,3,9], the first step to be made is the development of a good refutation system for intuitionistic logic.

The present paper is a starting point to work in this direction: in the next three paragraphs we will explain the refutation calculus; in the last paragraph we will make a short discussion about the potential interest of the calculus for Computer Science.

2. The intuitionistic model theory (Kripke models).

- First of all, we assume to deal with a first order language built up starting from a countable set of individual variables x,y,z,..., a countable set of predicative variables $P^i, Q^i, \ldots$, (for every arity i) and the connectives $\wedge, \vee, \supset, \neg$, and the quantifiers $\exists$, $\forall$.

– By a Kripke model [6,11,17] we mean a quadruple $\underline{K}=\langle K,\leq,D,\Vdash\rangle$ where $(K,\leq)$ is a non empty partially ordered set, D is a non decreasing function associating elements of K with non empty sets, and $\Vdash$ is a relation between elements of K and closed formulas (i.e., formulas without free variables but possibly containing constants a,b,c,... denoting elements of the domains associated by D to the elements of K) which satisfies the following (where the greek letters denote elements of K):

1) for $A(x_1,.,x_n)$ atomic, $\beta\geq\alpha$, $a_1,.,a_n\in D(\alpha)$, if $\alpha\Vdash A(a_1,.,a_n)$, then $\beta\Vdash A(a_1,.,a_n)$;
2) $\alpha\Vdash A\wedge B$ iff $\alpha\Vdash A$ and $\alpha\Vdash B$;
3) $\alpha\Vdash A\vee B$ iff $\alpha\Vdash A$ or $\alpha\Vdash B$;
4) $\alpha\Vdash A\supset B$ iff, for every $\beta\geq\alpha$, not $\beta\Vdash A$ or $\beta\Vdash B$;
5) $\alpha\Vdash\neg A$ iff, for every $\beta\geq\alpha$, not $\beta\Vdash A$;
6) $\alpha\Vdash\exists x\,A(x)$ iff there is $a\in D(\alpha)$ such that $\alpha\Vdash A(a)$;
7) $\alpha\Vdash\forall x\,A(x)$ iff, for every $\beta\geq\alpha$ and for every $a\in D(\beta)$, $\beta\Vdash A(a)$.

As it is well known, condition 1) of the above definition can be proved for formulas of any complexity [6,11,17]. To better understand the meaning of the above definitions we give two examples (for further details see, e.g.,[6,11,17]).

Example 1

Let us consider the following tree-model

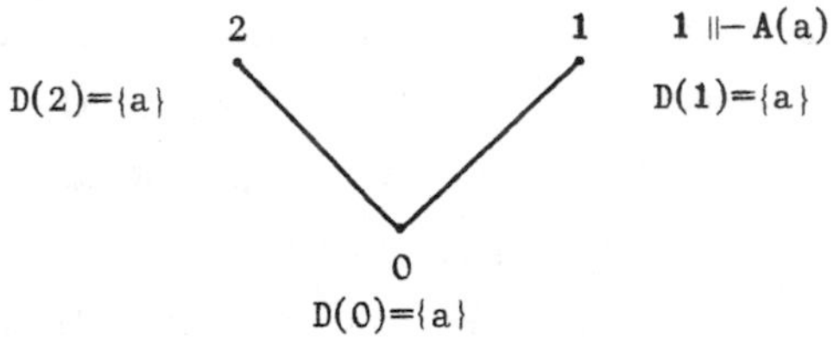

where $D(0)=D(1)=D(2)=\{a\}$ and where only the state 1 forces the atomic formula A(a). In the above model we have that both not $0\Vdash A(a)\vee\neg A(a)$ and not $0\Vdash\neg A(a)\vee\neg\neg A(a)$.

Example 2

Let us consider the infinite linear model

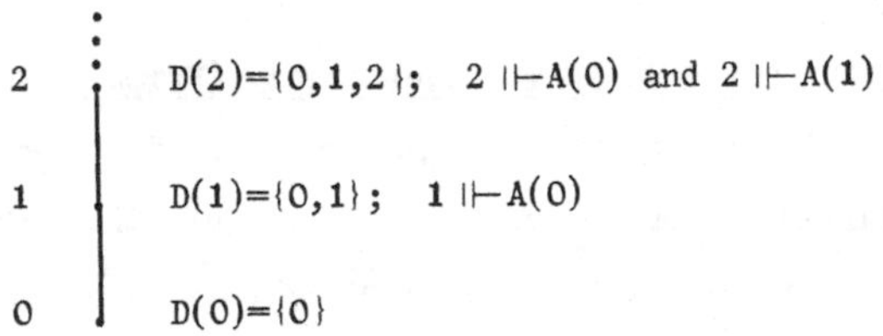

We have that not $0\Vdash\forall x\,(A(x)\vee\neg A(x))$.

– We say that a closed formula H holds in $\beta \in K$ of a model $\underline{K}$ iff $\beta \Vdash H$; we say that H holds in the whole model $\underline{K}$ iff H holds in every $\beta \in K$; finally, we say that H is <u>intuitionistically valid</u> iff, for every model $\underline{K}$, H holds in $\underline{K}$.

The above examples show that the classically valid formulas $A(a) \vee \neg A(a)$, $\neg A(a) \vee \neg\neg A(a)$, $\neg\neg \forall x(A(x) \vee \neg A(x))$ are not intuitionistically valid.

– We introduce the notion of <u>signed</u> formula: this will be any formula of the kind F H, $F_c H$, T H, H being any unsigned formula.

To explain the meaning of the signed formulas, we give the following definition: a set $S = F H_1,..,F H_n, F_c K_1,..,F_c K_m, T Z_1,..,T Z_k$ is <u>realized</u> in $\beta \in K$ of (a Kripke model) $\underline{K}$ iff:

a) for every i, $1 \leqq i \leqq n$, <u>not</u> $\beta \Vdash H_i$;

b) for every j, $1 \leqq j \leqq m$, $\beta \Vdash \neg K_j$;

c) for every h, $1 \leqq h \leqq k$, $\beta \Vdash Z_h$;

Remark that condition b) is much stronger than a). For, not $\beta \Vdash H$ doesn't exclude that, for some $\eta \geq \beta$, $\eta \Vdash H$ holds; otherwise, $\beta \Vdash \neg H$ implies not $\beta \Vdash H$ and $\eta \Vdash \neg H$ for every $\eta \geq \beta$, from which $\eta \Vdash H$ is excluded. In this line, "F_c" means definitive or "certain falsehood", while "F" means only "local falsehood"; of course, "T" means "truth" (local and global truth coincide according to the definition of $\Vdash$).

– We will say that a set S of signed formulas is <u>realizable</u> iff there is some $\beta \in K$ of some model $\underline{K}$ such that β realizes S.

3. Our refutation system.

Our refutation system is characterized by the explicit introduction of the sign F_c (see |15|). In this sense, it differs from the one of |6|, which uses only T and F with the above specified meaning. The introduction of this new sign not only provides a more adequate basis to capture in a calculus the semantics of Kripke models, but also makes considerably more efficient the calculus itself: more precisely, it lowers the non determinism involved in the refutation-trees, since it <u>exactly</u> states which rules may require "duplications of formulas" (i.e., which formulas "already used" in a step of a proof may be necessary also in successive steps). The reader will better appreciate the advantages of our calculus in the examples at the end of this paragraph.

Now we explain our calculus. Since the emphasis is on the possibility of avoiding duplications, we will put into clear evidence the rules possibly requiring duplications. The rules of the calculus are given in Table 1.

TABLE 1

Propositional calculus

$$T\wedge:\ \frac{S,\ T\ A\wedge B}{S,\ T\ A,\ T\ B} \qquad F\wedge:\ \frac{S,\ F\ A\wedge B}{S,\ F\ A/S,\ F\ B} \qquad F_c\wedge:\ \begin{array}{ccc} & S,\ F_c A\wedge B & \\ / & & \backslash \\ S_c, F_c A & & S_c, F_c B \end{array}$$

$$T\vee:\ \frac{S,\ T\ A\vee B}{S,\ T\ A/S,\ T\ B} \qquad F\vee:\ \frac{S,\ F\ A\vee B}{S,\ F\ A,\ F\ B} \qquad F_c\vee:\ \frac{S,\ F_c A\vee B}{S,\ F_c A, F_c B}$$

$$T\neg:\ \frac{S,\ T\neg A}{S,\ F_c A} \qquad F\neg:\ \begin{array}{c} S,\ F\neg A \\ | \\ S_c,\ T\ A \end{array} \qquad F_c\neg:\ \begin{array}{c} S,\ F_c\neg A \\ | \\ S_c,\ T\ A \end{array}$$

$$T\supset:\ \frac{S,\ T\ A\supset B}{S, F\ A, T\ A\supset B/S,\ T\ B} \qquad F\supset:\ \begin{array}{c} S,\ F\ A\supset B \\ | \\ S_c, T\ A, F\ B \end{array} \qquad F_c\supset:\ \begin{array}{c} S,\ F_c A\supset B \\ | \\ S_c, T\ A, F_c B \end{array}$$

Predicate calculus

$$T\forall:\ \frac{S,\ T\forall x\ A(x)}{S,\ T\ A(a),\ T\forall x\ A(x)} \qquad F\forall:\ \begin{array}{c} S,\ F\forall x\ A(x) \\ | \\ S_c,\ F\ A(a) \\ \text{with a new} \end{array} \qquad F_c\forall:\ \begin{array}{c} S,\ F_c\forall x\ A(x) \\ | \\ S_c, F\ A(a), F_c\forall x\ A(x) \\ \text{with a new} \end{array}$$

$$T\exists:\ \frac{S,\ T\exists x\ A(x)}{\begin{array}{c} S,\ T\ A(a) \\ \text{with a new} \end{array}} \qquad F\exists:\ \frac{S,\ F\exists x\ A(x)}{S,\ F\ A(a)} \qquad F_c\ :\ \frac{S,\ F_c\exists x\ A(x)}{S, F_c A(a), F_c\exists x\ A(x)}$$

Here:

- S means any finite set of signed formulas; for any given S, S_c denotes the subset of S containing only the formulas signed by T and F_c (the "certain part of S"). For sake of simplicity, our sets will be given in the form of sequences of signed formulas without repetitions and separated by commas.

- With every logical constant, three reduction rules are associated, one for T, one for F, one for F_c. Every rule is applied to a formula of a set S' consisting of a (possibly empty) set S of other formulas and of the formula in hand; the set S' is over the line, while under the line, according to the cases, one or two sets of signed formulas are generated, as indicated in Table 1.

- In Table 1 we have two kinds of rules, the first one with horizontal lines and the second one with non-horizontal lines. The non-horizontal lines involve the restriction of the set S to the "certain part" S_c of S.

As we will see in par.4, the calculus can be used not only to prove validity, but also to construct Kripke models realizing (realizable) sets of signed formulas; in this second case, the rules with horizontal lines give rise to signed formulas realized in the same node $\beta \in K$ (i.e., these rules don't change the node), while the rules with non-horizontal lines give rise to signed formulas realized in a node $\eta \geqq \beta$ (each non-horizontal line immediately "connects" η with β). For this reason, only the "certain part" S_c of S is preserved by these rules. For more details, see the proof of the Completeness Theorem in par.4.

- Our proofs are tree-like proofs, with a root (i.e., the starting set on the top) and with one or more leaves (i.e., the sets on the bottom). We say that a proof-tree is closed iff all the sets on the bottom are contradictory, i.e., they contain both T A, F A or T A, F_cA for some A. To prove the validity of a formula A, one starts with the set F A and applies a sequence of rules: if a closed proof-tree is reached, one has a proof of the validity of A; if there are no closed proof-trees with root F A, the formula A is unprovable (it may be satisfiable or unsatisfiable).

- The rules which may give rise to "duplications" are $T\supset$, $T\forall$, $F_c\forall$, $F_c\exists$; here, the signed formulas to which the rules are applied may be "duplicated", i.e., they may be repeated also under the lines. The other rules don't give rise to duplications: it isn't allowed to repeat the formulas to which they are applied.

- A classical (non intuitionistic) refutation system can be obtained from Table 1 by deleting all the F_c-rules and by preserving S also in the rules with non-horizontal lines; moreover, $T\supset$ doesn't require duplication, while $F\exists$ requires a duplication. Before giving some examples, let us briefly discuss the problem of duplications in our intuitionistic system. The reasons to duplicate are of two different kinds:

1) In $F_c\forall$ and in the left hand side of $T\supset$, a F_c or T signed formula respectively is reduced to a subformula signed by F. In other words, from a "certain information" we obtain an "uncertain information", i.e. an F signed formula which doesn't allow to recover the previous information, unless we duplicate the "certain formula", as we indeed do.

2) The rules $T\forall$ and $F_c\exists$ require possible duplications, when we must iterate their application to introduce many different constants. For, $T\forall xA(x)$ means $TA(c)$ for <u>any</u> c and $F_c\exists xA(x)$ means $F_cA(c)$ for <u>any</u> c.

For the same reason, one expects a duplication also in the rule $F\exists$ both in the classical and in the intuitionistic system; but, quite surprisingly, our intuitionistic system doesn't require such a duplication.[1)]

Let us see some examples:

$$
\begin{array}{cl}
F\neg\neg(A\vee\neg A) & \\
| & (F\neg) \\
T\neg(A\vee\neg A) & \\
\hline & (T\neg) \\
F_c\, A\vee\neg A & \\
\hline & (F_c\vee) \\
F_cA,\ F_c\neg A & \\
| & (F_c\neg) \\
F_cA,\ TA &
\end{array}
$$

where A is any formula. This is an example where our system is more efficient than Fitting's one; for, his system hasn't the sign F_c and so also the $T\neg$-rule may need duplication. The previous proof in Fitting's system is:

1) We know that the duplication is not required since we <u>can prove</u> the Completeness Theorem for intuitionistic predicate calculus without requiring it: on the other hand, we are not able to provide an easy intuitive justification of this fact. We want to put into evidence that the need of the duplications related to $T\supset$, $T\forall$, $F_c\forall$ and $F_c\exists$ can be proved with counterexamples: hence, <u>our calculus excludes all unnecessary duplications</u>.

$$F\neg\neg(A \vee \neg A)$$

$$\Big| \quad (F\neg)$$

$$T\neg(A \vee \neg A)$$

$$\overline{} \quad (T\neg)$$

$$F\, A \vee \neg A,\ T\neg(A \vee \neg A)$$

$$\overline{} \quad (F\vee)$$

$$F\, A,\ F\neg A,\ T\neg(A \vee \neg A)$$

$$\Big| \quad (F\neg)$$

$$T\, A,\ T\neg(A \vee \neg A)$$

$$\overline{} \quad (T\neg)$$

$$T\, A,\ F\, A \vee \neg A$$

$$\overline{} \quad (F\vee)$$

$$T\, A,\ F\, A,\ F\neg A$$

Of course, it is important to minimize the number of duplications, since every rule allowing duplication implies a non deterministic choice. Moreover, duplicated formulas increase the size of the sets S involved in a proof and of the proof itself. Our system is optimal with respect to duplications.

Let us consider another example which requires a duplication in the rule $T\supset$:

$$F\, ((\neg\neg A \vee \neg A) \supset A) \supset A$$

$$\overline{} \quad (F\supset)$$

$$T\, (\neg\neg A \vee \neg A) \supset A,\ F\, A$$

$$\overline{} \quad (T\supset)$$

$$F\neg\neg A \vee \neg A,\ F\, A,\ T\, (\neg\neg A \vee \neg A) \supset A \ /\ T\, A,\ F\, A$$

$$\overline{} \quad (F\vee)$$

$$F\neg\neg A,\ F\neg A,\ F\, A,\ T\, (\neg\neg A \vee \neg A) \supset A \ /\ T\, A,\ F\, A$$

$$\Big| \quad (F\neg)$$

$$T\neg A,\ T\, (\neg\neg A \vee \neg A) \supset A \ /\ T\, A,\ F\, A$$

$$\overline{} \quad (T\neg)$$

$$F_c A,\ T\, (\neg\neg A \vee \neg A) \supset A \ /\ T\, A,\ F\, A$$

$$\overline{} \quad (T\supset)$$

$$F_c A,\ F\neg\neg A \vee \neg A \ /\ F_c A,\ T\, A \ /\ T\, A,\ F\, A$$

$$\overline{} \quad (F\vee)$$

$$F_c A,\ F\neg\neg A,\ F\neg A \ /\ F_c A,\ T\, A \ /\ T\, A,\ F\, A$$

$$\Big| \quad (F\neg)$$

$$F_c A,\ T\, A \ /\ F_c A,\ T\, A \ /\ T\, A,\ F\, A$$

In the application of the rule $F\supset$ we have drawn an horizontal line instead a non-horizontal one; we have done so, because there was no F signed formula "to throw away" (in order to pass from S to the restriction S_c). Till now we make the following convention:

<u>Convention 1.</u> Whenever S and S_c coincide, the non-horizontal lines become horizontal. This convention will be used in the next paragraph, as we will see.

4. The Completeness Theorem and the use of our system to construct Kripke models.

The refutation system given in Table 1 is correct; we omit the proof, for sake of conciseness (the proof is based on the fact that our rules preserve the realizability of the involved sets of signed formulas). It is more interesting to prove that our system is indeed complete (i.e. that any intuitionistically valid formula is provable in it); for, such a proof contains also a method to construct "countermodels" of intuitionistically unprovable formulas, i.e., models realizing the starting formulas F A (and hence showing the unprovability of A).

For sake of conciseness, in this paragraph we will restrict our considerations to the propositional calculus; the completeness theorem for the predicate calculus requires a more sophisticated treatment, which we cannot give for lack of space.

Also, in the case of propositional intuitionistic logic, in the definition of Kripke models we omit the domain function D and the domains $D(\beta)$; i.e., a Kripke model is given simply by a triple $<K, \leq, \Vdash>$ (of course, in the considered case there is no loss of generality, as one easily sees).

Well, we prove the Completeness Theorem by showing that every consistent set S of signed formulas can be realized, where S is said to be consistent iff every proof-tree one can build by starting from S is not closed (in other words, at least one of the terminal leaves of any proof-tree is not a contradictory set containing T A and F A, or T A and $F_c A$, for some A). Of course, this immediately implies completeness: for, if A is intuitionistically valid but not provable, then on one hand F A is not realizable, on the other hand the set F A is consistent, a contradiction.

Thus, our Completeness Theorem consists in defining, for every consistent set S, a Kripke model $\underline{K}_S$ and showing that $\underline{K}_S$ indeed realizes S.

We explain the definition of a $\underline{K}_S$, for a consistent S, in the following points:

(a) A branching rule in Table 1 is any rule where from a set of signed formulas two sets are generated: these rules are $F\wedge$, $T\vee$, $F_c\wedge$ (involving two non-horizontal lines) and $T\supset$.

(b) If a branching rule is applied to a consistent set S, at least one of the two generated sets is consistent: we say that such a set is a consistent branching of S.

(c) If S' is obtained from a consistent S by applying a non branching rule with horizontal line, then S' is a consistent horizontal continuation of S; if S' is a consistent branching of S obtained by applying a rule with horizontal line, then S' is a consistent horizontal continuation of S; nothing else is a consistent horizontal continuation of S.

(d) By a <u>maximal consistent horizontal sequence</u> generated by a consistent S we mean a sequence $S_0,..,S_h$ (possibly h=0) such that:

1) S_0 is S;

2) for every $1 \leq j \leq h-1$, S_{j+1} is a consistent horizontal continuation of S_j;

3) either no rule with horizontal line can be applied to S_h, or such a rule provides S_h again.

By S* we will indicate the set of all formulas occurring in some set S_i, $1 \leq i \leq h$, of the sequence $S_0,.,S_h$; by $\bar{S}$ we will indicate the set S_h (of course, $S \cup \bar{S} \subseteq S^*$).

(e) Let a consistent S be such that $S=\bar{S}$; by an <u>associated set</u> of S we will mean any set S' such that: either S' is obtained from S by applying a non branching rule with non-horizontal line or S' is a consistent branching of S obtained by applying a non-horizontal line.

(f) Now, given a consistent S, we define a Kripke model $\underline{K}_S = \langle K_S, \leq, \Vdash \rangle$ in the following way:

f_1) K_S has the least element (the root of a tree) consisting of the set $\bar{S}$.

f_2) Having defined $\bar{S}'$ as an element of K_S, consider any associated set S'' of $\bar{S}'$ such that $\bar{S}'' \neq \bar{S}'$: then $\bar{S}''$ is an element of K_S immediately connected with $\bar{S}'$.

f_3) $\leq$ is the transitive and reflexive closure of the relation of immediate connectedness defined in the above point.

f_4) For every $\bar{S}' \in K_S$ and every atomic formula A, $\bar{S}' \Vdash A$ iff $T\ A \in \bar{S}'$; this is well given (if $T\ A \in \bar{S}'$ and $\bar{S}' \leq \bar{S}''$, then $T\ A \in \bar{S}''$) and automatically induces the extension of the relation to all well formed formulas.

Of course, our $\underline{K}_S$ is finite (we are dealing with propositional logic) and can be given in the form of a tree.

Now, by induction on the complexity of the formulas and analyzing the various cases and the various signs, one can prove the Completeness Theorem in the following form:

<u>Completeness Theorem</u>. Let S be a consistent set of signed formulas and let $\underline{K}_S$ be defined according to the previous points: then, for every $\bar{S}'$ of $\underline{K}_S$, $\bar{S}'$ realizes the set S'* (from which, being $S \cup \bar{S} \subseteq S^*$, $\bar{S}$ realizes S).

We conclude this paragraph by two simple examples of model constructions.

<u>Example 1</u>

To realize $F \neg\neg A \vee \neg A$, i.e., to find a counter-model for $\neg\neg A \vee \neg A$, we construct $K_{\{F\,\neg\neg A \vee \neg A\}}$ as seen above.

$$\frac{F\,\neg\neg A \vee \neg A}{F\,\neg\neg A,\ F\,\neg A} \qquad \left\{ \begin{array}{l} S^*_o = \{F\,\neg\neg A \vee \neg A,\ F\,\neg\neg A,\ F\,\neg A\} \\ \bar{S}_o = \{F\,\neg\neg A,\ F\,\neg A\} \end{array} \right\}$$

$$\left\{ \begin{array}{l} S^*_{1,1} = \{T\,\neg A,\ F_c A\} \\ \bar{S}_{1,1} = \{F_c A\} \end{array} \right\} \qquad \frac{T\,\neg A}{F_c\ A} \qquad\qquad T\ A \qquad \left\{ \begin{array}{l} S^*_{1,2} = \{T\ A\} \\ \bar{S}_{1,2} = \{T\ A\} \end{array} \right\}$$

The counter-model is just this one:

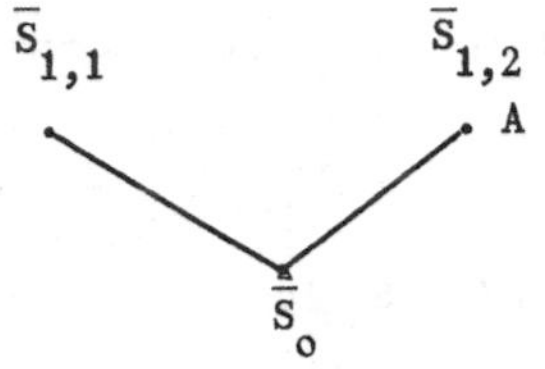

<u>Example 2</u>

To find a counter-model for $H = \neg\neg A \supset ((B \supset A) \vee \neg(A \supset B))$ we construct $K_{\{F\ H\}}$ as follows:

$$\frac{\dfrac{\dfrac{F\ (\neg\neg A \supset ((B \supset A) \vee \neg(A \supset B)))}{T\ \neg\neg A,\ F\ ((B \supset A) \vee \neg(A \supset B))}}{F_c\ \neg A,\ F\ ((B \supset A) \vee \neg(A \supset B))}}{F_c\ \neg A,\ F\ B \supset A,\ F\ \neg(A \supset B)} \qquad \bar{S}_o$$

$\bar{S}_{1,2}$ $\quad F_c\ \neg A,\ T\ B,\ F\ A$ $\qquad\qquad$ $\dfrac{\dfrac{F_c\ \neg A,\ T\ A \supset B}{F_c\ \neg A,\ T\ B}}{T\ A,\ T\ B}$ $\quad \bar{S}_{1,1}$

$\bar{S}_{1,3}$ $\quad T\ A,\ T\ B$

The counter-model is just this one:

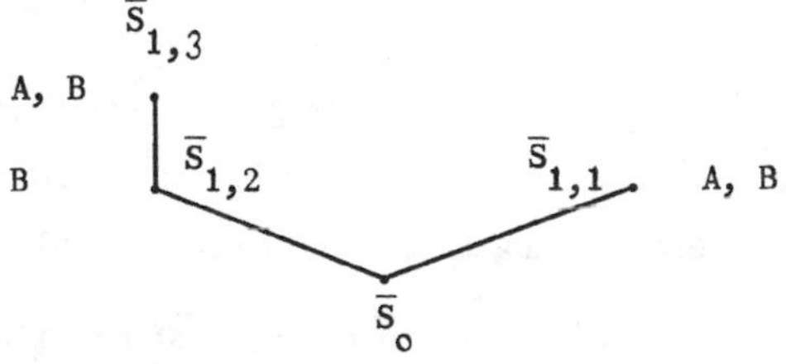

Of course, one can always verify whether or not a finite set S for propositional logic is consistent, in which case one can construct a model realizing it[1]. In the predicative logic, as it is known, this problem is not recursively solvable.

5. Concluding remarks about a possible use of intuitionistic refutation systems in Computer Science.

As mentioned in par.1, intuitionistic direct proofs can be interpreted as programs and, in this sense, the theory of direct proofs can be looked at as a kind of "theory of program-construction" (see also [4,5,9,10]). Moreover, direct proofs typically go to the proved formula by previously proving appropriate sub-formulas (this is particularly clear in Prawitz's normalized proofs or in cut-free Gentzen's proofs): in this sense the proof procedure is a typical "bottom up" technique.

Now we want to look at the mechanism involved in intuitionistic indirect proofs. The typical feature is that they start from the formula A to be proved (more precisely, from F A) and successively they analyze the sub-formulas of A: in this sense, the involved technique is a "top down" one, i.e., a complex problem is broken into simpler sub-problems. In this line, we want to give a few suggestions in order to look at our refutation system as a tool to outline algoritms in a top down way, by indicating appropriate sub-goals; we hope that these insights can be successively developed in a systematical research.

Our discussion is made through the following remarks.

1. A "constuctive content" can be devised even in classical refutation systems. In particular, we refer to the systems which are based on Herbrand's theorem (see the Resolution Method, e.g. [3,9]); these systems have been widely developed in Artificial Intelligence and have been proposed (see e.g. [3]) for program synthesis, i.e., to handle also general constructive facts.

2. Of course, there is a set of classical refutation trees which are included in our intuitionistic refutation system (it sufficies to consider, e.g., the proofs of our system where no F-signed formula is thrown away). The problem then arises of

---

1) Statman has shown that intuitionistic propositional logic is polynomial-space complete. See [19].

devising a corrispondence between these refutation trees and the classical ones which can be interpreted constructively according to the above point 1.

3. We can approximately assume that every classical refutation tree with a constructive content can be interpreted as an intuitionistic one[1]. From this point of view, the use of an intuitionistic logic directly contains (in the system itself) those constructive features which in the classical frame are extracted by introducing in the formal language metalinguistic tools (see e.g. the use of Green's predicate "ANS" quoted in[3]; see also remark 6. below).

4. The constructive features of intuitionism with respect to classical logic are better put into evidence by considering classical tautologies which are not intuitionistically provable. Let us consider the most typical example, i.e., the excluded third principle:

TABLE 2

$$\frac{F\ A \lor \neg A}{\begin{array}{c} F\ A,\ F\,\neg A \\ | \\ ?\quad T\ A \end{array}}$$

In the classical case F A is preserved, while the intuitionistic case puts into evidence that F A is thrown away. We want to interpret this circumstance as a way of looking at the formula $A \vee \neg A$ as the "problem of deciding the truth of A"; this is the intended meaning of the question mark under F A.

To better explain, let us assume that A is not atomic, i.e., it is a complex formula representing a problem to be solved in some formal frame $\mathcal{F}$. If in Table 2 one <u>delays</u> the application of the rule F$\neg$, one has to first apply rules to the branch F A which give rise to the "decomposition" of A into "sub-goals". At the end of this process, when one applies to F$\neg$A the rule F$\neg$, some subformulas coming from F A are thrown away and put into evidence by question marks; if in the considered frame $\mathcal{F}$ the latter subformulas can be refuted, one can automatically

---

1. A complete identification of the "constructive fragment" of classical logic with intuitionistic logic is indeed problematic. In this sense, there is some evidence of constructive features of classical logic interesting for Computer Science which cannot be captured in an intuitionistic frame. The authors are now developing a "logic of effectiveness", intended as an attempt of capturing the kind of constructivism involved in Computer Science: this logic, in a suitable sense, properly includes intuitionistic logic.

recover a solution of the problem in hand.

5. A very simple illustration of the above is provided by the following example. Let a problem be: "find, if it exists, a x such that $A(x) \vee B(x)$ holds; in this case, decide whether $A(x)$ holds or $B(x)$ holds". We relate this problem to an analysis in our system of the signed formula $F\ \exists x(A(x) \vee B(x)) \vee \neg \exists x(A(x) \vee B(x))$. We obtain the following Table 3 with question marks, along the lines discussed in remark 4.:

TABLE 3

$$\dfrac{\dfrac{\dfrac{F\ \exists x(A(x) \vee B(x)) \vee \neg\exists x(A(x) \vee B(x))}{F\ \exists x(A(x) \vee B(x)),\ F\ \neg\exists x(A(x) \vee B(x))}}{F\ A(p) \vee B(p),\ F\ \neg\exists x(A(x) \vee B(x))}}{F\ A(p),\ F\ B(p),\ F\ \neg\exists x(A(x) \vee B(x))}$$

$$\begin{array}{ccc} ? & ? & \dfrac{\dfrac{T\ \exists x(A(x) \vee B(x))}{T\ A(p) \vee B(p)}}{T\ A(p)\ /\ T\ B(p)} \end{array}$$

We associate the marked formula $F\ A(p)$? with $T\ A(p)$ and $F\ B(p)$? with $T\ B(p)$. Assuming $F\ A(p)$ can be independently refuted in our frame $\mathcal{F}$ (e.g. by exhibition of a value $\overline{p}$), then $A(\overline{p})$ is true, i.e. $T\ A(\overline{p})$ holds, and the sub-tree

$$\dfrac{\dfrac{T\ A(\overline{p})}{T\ A(\overline{p}) \vee B(\overline{p})}}{T\ \exists x(A(x) \vee B(x))}$$

coming from the branch $F\ \neg\exists x(A(x) \vee B(x))$ can be seen as a <u>direct proof</u> of $\exists x(A(x) \vee B(x))$ completely solving our problem; likewise if one has a refutation of $F\ B(\overline{p})$, one obtains a simmetric direct proof

$$\dfrac{\dfrac{T\ B(\overline{p})}{T\ A(\overline{p}) \vee B(\overline{p})}}{T\ \exists x(A(x) \vee B(x))}$$

6. The above examples correspond to a way of constructively interpreting, in a refutation system, the intuitionistic lack of the excluded third. The reader acquainted with the classical Resolution Method used in Artificial Intelligence will find an analogy between our interpretation and the use of the <u>metalinguistic</u> predicate ANS introduced by Green in order to "trace" the relevant information to be extracted from a refutation to solve the intended problem.

7. The above remarks are only a very rough sketch of our intuitive feeling that intuitionistic refutation systems can be used (as the direct ones) in program synthesis. Of course, our research is only at the beginning: this research should include also an analysis of recent tools such as the intuitionistic version of Herbrandt's theorem [20].

REFERENCES

1 E.W. Beth, Semantic Construction of Intuitionistic Logic. Mededelingen der Koninklijke Nederlandse Akademie van Wetenschappen, Afb. Letterkunde, Nieuve Reeks, Deel 19, No. 11.

2 W. Bibel, Predicative Programming, Report of the Institut für Informatik, Technische Universität München, Munich, 1974.

3 C.L. Chang, R.C.T. Lee, Symbolic Logic and mechanical theorem proving (book), Academic Press Inc., New-York 1973.

4 R. Constable, Constructive Mathematics and automatic program writers, Proc. IFIP Congress 1971 (Lublijana), TA-2, North-Holland 1971.

5 G. Degli Antoni, P. Miglioli, M. Ornaghi, The synthesis of programs as an approach to the construction of reliable programs, Proc. of the International Conf. on proving and improving programs, Arc et Senans, July 1975, Huet, Kahn eds.

6 M.C. Fitting, Intuitionistic Logic, Model Theory and Forcing, (book), North-Holland 1969.

7 C. Goad, Computational uses of the manipulation of formal proofs, Stanford Dept. of Comp. Sci., Rep. n. STAN-CS-80-819, August 1980.

8 S. Goto, Program Synthesis from Natural Deduction Proofs, Int. Joint Conf. on Artificial Intelligence, Tokyo, 1979.

9 R. Kowalski, Logic for Problem Solving, (book), Elsevier North-Holland, 1979.

10 G. Kreisel, Some uses of proof-theory for finding computer programs, Notes for a talk in the Logical Symp. of Clermond-Ferrand, 1975, available as a manuscript.

11 S. Kripke, Semantical Analysis of Intuitionistic Logic I, in Formal Systems and Recursive Function, North-Holland, 1975.

12 P. Miglioli, M. Ornaghi, A purely logical computing model: the open proofs as programs, Istituto di Cibernetica dell'Università di Milano, internal rep. MIG 7, Sept. 1979.

13 P. Miglioli, M. Ornaghi, A logically justified model for non deterministic and parallel computations, Istituto di Cibernetica dell'Università di Milano, int. rep. MIG 8, 1978.

14 P. Miglioli, M. Ornaghi, Some models of computation from a unified proof-theoretical point of view (to appear).

15 U. Moscato, Intuizionismo e teoria della dimostrazione. Su un'estensione dei sistemi di refutazione aderente alla semantica dell'intuizionismo. Tesi di laurea, Università di Milano, A.A.1979/1980.

16 D. Prawitz, Natural deduction : a proof theoretical study, (book), Almqvist--Wiksell,1965.

17 C.A. Smorynski, Applications of Kripke models, in S.A. Troelstra, Metamathe-

matical investigation of Intuitionistic Arithmetic and Analysis, (book), Lect. Notes in Mathematics n.344, Springer-Verlag, 1973.

18 R. Smullyan, First order logic, (book), Springer, New York, 1968.

19 R. Statman, Intuitionistic propositional logic is polynomial-space complete, Theoretical Computer Science 9 (1979), North-Holland.

20 G. Bellin, Herbrand's theorem for calculi of sequents LK and LJ, Rep. of the University of Stockholm.

This research has been supported by the italian Ministero della Pubblica Istruzione and by the CP Project of Honeywell Information System Italia.

EFFICIENT OPTIMIZATION OF MONOTONIC FUNCTIONS ON TREES

Yehoshua Perl
Dept. of Mathematics and Computer Science
Bar-Ilan University
Ramat-Gan, Israel

Yossi Shiloach
IBM Israel Scientific Center
Technion City
Haifa, Israel

ABSTRACT

The problem of optimizing weighting functions over all the k-subtrees (subtrees with k vertices) of a given tree, is considered. A general algorithm is presented, that finds an optimal k-subtree of a given tree whenever the weighting function is what we call 'monotonic'. Monotonicity is a very natural property, satisfied by most of the functions that one can think of.

The problem is solved for both cases of rooted and undirected trees. On the ohter hand, even simple extensions of it to general graphs are NP-hard.

1. *INTRODUCTION*

Let T be a rooted tree. Henceforth a SUBTREE of T will always mean a rooted subtree of T and a k-SUBTREE will mean a subtree with k vertices. A COMPLETE subtree of T is one that contains all the descendants of its root.

Let W be a weighting function that assigns a real number (called weight) to every subtree of T.

A k-subtree T' is MAXIMAL (MINIMAL) with respect to W if it is the heaviest (lightest) k-subtree of T. The word OPTIMAL will later be used for both maximal and minimal. Let T, T_1 and T_2 be rooted trees and let v be a vertex of T. Let $T_1'(T_2')$ be the tree obtained by hooking $T_1(T_2)$ on T at v. We say that W is a MONOTONIC weighting function if for any such triple T, T_1, T_2 and any vertex v of T:

$$W(T_1) \leq W(T_2) \text{ implies } W(T_1') \leq W(T_2') \quad .$$

In this paper we present a general and efficient algorithm for finding optimal k-subtrees of a given tree, which is good for any monotonic weighting function. Since most of the weighting functions that one can think of, including the following several examples, are monotonic - this algorithm might be proved very applicable. Moreover, in many cases this algorithm can be directly applied to corresponding problems in undirected trees. Furthermore, it will be shown that even some non-monotonic functions like the diameter can still be optimized if monotonic auxiliary functions are properly used.

Examples of several natural weighting functions:

Let $w: V \to R$ and $\ell: E \to R$ be two real-valued functions that assign weights to the vertices and lengths to the edges of T respectively.

Let $T' = (V',E')$ be a given subtree of T rooted at r'. Consider the following weighting functions:

1. $W1(T') = \sum_{v \in V'} w(v)$
2. $W2(T') = \sum_{e \in E'} \ell(e)$
3. $W3(T') = \min_{v \in V'} w(v)$
4. $W4(T') = \min_{e \in E'} \ell(e)$
5. $W5(T')$ = Number of terminal vertices in T'.

Defining the DISTANCE d(u,v) from a vertex u to its descendant v as the sum of lengths of the edges along the path connecting them, we can further define:

6. Internal Path Length: $W6(T') = \sum_{v \in V'} d(r',v)$

7. Height: $W7(T') = \max_{v \in V'} d(r',v)$.

Similar functions can be defined for undirected trees and undirected graphs as well. For almost all these functions, though, the corresponding problems of finding optimal k-subgraphs or even optimal k-subtrees in general graphs are NP-hard.

In the next section we describe the algorithm and analyze its complexity.

In Section 3, the corresponding problem for undirected trees is defined and solved with the aid of the algorithm below. It is also shown how the non-monotonic diameter function is optimized by utilizing an auxiliary monotonic function.

Section 4 contains a brief discussion on the NP-hardness of the problems of optimizing the weighting functions above over k-subgraphs and k-subtrees of general graphs.

2. *THE ALGORITHM*

The algorithm is based on dynamic programming. The common bottom-up approach requires exponential time and therefore an efficient variation of it is used.

Let T be a tree rooted at r and let $r_1,\dots,r_s$ be r's sons. Let $T_1,\dots,T_s$ be T's complete subtrees rooted at $r_1,\dots,r_s$ respectively. Given a monotonic weighting function W and an integer k, assume that optimal k-subtrees $T_1',\dots,T_s'$ of $T_1,\dots,T_s$ respectively, have already been found. The optimal k-subtree of T is either the optimal subtree among $T_1',\dots,T_s'$ or another k-subtree rooted at r. Thus, it remains to find an optimal k-subtree among those rooted at r. Such a subtree is called an r-OPTIMAL subtree. It turns out, however, that in order to find an r-optimal k-subtree, one has to find r_j-optimal i-subtrees for all $1 \leq i \leq k$ in each of the subtrees T_j, $1 \leq j \leq s$. Let $T_j'(i)$ denote an r_j-optimal i-subtree of T_j, for all $1 \leq j \leq s$ and $1 \leq i \leq k$. Adopting a well-known dynamic programming technique, we assume that these trees are already known and proceed to find r-optimal i-subtrees for all $1 \leq i \leq k$. At this point, the straightforward approach requires that for each sequence $i_1,\dots,i_s$ such that $i_1 + \dots + i_s = k$, a corresponding sequence of $T_j'(i_j)$, $1 \leq j \leq s$, will be considered. Since the number of such sequences is $O(k^{\deg(r)})$, this approach is intractable when r has many sons. It turns out, however, that a left-to-right propagation will save us this time. To this end let $\underline{T}_j$, $1 \leq j \leq s$, be the r-rooted subtree of T, spanned by $T_1,\dots,T_j$ and r. As expected, let $\underline{T}_j'(i)$ denote an r-optimal i-subtree of $\underline{T}_j$ for all $1 \leq j \leq s$ and $1 \leq i \leq k$. Since $\underline{T}_1$ is the subtree spanned by T_1 and r, it follows from the monotonicity that $\underline{T}_1'(i)$ can be taken as the subtree spanned by $T_1'(i-1)$ and r, $1 \leq i \leq k$. It should now be shown that the trees $\underline{T}_{j+1}'(i)$, $1 \leq i \leq k$, can be efficiently obtained from $\underline{T}_j'(i)$ and $T_{j+1}'(i)$, $1 \leq i \leq k$. Let i_o be a fixed integer between 1 and k. For all i, $1 \leq i \leq i_o$, denote by $\underline{TT}_{j+1}(i,i_o)$ the

i_o-subtree rooted at r that is spanned by $\underline{T}'_j(i)$ and $T'_{j+1}(i_o-i)$, (see Fig. 1).

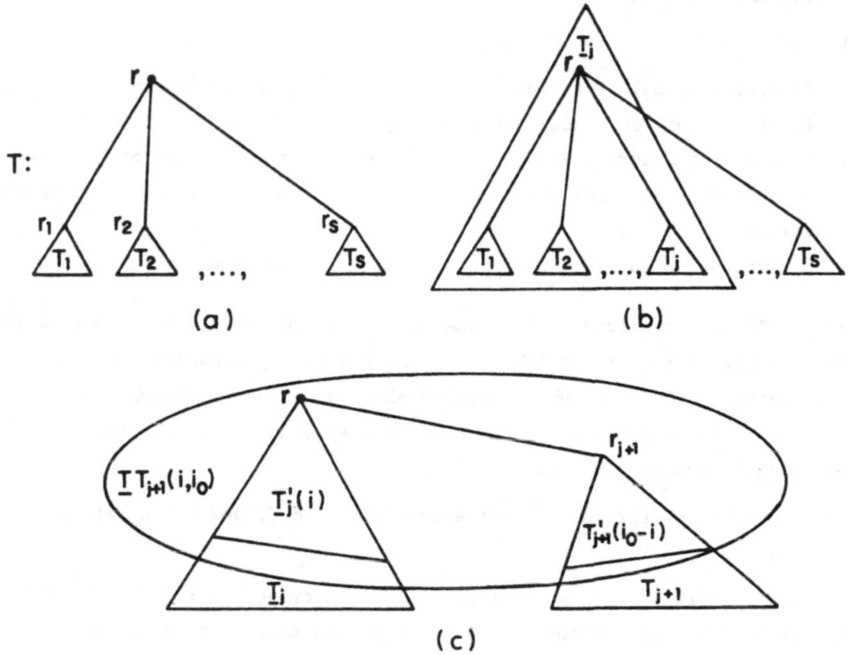

<u>Figure 1</u>

The following simple theorem both motivates the algorithm below and establishes its validity.

Theorem: $\underline{T}'_{j+1}(i_o)$ can be chosen as one of the subtrees $\underline{T}T_{j+1}(i,i_o)$, $1 \leq i \leq i_o$, that optimizes W.

Proof: Follows immediately from the monotonicity of W. □

In order to translate the ideas above into a more formal algorithm, let us associate a sequence $S(v)$ of length k with each vertex $v \in T$. $S(v)[i]$, the i-th element of $S(v)$, should be by the end of the algorithm, a v-optimal i-subtree.

Let v be given, let $v_1,\ldots,v_s$ be its sons, and let $\underline{T}_j$ be defined as above for $1 \leq j \leq s$. Denote by $S_j(v)$, a sequence whose i-th element is a v-optimal i-subtree of $\underline{T}_j$ rather than T. If the sequences $S(v_1),\ldots,S(v_s)$ have already been constructed then v's sequence can be obtained by:

CONSTRUCT $S(v)$

Step 1: $S_1(v)[i] \leftarrow$ The subtree spanned by $S(v_1)[i-1]$ and r; $j \leftarrow 1$.

Step 2: While $j < s$, obtain $S_{j+1}(v)$ from $S_j(v)$ and $S(v_{j+1})$ as described in the theorem above; $j \leftarrow j+1$.

Step 3: $S(v) \leftarrow S_s(v)$.

Finding an r-optimal k-subtree of T may now look like:

OPT(r,k)

Step 1: (Initialization) For each terminal vertex $v_t \in T$, $S(v_t) \leftarrow (\{v_t\},\phi,\ldots,\phi)$. The i-th empty set indicates that v_t has no i-subtree, $2 \leqslant i \leqslant k$.)

Step 2: Process the internal vertices of T in end-order (or any other order in which each vertex follows its sons) constructing S(v) for each internal vertex v of T.

Step 3: Output the k-th element of S(r) as an r-optimal k-subtree of T.

Since OPT(r,k) constructs the sequences S(v) for all $v \in V$, we know the v-optimal k-subtrees for all $v \in V$. Thus one can easily obtain an optimal k-subtree, simply by taking the best k-subtree out of all the v-optimal k-subtrees. Thus, substituting the following Step 3' instead of Step 3 in OPT(r,k), yields OPT(k) that finds an optimal k-subtree of T.

Step 3': Find a best v-optimal k-subtree over all $v \in V$, and output it as an optimal k-subtree of T.

Note that OPT(r,k) yields the r-optimal i-subtrees for all i, $1 \leqslant i \leqslant k$. Hence the optimal i-subtrees for all $1 \leqslant i \leqslant k$ can also be easily found.

Complexity.

The complexity of the algorithm strongly depends on the time required to compute W(T) for a given tree T. In order to eliminate this factor from our evaluation, it is taken as O(1). In any other case, our complexity should be multiplied by the appropriate factor in order to obtain the right figure.

Assume that T has n vertices and that merging several disjoint subtrees into one tree takes O(n) time. Let us first analyze the complexity of CONSTRUCT S(v). Obviously, Step 2 is the most time consuming. For a fixed j, obtaining $S_{j+1}(v)$ from $S_j(v)$ and $S(v_j)$ requires $O(k^2n)$ time. This yields time of $O(\deg(v) \cdot k^2n)$ for the whole Step 2 and thus for CONSTRUCT S(v) too. Summing the last expression over all $v \in T$ yields a total time of $O(k^2n^2)$ for the entire OPT(r,k) algorithm. The same time bound applies to OPT(k) too.

It should be noted that many weighting functions, including all the examples above, can be computed without a complete information on the tree's structure. In most of the cases, only the weights of the optimal subtrees $\underline{T}'_j(i)$ and $T'_{j+1}(i)$, $1 \leqslant i \leqslant k$, together with some additional information on the root and the edges connecting it to its sons, is really required. In such cases, one can store in S(v) just the weights of the appropriate optimal subtrees. Additional pointers that would enable us to recover the desired final tree, should also be maintained. Both time and space can be reduced in these cases by a factor of n. This yields an $O(k^2n)$ time bound for all the examples above and for many other functions as well.

For example let us take the Internal Path Length function W6. The basic step in the construction of S(v), namely Step 2 of CONSTRUCT S(v) has in this case the form:

Choose $\underline{T}'_{j+1}(i_o)$ as one of the subtrees of the form $\underline{T}T_{j+1}(i,i_o)$ for which $W6(\underline{T}'_j(i)) + W6(T'_{j+1}(i_o-i)) + (i_o-i) \cdot \ell(v,v_j)$ is optimized. This formula shows that in this case, as in many others, we need the weights of the appropriate optimal trees rather than then trees themselves.

3. *APPLICATIONS TO OPTIMIZATION PROBLEMS IN UNDIRECTED TREES*

In this section, T is an undirected tree and 'subtrees' are undirected subtrees unless otherwise specified. Let v be a vertex of T and let T(v) denote the rooted tree obtained by hooking T on v. In order to apply the algorithm above to undirected trees we would first like to extend the notion of a monotonic function to weighting functions that are defined on undirected trees. Fortunately, there is a natural way to do so. If W is innitially defined only for undirected trees, it can easily be extended to rooted trees by defining the weight of a rooted tree as that of its underlying tree. W is UD-MONOTONIC if its extension to rooted trees is monotonic according to the first definition. Note that if T' is any subtree of T, then rooting T at any vertex v, turns T' to a rooted subtree of T, say T'(u), for some u in T'. This observation yields the following undirected modification of OPT(k) for ud-monotonic functions.

UD-OPT(k)

Step 1: Choose a vertex v of T.
Step 2: Apply OPT(k) to T(v) yielding a k-optimal rooted subtree T'(u) to T(v).
Step 3: Return T', the underlying undirected tree of T'(u), as an optimal k-subtree.

It turns out that optimization of non-monotonic functions can sometimes be carried out by the aid of monotonic 'middle' functions. This is the case of the following two functions:

The Diameter: $W8(T') = \max_{u,v \in V} d(u,v)$, and

The Radius: $W9(T') = \min_{c \in V} \max_{v \in V} d(c,v)$.

Both functions are not ud-monotonic as one can easily verify.

Since UD-OPT(k) calls for OPT(k) and OPT(k) is solved via OPT(r,k), we just have to hook T on an arbitrary root r and show that OPT(r,k) can be solved for T(r). The 'middle' function in both cases is the monotonic Height function W7. Since both cases are quite similar, only W8 will be discussed. Again, we restrict ourselves to Step 2 of CONSTRUCT S(v) which is the heart of the algorithm.

As before, let $\underline{T}_j'(i)$ denote an r-optimal i-subtree of $\underline{T}_j$ and let $T_{j+1}'(i)$ be an r-optimal i-subtree of T_{j+1}, $1 \leq i \leq k$, where optimality is taken with respect to the height. Similarly, let $\underline{T}_j''(i)$ be an r-optimal i-subtree of $\underline{T}_j$ with respect to the diameter. Assuming that the weights of all these trees are already known, and an integer i_o, $1 \leq i_o \leq k$, is given - the next tree to be computed, namely $\underline{T}_{j+1}''(i_o)$ is either $\underline{TT}_{j+1}(i,i_o)$ for some i, $1 \leq i \leq i_o$, that optimizes

$$W7(\underline{T}_j'(i)) + W7(T_{j+1}'(i_o-i)) + \ell(r,r_{j+1})$$

or

$\underline{T}_j''(i_o)$ if $W8(\underline{T}_j''(i_o))$ is even better.

The validity proof for this way of choosing $\underline{T}_{j+1}''(i_o)$ is straightforward.

4. *COMPLEXITY OF SIMILAR OPTIMIZATION PROBLEMS IN GENERAL GRAPHS*

A k-SUBGRAPH of a given undirected graph is a connected subgraph with k vertices.

In this section we consider the complexity of finding optimal k-subgraphs and k-subtrees of a given graph. For most of the weighting functions mentioned above, the corresponding decision problems are NP-complete even in their simple 0-1 forms. For each problem claimed to be NP-complete, a corresponding NP-complete problem which is reducible to it, is listed. For the exact definitions of the source NP-complete problems, see [GJ].

1. Maximizing W over k-subgraphs.

W1: The unit length Steiner tree problem.

W2, W4: The maximum clique problem.
W5: The maximum leaf spanning tree problem.
W8, W9: The longest path problem with unit lengths.
W3: Can be solved in polynomial time.

2. Minimizing W over k-subgraphs.

W1, W2: The unit length Steiner tree problem.
W7: The longest path problem with unit lengths.
W8: The maximum clique problem.
W3, W4, W9: Have polynomial solutions.

3. Maximizing W over k-subtrees.

W1, W7, W8, W9: Same reductions as for k-subgraphs.
W2: The unit length Steiner tree problem.

W3, W4: Have polynomial solutions.

4. Minimizing W over k-subtrees.

W1, W2, W7: Same reductions as for k-subgraphs.

W3, W4, W8, W9: Can be solved in polynomial time.

REFERENCES

[GJ] Garey, M.R. and Johnson, D.S., COMPUTERS AND INTRACTABILITY; A GUIDE TO THE THEORY OF NP-COMPLETENESS, W.H. Freeman and Co., San Francisco, 1979.

DIFFERENTS TYPES DE DERIVATIONS INFINIES DANS LES GRAMMAIRES ALGEBRIQUES D'ARBRES

N. Pōlian

Laboratoire d'Informatique
Université de Poitiers
40, av. du Recteur Pineau 86022 POITIERS
(France)

Introduction

Dans l'étude des grammaires algébriques d'arbres comme modèles de schémas de programmes (non déterministes) on est amené à introduire les dérivations infinies et à définir les résultats de ces dérivations infinies. On sait qu'il existe plusieurs types de dérivations finies (quelconques, descendantes, ascendantes, parallèles descendantes etc.) et l'on connait des résultats comparant les langages d'arbres engendrés par une même grammaire en utilisant différents types de dérivations.

Le but de cet article est de généraliser de tels résultats aux dérivations infinies. Ceci nous amène :

1°) A introduire de nouveaux types de dérivations :

- une dérivation est dite *faiblement descendante* si aucun symbole dérivé ne figure en dessous d'un symbole dérivé ultérieurement.
- une dérivation est dite *faiblement par niveaux* si la profondeur des symboles dérivés est croissante. Elle est dite *par niveaux* si elle est de plus descendante.

2°) A étudier le problème de l'équivalence de ces dérivations, lorsqu'on ne considère plus seulement le langage engendré formé uniquement d'arbres terminaux mais aussi lorsque l'on considère tous les "résultats" (terminaux et non terminaux) de ces dérivations : on montre que toute dérivation finie est équivalente à une dérivation faiblement descendante.

3°) A introduire de nouvelles définitions des résultats de dérivations infinies qui permettent de généraliser le concept de résultat non terminal d'une dérivation finie ;

- une dérivation infinie est dite *réussie* si elle converge vers un élément de $M^{\infty}(F)$, (F = ensemble des symboles terminaux).
- une dérivation infinie est dite *faiblement réussie* si elle converge vers un élément de $M^{\infty}(F \cup \Phi)$ (Φ = ensemble des symboles non terminaux).

4°) A comparer les langages engendrés par les dérivations infinies :

- deux dérivations à partir du même arbre sont dites C-équivalentes si elles convergent vers le même arbre.

On montre alors que toute dérivation réussie est C-équivalente à une dérivation descendante.

Les résultats obtenus font apparaître qu'à cet égard aussi les grammaires de Greibach jouissent de propriétés très particulières :

Si G est de Greibach : toute dérivation faiblement réussie est C-équivalente à une dérivation faiblement descendante.

Ce résultat n'est pas vrai dans le cas général.

I - NOTATIONS ET DEFINITIONS

1) Soit F un ensemble fini de symboles de fonctions, à chaque $f \in F$ on associe son arité $ar(f) \in \mathbb{N}$. On note $F_i = \{f \in F, ar(f) = i\}$.

Soit M(F) l'ensemble des arbres finis sur F, pour $t \in M(F)$ on note $Dom(t) \subset \mathbb{N}_+^*$ le domaine de l'arbre t([7]).

L'arbre t peut alors être représenté comme l'application : $Dom(t) \to F$, qui à chaque élément du domaine associe un symbole de F.

- On appellera *noeud* de l'arbre $t(I_f, f)$ le couple formé d'un élément I_f du domaine et du symbole $f \in F$ qui lui est associé.
- Soit (I_f, f) un noeud de t, on appelle profondeur du noeud la longueur $|I_f|$
- On appelle profondeur de l'arbre t le Sup des profondeurs des noeuds de t.
- On dira qu'un noeud (I,ϕ) est "au-dessus" du noeud (I,ψ) si I est un facteur gauche propre de J : $I \in FGP(J)$.
- On définit classiquement la distance d sur M(F) :

$$d(t_1,t_2) = \begin{cases} 0 & \text{si } t_1 = t_2 \\ 2^{-\alpha(t_1,t_2)} & \text{où } \alpha(t_1,t_2) \text{ est la profondeur du premier noeud où } t_1 \text{ et } t_2 \text{ diffèrent.} \end{cases}$$

En complétant l'espace métrique M(F) on obtient alors $M^\infty(F) = M(F) \cup M^\omega(F)$ l'espace des arbres finis ou infinis.

- Soit Ω un symbole supplémentaire d'arité zéro on définit alors sur $M_\Omega^\infty(F) = M^\infty(F \cup \Omega)$ l'ordre syntaxique habituel.

2) Une grammaire algébrique d'arbre [1,2] est un quadruplet $< F,\phi,X_o,\mathcal{R}>$ où
   - F est un ensemble fini de fonctions graduées appelé ensemble des terminaux.
   - Φ est un ensemble fini de fonctions graduées appelé ensemble des non-terminaux.
   - $X_o \in F$ est l'axiome.
   - $\mathcal{R}$ est l'ensemble des règles de dérivations de la forme :

     $\phi(x_1,\ldots,x_n) \to \theta \in M(F \cup \Phi, x_1,\ldots,x_n)$

Soit $G = < F,\Phi,X_o,\mathcal{R} >$ une grammaire algébrique d'arbre. De même que dans le cas des langages de mots [7] on définit le langage infini engendré par G.

A. Arnold et M. Nivat [3] ont proposé la définition suivante :
soit ω l'application de $M^{\infty}(F \cup \Phi) \to M^{\infty}_{\Omega}(F)$ qui consiste à remplacer les sous-arbres de racine non terminale par le symbole "indéterminé" Ω c'est-à-dire :

$$\omega(a) = a \quad \text{si } a \in F_o$$
$$\omega(f(t_1,\dots,t_n)) = f(\omega(t_1),\dots,\omega(t_n)) \quad \text{si } f \in F_n$$
$$\omega(\phi(t_1,\dots,t_n)) = \Omega$$

<u>Définition 1</u> : Soit $\delta : t_o \to \dots \to t_n \to \dots$ une dérivation infinie dans G, δ est dite réussie si $\sup_{i \in \mathbb{N}} (\omega(t_i))$ est un arbre infini maximal.

En nous appuyant sur la propriété bien connue : voir par exemple [4]

<u>Propriété</u> : pour toute suite infinie croissante dans $M^{\infty}_{\Omega}(F \cup \Phi)$ la limite au sens de l'ordre est la même que la limite au sens de la topologie.

Nous proposons la définition équivalente suivante :

<u>Définition 2</u> : Une dérivation infinie $\delta : t_o \to t_1 \dots \to t_n \to \dots$ dans G, est réussie si la suite (t_n) des arbres dérivés converge vers un arbre $T \in M^{\infty}(F)$ au sens de la topologie définie par la distance d.

ou encore :

il existe $T \in M(F)$ tel que pour tout $n \in \mathbb{N}$ il existe $m \in \mathbb{N}$ tel que $d(T,t_m) \leq 2^{-n}$.

On dira alors que T est le résultat de la dérivation réussie.

<u>Exemple 1</u>

$F = F_2 \cup F_1 \cup F_o \qquad F_2 = \{f\} \qquad F_1 = \{g\} \qquad F_o = \{a\}$

$\Phi = \Phi_1 = \{\phi\}$

$: \ \phi(x) \to g(\phi(x))$

La dérivation infinie δ_1 :

δ_1 :

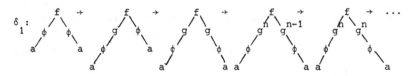

est réussie de résultat : $T_1 = f(g^{\omega},g^{\omega})$

La dérivation infinie δ_2 :

δ_2:
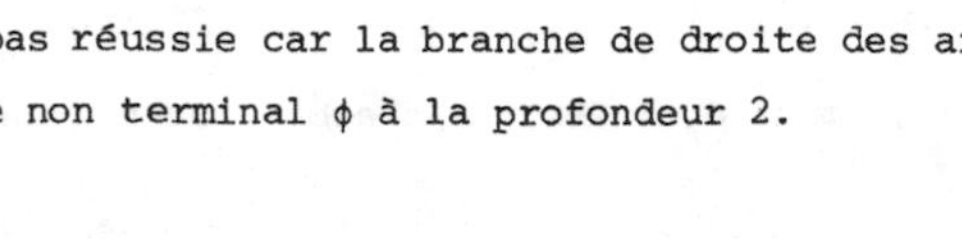

n'est pas réussie car la branche de droite des arbres dérivés contient toujours le symbole non terminal ϕ à la profondeur 2.

<u>Définition 3</u> On appelle langage infini $L^{\omega}(G,X_o)$ engendré par G à partir de X_o l'ensemble des résultats des dérivations infinies réussies à partir de X_o.

On remarque que dans le cas fini on a d'une part les dérivations terminales : $X \overset{*}{\rightarrow} t \in M(F)$ dont les résultats définissent le langage engendré par G, d'autre part les dérivations non terminales $X \overset{*}{\rightarrow} t \in M(F \cup \Phi)$. Les dérivations infinies réussies correspondent aux dérivations finies terminales. On peut définir les dérivations infinies non réussies mais cependant convergentes qui correspondent aux dérivations finies non terminales :

<u>Définition 4</u> Une dérivation infinie $\delta : X_o \rightarrow t_1 \rightarrow \ldots \rightarrow t_n \rightarrow \ldots$ est faiblement réussie si la suite (t_n) converge vers un arbre $T \in M^{\infty}(F \cup \Phi)$ au sens de la topologie définie par d.

Dans l'exemple 1, la dérivation δ_2 est faiblement réussie de résultat $T_2 = f(g^{\omega},\phi(a))$

<u>Définition 5</u> On dira que deux dérivations infinies faiblement réussies à partir du même arbre t_o sont C-équivalentes si elles ont le même résultat.

II - RESULTATS

On sait [5,8] que dans le cas fini tout élément du langage L(G,X), peut être obtenu comme résultat d'une dérivation descendante : $L(G,X) = L_{OI}(G,X)$. On va demontrer que ce résultat est encore vrai dans le cas infini. Pour faire cette démonstration nous avons besoin de quelques définitions supplémentaires.

<u>Définition 6</u> Soit $t \in M(F \cup \Phi)$, un noeud (I_A,A) de t est dit externe si pour tout noeud (I_B,B) "au-dessus" de lui alors $B \in F$. Un noeud non externe est dit interne.

<u>Définition 7</u> Une dérivation $t_1 \overset{*}{\rightarrow} t_2$ dans G est dite externe ou descendante (resp. interne) si elle est composée de dérivations élémentaires qui dérivent des noeuds externes (resp. interne). On note :

$$t_1 \overset{*}{\underset{D}{\longrightarrow}} t_2 \quad (\text{resp. } t_1 \overset{*}{\underset{I}{\longrightarrow}} t_2)$$

<u>Définition 8</u> Une dérivation δ est dite faiblement descendante si pour tout noeud (I_{ϕ},ϕ) dérivé dans δ aucun noeud (I_{ψ},ψ) "au-dessus" de lui n'est dérivé ultérieurement dans δ. On note $t_1 \overset{*}{\underset{F}{\longrightarrow}} t_2$.

Autrement dit si on dérive un noeud non externe, les noeuds non terminaux "au-dessus" de lui ne seront plus dérivés.

Remarque : il est clair que toute dérivation descendante est faiblement descendante.

<u>Définition 9</u> Une dérivation $\delta : t_1 \xrightarrow{*} t_2$ est dite par niveaux (de profondeur n) si les conditions suivantes sont réalisées :

(i) la profondeur des noeuds dérivés est croissante (de profondeur $\leq n$)

(ii) δ est descendante

On note : $t_1 \overset{*}{N\!\!\rightarrow} t_2$.

δ est dite faiblement par niveaux si seule la condition (i) est réalisée.

Autrement dit on dérive "niveau par niveau" : les noeuds de profondeur 1, puis les noeuds de profondeur 2,...

<u>Définition 10</u> Deux noeuds (I_ϕ, ϕ) et (I_ψ, ψ) sont dits indépendants si ils sont différents et si aucun d'entre eux n'est "au-dessus" de l'autre.

Deux dérivations élémentaires sont indépendantes si elles dérivent deux noeuds indépendants.

Les deux lemmes suivants découlent immédiatement des définitions :

<u>Lemme 1</u> Si au cours d'une dérivation faiblement descendante une dérivation élémentaire interne d_1 précède une dérivation élémentaire externe d_2, alors d_1 et d_2 sont indépendantes.

<u>Lemme 2</u> Deux dérivations successives indépendantes sont permutables.

<u>Proposition 1</u> Soit $\delta : t_1 \overset{*}{F\!\!\rightarrow} t_2$ une dérivation faiblement descendante, on peut réordonner les dérivations élémentaires de δ de telle sorte qu'il existe t' et une dérivation :

$$t_1 \overset{*}{D\!\!\rightarrow} t' \overset{*}{I\!\!\rightarrow} t_2$$

Démonstration : Par récurrence sur la longueur n de la dérivation δ.

- vrai pour n = 1

- supposons la propriété vérifiée pour n-1. Soit $\delta : t_1 F\!\!\rightarrow \ldots F\!\!\rightarrow t_n$ de longueur n. En appliquant l'hypothèse de récurrence à $t_1 \overset{*}{F\!\!\rightarrow} t_{n-1}$ on obtient :

$$\delta' : t_1 \overset{d_1}{\overset{*}{D\!\!\longrightarrow}} t' \overset{d_2}{\overset{*}{I\!\!\longrightarrow}} t_{n-1}$$

Deux cas se présentent alors :

- si $t_{n-1} \rightarrow t_n$ est interne il suffit de prolonger δ'

- si $t_{n-1} \rightarrow t_n$ est externe, on applique alors le lemme 1, elle est donc indépendante des dérivations composant d_2 on peut donc la permuter avec celles-ci, et on obtient :

$$t_1 \overset{d_1}{\overset{*}{D\!\!\longrightarrow}} t' D\!\!\longrightarrow t'' \overset{*}{I\!\!\longrightarrow} t_n$$

<u>Proposition 2 [5,8]</u> Soit $t_1 \xrightarrow{*} t_2$ une dérivation dans G avec $t_2 \in M(F)$, il existe une dérivation descendante : $t_1 \overset{*}{D\!\!\rightarrow} t_2$.

<u>Lemme 3 [6]</u> Soit $t_1 \xrightarrow{*} t_2$ une dérivation dans G. On peut toujours supposer que les dérivations par des règles linéaires et terminales sont en dernier. (une règle

$\phi(x_1,\ldots,x_n) \to \tau$ est dite linéaire si chaque variable x_i figure une fois et une seule dans τ, elle est dite terminale si τ ne contient pas de symboles non terminaux.

Ce lemme permet alors de démontrer une propriété analogue à la Proposition 2 pour les dérivations finies non terminales :

<u>Proposition 3</u> Soit d : $t_1 \xrightarrow{*} t_2$ une dérivation dans G, il existe une dérivation faiblement descendante de t_1 en t_2.

Démonstration :

Considérons la grammaire $\bar{G} = < F \cup \bar{\Phi}, \Phi, X_o, \mathcal{R} \cup \bar{\mathcal{R}} >$
où $\bar{\Phi} = \{\bar{\phi}, \phi \in \Phi\}$ et $\bar{\mathcal{R}} = \{\phi(x_1,\ldots,x_n) \to \bar{\phi}(x_1,\ldots,x_n), \phi \in \Phi\}$

Soit ω l'application définie par :

- $\omega(\phi(t_1,\ldots,t_n)) = \bar{\phi}(\omega(t_1),\ldots,\omega(t_n))$ si $\phi \subset \Phi$

- $\omega(f(t_1,\ldots,t_n)) = f(\omega(t_1),\ldots\omega(t_n))$ si $f \in F$

qui consiste à "barrer" dans un arbre tous les symboles non terminaux.

Soit $\bar{\delta}$: $t_1 \xrightarrow{d} t_2 \longrightarrow \omega(t_2)$

en appliquant la prop. 2 il existe une dérivation $\bar{\delta}_D$ descendante dans $\bar{G}$

$\bar{\delta}_D$: $t_1 \; D\xrightarrow[\bar{G}]{*} \omega(t_2)$

en utilisant le lemme 3 on peut permuter les dérivations de telles sortes que les règles linéaires terminales $\phi(x_1,\ldots,x_n) \to \bar{\phi}(x_1,\ldots,x_n)$ soient à la fin, on obtient donc :

$$t_1 \xrightarrow[*]{d_1} t' \xrightarrow[*]{d_2} \omega(t_2)$$

où d_1 (resp.d_2) est composée de dérivations utilisant des règles de $\mathcal{R}$ (resp. de $\bar{\mathcal{R}}$)
On a donc $t'=t_2$. D'autre part il est clair que d_1 est faiblement descendante.

<u>Lemme 4</u> Soit δ: $t_1 \xrightarrow{*} t_2$ une dérivation dans G, il existe δ_1 dérivation descendante et δ_2 dérivation interne telle que :

$$\delta : t_1 \; D\xrightarrow[*]{\delta_1} t' \; I\xrightarrow[*]{\delta_2} t_2$$

Démonstration : Par application de la proposition 3 puis de la proposition 1.

<u>Lemme 5</u> Soit δ: $t_1 \; I\xrightarrow{*} t_2$ une dérivation interne dans G . On a $\omega(t_1) = \omega(t_2)$.

<u>Théorème 1</u> Soit δ : $t_o \to \ldots \to t_n \to \ldots$ une dérivation infinie réussie il existe une dérivation infinie δ' descendante, C-équivalente à δ.

Démonstration : On construit par induction une dérivation

δ' : $t_o \; D\xrightarrow{*} t'_1 \;\; \ldots D\xrightarrow{*} t'_n \; D\rightarrow \; \ldots$ telle que

Pour tout i il existe : $t'_i \; I\xrightarrow{*} t_i$.

Supposons la construction réalisée jusqu'à l'étape n

On a : $t_o \; D\xrightarrow{*} t'_n \; I\xrightarrow{*} t_n \rightarrow t_{n+1}$

d'après le lemme 4, il existe $t'_n \xrightarrow{*}_D t'_{n+1} \xrightarrow{*}_I t_{n+1}$. On a donc prolongé la construction à l'étape n+1.

La construction étant trivialement réalisée à l'étape n=0, on peut donc construire δ'.

De plus d'après le lemme 5 on a $\omega(t'_i) = \omega(t_i)$ pour tout $i \in \mathbb{N}$. Comme δ est réussie, son résultat T est $T = \sup_i \omega(t_i) = \sup_i \omega(t'_i)$, donc δ' est aussi réussie de même résultat.

<u>Lemme 6</u> Soit δ une dérivation descendante dans G. $\delta : t_1 \xrightarrow{*}_D t_2$ où t_2 est terminal jusqu'à la profondeur n (i.e. tous les noeuds de profondeur inférieure ou égale à n sont terminaux) il existe t' terminal jusqu'à la profondeur n tel que :

$$t_1 \underset{N}{\xrightarrow{d_1 \; *}} t' \underset{D}{\xrightarrow{d_2 \; *}} t_2$$

où d_1 est une dérivation par niveaux de profondeur n.

Démonstration : par récurrence sur n.

<u>Corollaire</u> : Soit $\delta : t_o \to \ldots \to t_n \to$ une dérivation infinie réussie à partir de t_o, de résultat T, il existe une dérivation infinie par niveaux à partir de t_o, de résultat T.

Démonstration : Soit $\delta' : t_o \to \ldots \to t'_n \to \ldots$ une dérivation infinie descendante de résultat T, on peut alors construire en utilisant le lemme 6 une dérivation infinie par niveaux δ'' :

$$\delta'' : t_o \xrightarrow{N} t''_1 \xrightarrow{N} t''_2 \xrightarrow{N} \ldots t''_n \xrightarrow{N} \ldots \quad \text{telle que } t''_i \xrightarrow{N} t''_{i+1} \xrightarrow{D} t'_{j_{i+1}}$$

où t''_{i+1} et $t'_{j_{i+1}}$ sont tous deux terminaux jusqu'à la profondeur i+1, d'où on déduit que :

$$\lim_{i\to\infty}(t''_i) = \lim_{i\to\infty}(t'_{j_i}) = \lim_{j\to\infty}(t'_j) = T$$

<u>Remarque</u> :

Toute dérivation infinie faiblement réussie n'est pas C-équivalente à une dérivation faiblement descendante.

Contre exemple :

Soit G la grammaire définie par :

$F_1 = \{A,B\}$, $F_o = \{a_o,b_o\}$

$\Phi_2 = \{\phi\}$ $\Phi_1 = \{\phi\}$ $\Phi_o = \{X_o\}$

$$\mathcal{R} \begin{cases} X_o \to \phi(a_o,\psi(b_o)) \\ \phi(x,y) \to \phi(A(x),y) \\ \psi(x) \to \psi(B(x)) \end{cases}$$

et δ la dérivation infinie :

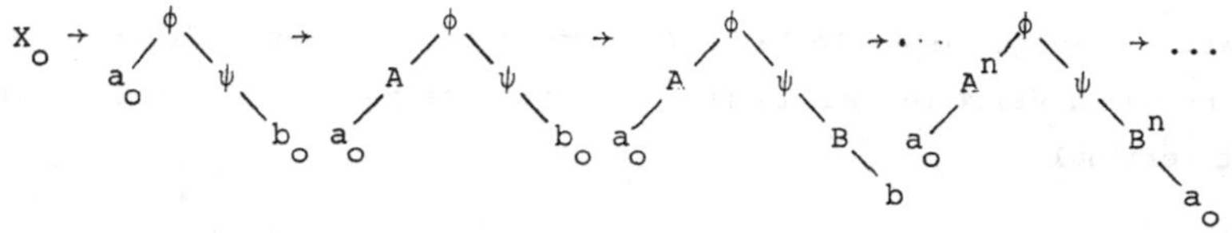

obtenue en dérivant alternativement ϕ et ψ, δ n'est donc pas faiblement descendante, δ est faiblement réussie de résultat $T = \phi(A^{\omega},\psi(B^{\omega}))$

Soit $L'^{\omega}_{FD}(G,X_o)$ l'ensemble des résultats des dérivations infinies, faiblement descendantes, faiblement réussies dans G à partir de X_o.

On a : $L'^{\omega}_{FD}(G,X_o) = \phi(A^{\omega},\psi(b_o)) \cup \{\phi(A^n(a_o),\psi(B^{\omega}))\ n \geq 0\}$

T n'appartient pas à $L'^{\omega}_{FD}(G,X_o)$.

III - ETUDE DANS LE CAS OU G EST UNE GRAMMAIRE DE GREIBACH

Définition 11 Une grammaire $G = < F,\Phi,X_o,\mathfrak{R} >$ est dite de Greibach si les règles de dérivation sont de la forme (i) ou (ii) :

(i) $\phi(x_1,\ldots,x_n) \to t$ où la racine de t est dans F

(règles normales)

(ii) $\phi(x_1,\ldots,x_n) \to x_i$

(règles effaçantes)

On supposera toujours dans la suite que les grammaires considérées sont de Greibach.

Lemme 7 Soit δ: $t_o \to t_n \to \ldots$ une dérivation infinie, faiblement descendante dans G. Soit (I_{ϕ},ϕ) un noeud externe de l'arbre t_i, il y a un nombre fini de dérivations élémentaires dans δ qui devient un noeud de position I_{ϕ}.

Démonstration :

Les règles de dérivations appliquées sont soit normales de la forme :

$\phi(x_1,\ldots,x_n) \to A(t_1,\ldots,t_i)$ où $A \in F$

soit effaçantes :

$\phi(x_1,\ldots,x_n) \to x_i$

Il y a donc au plus une dérivation d'un noeud de position I_{ϕ} utilisant une règle normale.

Soit K la profondeur du sous-arbre de sommet (I_{ϕ},ϕ) dans t_i. Le nombre de dérivations ultérieures à t_i d'un noeud de position I_{ϕ} utilisant une règle effaçante est borné par K.

<u>Proposition 4</u> Toute dérivation infinie par niveaux dans G est réussie.

Démonstration : En utilisant le lemme 7 on démontre qu'au bout d'un nombre fini de dérivations par niveaux le symbole figurant dans une position I donnée devient nécessairement terminal.

<u>Proposition 5</u> Toute dérivation δ: $t_o \to t_1 \to \ldots \to t_n \to \ldots$, infinie, faiblement descendante dans G est faiblement réussie.

Démonstration : La démonstration est analogue à la précédente mais repose sur une extension du lemme 7 au cas où (I_ϕ, ϕ) c'est pas un noeud externe mais un noeud tel que tout noeud non-terminal "au-dessus" de lui ne sera plus jamais dérivé dans δ.

<u>Remarque</u> :

L'hypothèse "par niveaux" dans la proposition 5 ne peut pas être remplacée par l'hypothèse plus faible "descendante".

Contre exemple : Soit $G = < F, \phi, \theta, \mathcal{R} >$ la grammaire de Greibach suivante :

$F_2 = \{f\} \quad F_1 = \{g\} \quad F_o = \{a\}$

$\Phi_1 = \{\phi, \psi\}$

$$\mathcal{R} \begin{cases} \theta \to f(\psi(a), \phi(a)) \\ \phi(x) \to g(\phi(x)) \end{cases}$$

la dérivation δ :

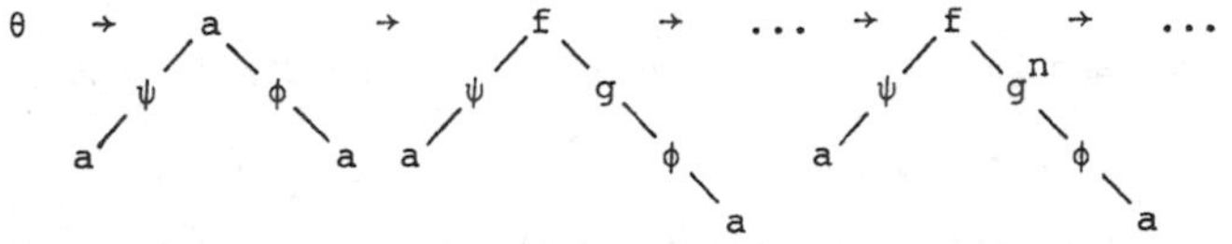

est descendante mais non réussie.

<u>Proposition 6</u> Soit δ : $t_o \to \ldots \to t_n \to \ldots$ une dérivation infinie faiblement réussie dans G. Il existe une dérivation faiblement descendante dans G, C-équivalente à δ.

Démonstration : Soit $\bar{G}$ la grammaire obtenue à partir de G en rajoutant l'ensemble des terminaux $\{\bar{\phi}, \phi \in \Phi\}$ et les règles : $\phi(x_1,\ldots,x_n) \to \bar{\phi}(x_1,\ldots,x_n)$ et soit ρ (resp ρ_n) l'application de $M^\infty(F \cup \Phi) \to M^\infty(F \cup \bar{\Phi})$ (resp de $M^\infty(F \cup \Phi) \to M^\infty(F \cup \Phi \cup \bar{\Phi})$) qui consiste à barrer tous les symboles non terminaux (resp jusqu'à la profondeur n).

ρ_o est donc l'identité, de plus $d(\rho(T), \rho_n(T)) \le 2^{-n}$.

On va construire une dérivation infinie réussie dans $\bar{G}$ à partir de t_o qui converge vers $\rho(T)$.

On pose $\delta_o = \delta$ dérivation qui converge vers $\rho_o(T) = T$.

Supposons par hypothèse de récurrence que l'on a construit une dérivation $\delta_n : t_o \to t_1^{(n)} \to \ldots \to t_m^{(n)} \to \ldots$ qui converge vers $\rho_n(T)$.

Soit i_n l'indice tel que $d(t_{i_n}^{(n)}, \rho_n(T)) \leq 2^{-n}$ et donc $d(t_{i_n}^{(n)}, \rho(T)) \leq 2^{-n}$.

Par application du lemme 7 soit $i_{n+1} \geq i_n$ l'indice à partir duquel on ne dérive plus aucun noeud de profondeur $\leq n+1$ dans δ_n.

En dérivant par les règles $\phi \to \bar{\phi}$ tous les noeuds non terminaux de profondeur n+1 dans $t_{i_{n+1}}^{(n)}$ on obtient la dérivation δ_{n+1} :

$$\delta_{n+1} : t_o \xrightarrow{*} t_{i_{n+1}}^{(n)} \xrightarrow{*} \rho_{n+1}(t_{i_{n+1}}^{(n)}) \to \ldots \rho_{n+1}(t_j^{(n)}) \to \ldots$$

δ_{n+1} converge donc vers $\rho_{n+1}(T)$

On en déduit donc une dérivation

$\delta' : t_o \to t_1' \to \ldots \to t_n' \to \ldots$

convergent vers $\rho(T)$, donc réussie.

Il existe alors une dérivation descendante dans $\bar{G}$ qui converge vers $\rho(T)$.

En enlevant dans cette dérivation toutes les règles $\phi(x_1,\ldots,x_n) \to \bar{\phi}(x_1,\ldots,x$ on obtient alors une dérivation faiblement descendante dans G qui converge vers T.

REFERENCES

[1] A. Arnold, M. Dauchet. Un théorème de duplication pour les forêts algébriques. *J. Comput. System Sci.* 13 (1976) p. 223-244.

[2] A. Arnold, M. Dauchet. Forêts algébriques et homomorphismes inverses. *Information and control* 37 (1978) p. 182-196.

[3] A. Arnold, M. Nivat. Formal computations of non-deterministic recursive program schemes. *Math. Syst. Theory* 13 (1980) p. 219-236.

[4] S. Bloom, R. Tindell. Compatible orderings on the metric theory of trees. *SIAM J. Comput.* Vol. 9, n° 4 (Nov. 1980).

[5] J. Engelfriet, E.M. Schmidt. IO and OI. *J. Comput. System. Sci.* 15 (1977) p. 328-353.

[6] B. Leguy. Réductions, transformations et classification des grammaires algébriques d'arbres. Thèse 3è cycle, Lille (1980).

[7] M. Nivat. Mots infinis engendrés par une grammaire algébrique. *RAIRO Informatique théorique* 11 (1977) p. 311-327 et 12 (1978)p. 259-278.

[8] W.C. Rounds. Mapping and Grammars on trees. *Math. Syst. Theory* 4, p. 257-287.

Discriminability of infinite sets of terms in the D_∞ - models of the λ-calculus.

S. Ronchi della Rocca

Istituto di Scienza dell'Informazione - C.M. D'Azeglio, 42 - Torino.

1. Introduction

The first important result about the discriminability of terms is given by Böhm [2], who has shown that any two non congruent terms in β - η - normal form, M and N, can always be discriminated, i.e., there exists a context (in the sense of [13]), whose value can be interpreted as "true" when it is filled by M and "false" [(1)] when it is filled by N. An important consequence of this is that any proper extension of λ- calculus in which two different normal forms are equated is inconsistent (since we could immediately prove "true" is equal to "false"). In [1] and [6] each finite set $F = \{F_1, ..., F_n\}$ of β - η - normal forms has been proved be discriminable, i.e., there exists a context which maps F into a set of n distinguished variables. In [12] Hindley notes that the discrimination algorithm given in [6] holds also for the combinatory logic. Wadsworth [13] shows that any two terms, which have different values in Scott's D_∞- models of the λ-calculus, can be semi-discriminated (but they cannot, in general, be discriminated). I.e., there exists a context whose value can be interpreted as "true" or "undefined" according to it is filled by one or by the other of the two terms. From this result it follows that any system obtained from D_∞ by adjoining a new equality is inconsistent (i.e., D_∞ determines a maximal consistent extension of λ-calculus). In [8] necessary and sufficient conditions under which a set of n arbitrary terms different in D_∞ can be discriminated (or semi-discriminated) are given.

Moreover it is well known [3] that it does not exist a general discriminator of terms, i.e., given any term F, there is not a context $C_F[\ \ ]$ giving the value "true" or "false" according to it is filled by a term which is β-η-convertible to F or not. Then it is interesting to characterize discriminable sets of terms, and the extension of λ-calculus determined by D_∞ is, by the before consideration, the most interesting system for studying discriminability of terms.

(1) We can represent "true" and "false" by the terms $\mathbf{K} \equiv \lambda xy.x$ and $\mathbf{O} \equiv \lambda xy.y$.

There is no general characterization of infinite sets of terms which can be discriminated, although every numeral system is an example of a discriminable, infinite set of terms (not necessarily in normal form, as shown, by examples, in [1], chapter 6]).

In this paper the property of discriminability of an infinite set F of terms, different in D_∞-models of the λ-calculus, is studied. We consider two kinds of discriminability: first, a set $F = \{F_n \mid n \geq 0\}$ of different terms can be discriminated in respect of one of its elements, F_i, if there exists a term which, applied to F_i, gives the value "true", while it gives "false" if applied to each other element of F different from F_i. Second, we define F completely discriminable if there exists a term which, applied to two of its elements, gives the value "true" if these two elements are the same, "false" otherwise.

In this paper, some properties of the discriminable sets are proved. Moreover the discriminability of uniform sets is studied, where a set $F = \{F_n \mid n \geq 0\}$ is called uniform if there exists a term G such that $G\ \underline{n} \doteq F_n$ (2). The interest of the uniform sets, in respect of their discriminability, is given by the fact that all the numeral systems are uniform. We prove that it is not decidable if a uniform set is discriminable or not. Finally, some conditions on the discriminability of infinite sets of terms (not necessarily uniform) are given, which relate the discriminability of a set F to the existence of a discriminable set of approximants of the elements of F.

2. Discriminable sets and their properties.

Let $F = \{F_i \mid i \geq 0\}$ be a set of terms, $\doteq$ denote the α-β-η-convertibility, and $=_{D_\infty}$ the equality in D_∞. We will say that F is *proper* if, for $i \neq j$, $F_i \neq_{D_\infty} F_j$. We are interested to study the conditions under which a proper set F is discriminable.

It seems natural to define two different kinds of discriminability, in the following way.

Definition 1 - F is discriminable if there exists at least one element of F, say F_i, and a term δ_i such that for every $F_j \in F$ $(j \geq 0)$: $\delta_i F_j \doteq \begin{cases} \mathbf{K} & \text{if } i=j \\ \mathbf{O} & \text{otherwise.} \end{cases}$

(2) $\underline{n}$ is the term which represents the number n in the standard numeral system of Church [7], i.e., $\underline{n} \doteq \lambda xy.\underbrace{x(x(x \ldots (xy) \ldots)}_{n} \doteq \lambda xy.x^n y$.
($\doteq$ denotes the α-β-η-convertibility).

(we call δ_i *discriminator* for F, in respect of F_i).

Definition 2 - F is *completely discriminable* if there exists a term δ such that, for every F_i, $F_j \in F$ $(i, j \geqslant 0)$:

$$\delta\, F_i\, F_j \doteq \begin{cases} \mathbf{K} & \text{if } i = j \\ \mathbf{O} & \text{otherwise} \end{cases}$$

(we call δ a *complete discriminator* for F).

We notice that each completely discriminable set is discriminable, in respect of each of its elements. These two definitions of discriminability are not redundant: in fact in the example 1, at the end of this section, we exhibit a proper set which is discriminable but not completely discriminable.

It is well known [10] that a *numeral system* is a proper set of terms, $N = \{[\![n]\!] \mid n \geqslant 0\}$ such that there exist three terms $[\![\sigma]\!]$, $[\![\pi]\!]$ and $[\![\delta_0]\!]$ (respectively *successor, predecessor* and *discriminator* for zero) and the following relationships hold:

$$[\![\sigma]\!]\,[\![n]\!] \doteq [\![n+1]\!]$$

$$[\![\pi]\!]\,[\![n+1]\!] \doteq [\![n]\!]$$

$$[\![\delta_0]\!]\,[\![n]\!] \doteq \begin{cases} \mathbf{K} & \text{if } n = 0 \\ \mathbf{O} & \text{otherwise.} \end{cases}$$

Then in particular a numeral system must be discriminable in respect of the term $[\![0]\!]$. Really the discriminability of a numeral system is stronger; in fact the following property holds:

Property 1. A numeral system is always completely discriminable.

In fact if N is a numeral system we can build a complete discriminator δ for N in function of its predecessor $[\![\pi]\!]$ and discriminator for zero $[\![\delta_0]\!]$.

For reasons of clarity, here and in what follows, before showing the term F which λ-defines a function $\hat{F}$, we will give an informal description of $\hat{F}$ in a language Algol-like. Then:

$$\hat{\delta}\,[\![n]\!]\,[\![m]\!] = \textit{if } n = 0 \textit{ then}$$
$$\textit{if } m = 0 \textit{ then } \mathbf{K} \textit{ else } \mathbf{O}$$
$$\textit{else if } m = 0 \textit{ then } \mathbf{O}$$
$$\textit{else } \hat{\delta}\,([\![\pi]\!]\,[\![n]\!])\,([\![\pi]\!]\,[\![m]\!]). \qquad (n, m \geqslant 0).$$

and

$$\delta \doteq \mathbf{Y}\,(\lambda txy.\, [\![\delta_0]\!]\, x\, ([\![\delta_0]\!] y)\, ([\![\delta_0]\!] y\, \mathbf{O}\, (t([\![\pi]\!] x)\, ([\![\pi]\!] y))))\ ^{(3)}.$$ ■

We notice that the given (classical) definition of numeral system is not redundant, i.e. the following property holds:

Property 2. No one of the three terms $[\![\sigma]\!]$, $[\![\pi]\!]$ and $[\![\delta_0]\!]$ of a numeral system can be expressed as a function of the others.

In fact it is possible to exhibit some proper sets which have only two of these three terms, and then they cannot be numeral systems (see examples 2, 3 and 4 at the end of this section). ■

But it is possible to give an alternative definition of numeral system, based on the existence of the complete discriminator δ instead of the existence of the discriminator of zero.

Property 3. A proper set which is completely discriminable and has a successor, is a numeral system.

Let $F = \{F_i \mid i \geqslant 0\}$ be such a set and $\delta^{(F)}$ and $\sigma^{(F)}$ be its complete discriminator and successor (i.e., $\sigma^{(F)} F_i \doteq F_{i+1}$). Then $\delta^{(F)}$ implies the existence of $\delta_i^{(F)}$, for every $i \geqslant 0$, and a predecessor for F can be built in the following way:

$$\hat{\pi}^{(F)} F_n = \textit{if } n=0 \textit{ then } F_0 \textit{ else } \hat{H}\ F_n\ F_1\ F_0$$

where:

$$\hat{H}\ F_n\ F_m\ F_{m-1} = \textit{if } n=m \textit{ then } F_{m-1} \textit{ else } \hat{H}\ F_n\ F_{m+1}\ F_m$$

and $\pi^{(F)}$ can be the term:

$$\pi^{(F)} \doteq \lambda x.\, \delta_0^{(F)}\, x\, F_0 (H\, x\, (\sigma^{(F)} F_0)\, F_0)$$

where $$H \doteq \mathbf{Y}\,(\lambda tuvz.\, \delta^{(F)}\, u\, v\, z\, (t\, u\, (\sigma^{(F)} v)\, v)).$$

Then F is the numeral system: $[\![n]\!] \doteq F_n$. ■

Notice that the predecessor $\underline{\pi}$ of the standard numeral system of Church is built in this way.

The following (obvious) property will be useful in the next paragraph.

---

(3) $\mathbf{Y}$ is a term with the following reduction rule: $\mathbf{Y}z = z\,(\mathbf{Y}z)$, for all term z.

Property 4. A proper set F is completely discriminable iff there exists an injective, λ-definable map between F and G, where G is any proper and completely discriminable set.

It is well know [14] that, if G is a numeral system and the map is bijective, then F is a numeral system too.

Now, to show the promised examples, we need the following lemma:

Lemma 1. The proper set $F = \{\mathbf{K}^n \mathbf{I} \mid n \geqslant 0\}$ [4] is not (completely) discriminable.

Proof. See [1] and [14]. ∎

Example 1. The proper set $F = \{F_n \mid n \geqslant 0\}$, where:

$$F_0 \doteq \langle \mathbf{O}, \mathbf{I} \rangle^{(5)}$$

$$F_n \doteq \langle \underline{1}, \mathbf{K}^n \mathbf{I} \rangle \quad (n > 0)$$

is discriminable in respect of F_0: in fact, δ_0 can be built in the following way:

$$\hat{\delta}_0 \langle x_1, x_2 \rangle = \textit{if } x_1 = \mathbf{O} \textit{ then } \mathbf{K} \textit{ else } \mathbf{O}$$

and δ_0 is the term:

$$\delta_0 \doteq \lambda x . \underline{\delta_0} (x\, \mathbf{K})$$

By Lemma 1, F is not completely discriminable.

Example 2. (Jacopini) The proper set of the example 1 has a discriminator in respect of F_0 (given in the same example), a successor for it can be built in the following way:

$$\hat{\sigma} \langle x_1, \mathbf{K}^n \mathbf{I} \rangle = \textit{if } x_1 \doteq \underline{0} \textit{ then } \langle \underline{1}, \mathbf{KI} \rangle \textit{ else } \langle \underline{1}, \mathbf{K}(\mathbf{K}^n \mathbf{I}) \rangle$$

i.e.:

$$\sigma \doteq \lambda x . \underline{\delta_0}(x\mathbf{K}) \langle \underline{1}, \mathbf{KI} \rangle \langle \underline{1}, \mathbf{K}(x\mathbf{O}) \rangle .$$

But the existence of a predecessor for F would imply the existence of a discriminator for the set $\{\mathbf{K}^n \mathbf{I} \mid n \geqslant 0\}$, what is impossible, by Lemma 1.

Example 3. (Wadsworth) The proper set $F = \{\mathbf{K}^n \mathbf{I} \mid n \geqslant 0\}$ has a successor: $\sigma \doteq \mathbf{K}$

---

(4) $\mathbf{I} \doteq \lambda x.x$, $\mathbf{K}^n \doteq \lambda x_1 \ldots x_{n+1} . x_1$. Then $\mathbf{K}^n \mathbf{I} \doteq \lambda x_1 \ldots x_{n+1} . x_{n+1}$.

(5) We will represent the pair $\langle a, b \rangle$ by the term $\lambda x.xab$. Clearly $\langle a, b \rangle\, \mathbf{K} \doteq a$ and $\langle a, b \rangle\, \mathbf{O} \doteq b$.

and a predecessor: $\pi \doteq \lambda x.\, xA$ (where A is any term)

but, by Lemma 1, it does not have a discriminator.

Example 4. The proper set $F = \{F_n \mid n \geqslant 0\}$, where:

$$F_0 \doteq \langle \mathbf{O} \rangle^{(6)}$$

$$F_i \doteq \langle a_i, F_{i-1} \rangle \quad (i > 0)$$

and a_i $(i > 0)$ is a free variable, has as predecessor: $\pi \doteq \langle \mathbf{O} \rangle$

as discriminator in respect of F_0: $\delta_0 \doteq \langle U_3^3, U_3^3, \mathbf{K} \rangle^{(7)}$

but clearly it does not have a successor.

3. Uniform sets

Definition 4. [1] A proper set $F = \{F_i \mid i \geqslant 0\}$ is *uniform* iff there exists a term $G^{(F)}$ such that:

$$G^{(F)}\ \underline{n} \doteq F_n .$$

$G^{(F)}$ is a *generator* of F.

The interest to study the uniform sets is given by the fact that the numeral systems are uniform. In fact the following property holds (but not the inverse):

Property 5. A proper set with a successor is a uniform set.

In fact, let $F = \{F_i \mid i \geqslant 0\}$ be a proper set and let $\sigma^{(F)}$ be a successor for F, i.e, $\sigma^{(F)}\ F_i \doteq F_{i+1}$. Then a generator for F can be:

$$\hat{G}^{(F)}\ \underline{n} = \textit{if}\ n = 0\ \textit{then}\ F_0\ \textit{else}\ (\sigma^{(F)}\ (\hat{G}^{(F)}\ \underline{n-1}))$$

i.e., $G^{(F)}$ is the term:

$$G^{(F)} \doteq Y\,(\lambda t x.\, \underline{\delta_0}\ x\ F_0\ (\sigma^{(F)}\,(t\,(\underline{\pi}\, x)))). \blacksquare$$

If we limit ourselves to consider the uniform sets, the previous definition of a numeral system becomes redundant, for what follows.

---

(6) $\langle a_1, ..., a_n \rangle$ is the term $\lambda x.x a_1 \ldots a_n$ $(n > 0)$.
(7) U_i^n is the term $\lambda x_1 \ldots x_n . x_i$ $(i \leq n)$.

Property 6. A uniform set with a predecessor and a discriminator for zero has also a successor (i.e., it is a numeral system).

Let $F = \{F_i \mid i \geqslant 0\}$ be a set generated by $G^{(F)}$, and let $\pi^{(F)}$ and $\delta_0{}^{(F)}$ its predecessor and its discriminator in respect of F_0. A successor for F then can be:

$$\hat{G}^{(F)} = \hat{Q}\,\underline{1}\text{, where } \hat{Q}\,x\,F_n = \quad \textit{if}\; n=0 \;\textit{then}\; (G^{(F)}x) \;\textit{else}\; (\,\hat{Q}\;\; (x+1)\,F_{n-1})$$

i.e., $\sigma^{(F)}$ is the term:

$$\sigma^{(F)} \doteq \mathbf{Y}\,(\lambda txy.\underline{\delta_0}{}^{(F)}\; y\; (G^{(F)}x)\; (t\; (\underline{\sigma}\; x)\; (\pi^{(F)}\; y)))\; \underline{1}.$$

Then F is the numeral system $[\![n]\!] \doteq G^{(F)}\underline{n}$. ⊠

Property 7. If a uniform set $F = \{G^{(F)}\underline{n} \mid n \geqslant 0\}$ is discriminable in respect of each of its elements by a uniform set of discriminators, i.e., if there exists a term D such that:

$$D\underline{n}\; (G^{(F)}\; \underline{m}) \doteq \begin{cases} \mathbf{K} & \text{if } m=n \\ \mathbf{O} & \text{otherwise} \end{cases}$$

then F is a numeral system.

In fact, a predecessor for F can be built in the following way:

$$\hat{\pi}^{(F)}\; F_n = \textit{if}\; n=0 \;\textit{then}\;\; G^{(F)}\; \underline{0} \;\textit{else}\; (\hat{N}\; 0\; 1\; F_n)$$

where:

$$\hat{N}\; p\; q\; F_n = \textit{if}\;\; n=q \;\textit{then}\;\; p \;\;\textit{else}\;\; \hat{N}\; q\; (q+1)\; F_n$$

i.e., $\pi^{(F)}$ is the term:

$$\pi^{(F)} \doteq \lambda x.\; D\; \underline{0}\; x\; (G^{(F)}\; \underline{0})\; (N\; \underline{0}\; \underline{1}\; x)$$

where $N \doteq \mathbf{Y}\; (\lambda txyz.Dyzx\; (ty\; (\underline{\sigma}y)\; z))$.

For the property 6, F has a successor too, then it is the numeral $[\![n]\!] \doteq G^{(F)}\; \underline{n}$. ⊠

Property 8. If the generator G of a set F has a left inverse, then F is a numeral system.

If G^{-1} is a left inverse of G, then, by definition:

$$\mathbf{B}G^{-1}\; G \doteq \lambda z.G^{-1}\;\; (G\; z) \doteq \mathbf{I}$$ [8]

---

[8] $\mathbf{B} \doteq \lambda xyz.x(yz)$.

by applying to $\underline{n}$:

$$G^{-1}\ (G\underline{n}) \doteq \underline{n}.$$

Then there exists a bijective map between F and the standard numbers, and so F is a numeral system, by property 4. ∎

In [5] a characterization of terms possessing a left inverse in the λ-β-calculus is given; for the λ-β-η-calculus, only sufficient conditions are known (see [11]).

In the definition of a uniform set, we supposed that such a set is proper: unfortunately this property is **not** decidable [9] and moreover, even if we assume this to be known from an oracle, it is **not** decidable if a uniform set is discriminable or not. In order to prove this, let us introduce the following construction. It is well known that, given a finite alphabet $S = \{a_1, a_2, \ldots, a_p\}$ it is possible, in many ways, to construct a one-one correspondence between the set of the strings built on S and the set N^+ of positive integers [4]. For example, let $a_{i_1}\ a_{i_2} \ldots a_{i_t}$ be a string built on S; we can associate to it the integer:

$$\overline{a_{i_1} \ldots a_{i_t}} = (i_t + i_{t-1} \cdot p + i_{t-2} \cdot p^2 + \ldots + i_1 \cdot p^{t-1}).$$

where $p = |S|$[10]. This correspondence (which will be denoted by C_p) arranges the strings in lexicographical order. Let consider now the sequence of pairs $\Lambda = <\alpha_1, \beta_1>, \ldots, <\alpha_m, \beta_m>$ where α_i and β_i $(1 \leqslant i \leqslant m)$ are strings built on the alphabet $\{1, 2\}$, and let us construct:

$$A = \{\alpha_{i_1} \ldots \alpha_{i_r} \mid r \geqslant 1,\ 1 \leqslant i_j \leqslant m\}$$

$$B = \{\beta_{i_1} \ldots \beta_{i_r} \mid r \geqslant 1,\ 1 \leqslant i_j \leqslant m\}$$

where $\alpha_{i_1} \ldots \alpha_{i_r}$ $(\beta_{i_1} \ldots \beta_{i_r})$ is the string obtained by concatenating $\alpha_{i_1}, \ldots, \alpha_{i_r}$ $(\beta_{i_1}, \ldots, \beta_{i_r})$. Let $\overline{\alpha_{i_1} \ldots \alpha_{i_r}}$ $(\overline{\beta_{i_1} \ldots \beta_{i_r}})$ be the number corresponding to $\alpha_{i_1} \ldots \alpha_{i_r}$ $(\beta_{i_1} \ldots \beta_{i_r})$ in the correspondence C_2, and $\overline{i_1 \ldots i_r}$ be the number corresponding to $i_1 \ldots i_r$ in the correspondence C_m. Then we associate to the string $\alpha_{i_1} \ldots \alpha_{i_r}$ $(\beta_{i_1} \ldots \beta_{i_r})$ the pair:

$$< \overline{\alpha_{i_1} \ldots \alpha_{i_r}}\,,\ \overline{i_1 \ldots i_r} > \quad (< \overline{\beta_{i_1} \ldots \beta_{i_r}}\,,\ \overline{i_1 \ldots i_r} >).$$

---

(9) A property is **not** decidable if the set of objects which have such a property is not recursive.

(10) $|A|$ denotes the cardinality of the set A.

If we call $\alpha^{(n)}$ ($\beta^{(n)}$) the pair $<\overline{\alpha_{i_1} \dots \alpha_{i_r}}, n>$ ($<\overline{\beta_{i_1} \dots \beta_{i_r}}, n>$), we define:

$$A' = \{\alpha^{(n)} \mid n > 0\}$$

$$B' = \{\beta^{(n)} \mid n > 0\}.$$

Fact. $\exists n\, (\alpha^{(n)} \equiv \beta^{(n)})?$ [11] is not decidable.

In fact $\exists n\, (\alpha^{(n)} \equiv \beta^{(n)})?$, if $n = \overline{i_1 \dots i_r}$ means: $(\alpha_{i_1} \dots \alpha_{i_r} \equiv \beta_{i_1} \dots \beta_{i_r})?$ that is the not decidable problem of Post's correspondences.

Theorem 1. For a set built by a generator the property of being proper is not decidable.

Proof. Consider the function on the integers:

$$\hat{G}^A = \lambda x.\, \mathit{if}\ \alpha^{(x+1)} \not\equiv \beta^{(x+1)}\ \mathit{then}\ \underline{(x+1)}\ \mathit{else}\ 0$$

Clearly $\hat{G}^A$ is λ-definable, since all the construction given below can be λ-defined, using the standard numerals, for example, and the predicate $\alpha^{(y)} \equiv \beta^{(y)}$ is simply:

$$\underline{\delta}\ (\underline{\alpha^{(y)}}\ \mathbf{K})\ (\underline{\beta^{(y)}}\ \mathbf{K})$$

where $\underline{\alpha^{(y)}}$ ($\underline{\beta^{(y)}}$) is the pair of standard numerals corresponding to $\alpha^{(y)}$ ($\beta^{(y)}$). Let G^A be the λ-term which λ-defines $\hat{G}^A$: G^A is a generator of a set F, and two cases are possible:

1) if $\alpha^{(n)} \not\equiv \beta^{(n)}$ for every n, $F = \{\underline{n} \mid n > 0\}$, and then F is proper.

2) if, for some n_i, $\alpha^{(n_i)} \equiv \beta^{(n_i)}$, $F = \{F_n \mid n \geq 0\}$, where:

$$F_n \doteq \begin{cases} \underline{n+1} & \text{if } \alpha^{(n+1)} \not\equiv \beta^{(n+1)} \\ \underline{0} & \text{otherwise.} \end{cases}$$

Then F is not proper, since there exists an infinite number of index i such that $F_i \doteq \underline{0}$.

But, by the fact, it is not decidable if the generated set F belongs to the case 1) or 2), and then it is not decidable if it is proper. ∎

From now on, let us assume the existence of an oracle, which tell us if a set is or not proper.

---

(11) $\equiv$ here denotes the identity of strings.

Lemma 2 - If a proper set F is a union of two sets F_1 and F_2, and if F_1 or F_2 is not (completely) discriminable, then F is not (completely) discriminable.

Proof. Obvious. ■

Theorem 2. For a uniform set (we suppose that an oracle stated that such a set is proper) the property of being (completely) discriminable is **not** decidable.

Proof. Similarly as in the proof of theorem 1, we can built a function on the integers:

$$\hat{G}^A = \lambda x.\ \textit{if}\ \alpha^{(x+1)} \equiv \beta^{(x+1)}\ \textit{then}\ (K^{x+1}\ I)\ \textit{else}\ (x+1).$$

The term G^A which λ-define $\hat{G}^A$ is a generator of a set F, and two cases are possible:

1) if $\alpha^{(n)} \not\equiv \beta^{(n)}$, for every n, $F = \{\underline{n} \mid n > 0\}$ and then F is discriminable.

2) if, for some n_i, $\alpha^{(n_i)} \equiv \beta^{(n_i)}$, $F = F_1 \cup F_2$ where $F_1 \doteq \{K^m I \mid m \in M\}$ where M is the set $\{n \mid \alpha^{(n)} \equiv \beta^{(n)}, n > 0\}$ and $F_2 = \{\underline{n} \mid n \notin M\}$; but F_1, by lemma 1, is not (completely) discriminable and then, by Lemma 2, F is not (completely) discriminable too.

But, also in this case, as in Theorem 1, it is **not** decidable if F belongs to case 1) or 2), and then it is **not** decidable if it is (completely) discriminable. ■

4. Some conditions on the discriminability of proper sets of terms

In what follows, we will give some conditions for discriminability of proper sets. First we need the notion of *approximants.* Following [13] a term N is said to be a *direct approximant* of a term M, if N and M are identical (modulo Ω-reductions) excepts at components which are occurrences of Ω in N and moreover N is in β-normal form. The set $\mathcal{A}$(M) of *approximate normal forms* of M is then defined by:

$\mathcal{A}(M) = \{N \mid \exists\ M'$ such that M β-reduces to M' and N is a direct approximant of M'$\}$.

It is useful to extend to sets of terms the notion of approximants.

Definition 5. Let $F = \{F_i \mid i \geq 0\}$ be a set of terms, then $\mathcal{A}(F) = \{\ \{F_i' \mid i \geq 0\} \mid F_i' \in \mathcal{A}(F_i)\}$.

The next theorem relates the discriminability of a proper set F to the discriminability of the elements of $\mathcal{A}(F)$: in [8] this theorem is proved for finite sets of terms. First we need the following lemma, proved in [13].

Lemma 3. For all terms N and contexts C [ ], C [N] has an Ω - free normal form iff C [N'] has the same Ω -free normal form for some $N' \in \mathcal{A}(N)$.

Theorem 3. A proper set F is discriminable by a discriminator δ iff there exists $F' \in \mathcal{A}(F)$ such that F' is discriminable by δ itself.

Proof. Only if part. Let $F = \{F_n \mid n \geqslant 0\}$, and let exist a discriminator δ_i for F. For definition, δ_i is such that:

$$\delta_i F_j \doteq \begin{cases} K & \text{if } i=j \\ O & \text{otherwise} \qquad (i, j \geqslant 0). \end{cases}$$

Since O and K are Ω -free normal forms, by Lemma 3 there exists $F_j' \in \mathcal{A}(F_j)$ such that

$$\delta_i F_j' \doteq \begin{cases} K & \text{if } i=j \\ O & \text{otherwise}, \end{cases}$$

Then the set $F' = \{F_n' \mid n \geqslant 0\}$ is discriminated by δ_i. The *if part* is proved by a similar argument using Lemma 3 in the other direction. ∎

Moreover, we are able to give a sufficient condition for the discriminability of a set F, using a result proved in [8] for finite sets of terms. First, we need some definitions and lemmas.

It is well known [1] that, given a term F, its Böhm-tree B.T. (F) is a labelled oriented tree defined in the following way:

- if F is unsolvable, then B.T. (F) is a single root labelled Ω
- if $F \doteq \lambda x_1 \ldots x_n . \mathfrak{z} F_1 \ldots F_m$, then B.T. (F) is:

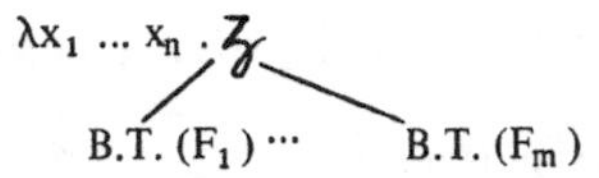

It is easy to see that an approximant has always a finite B.T.. B.T.s will be always considered modulo α-reduction. A *path* γ on a tree is a finite (possibly empty) list of integers $< i_1, \ldots, i_k >$. We say that γ *exists* in an ordered tree T iff there exists in T a sequence of nodes whose first node is the root of T and whose j-th node $(1 < j \leqslant k+1)$ is the i_{j-1}-th son of the (j - 1)-th node. The last node of this sequence is the *end-node*

of γ while k is the *lenght* of γ. Any path $\gamma' = < i_1, ..., i_h >$ with $h \leq k$ is a *subpath* of γ. We will say that two nodes in two trees are *correspondent* iff they are endnodes of the same path.

Definition 6. Given a term F and a path γ, γ is *defined* in F iff there exists F' such that $F' \doteq F$, γ exists in B.T. (F') and its endnode is not labelled Ω.

A path γ is undefined in F iff in B.T. (F) (and in B.T. (F') forall $F' \doteq F$) the endnode of some proper subpath of γ is labelled Ω.

Definition 7. Two terms F and H are *γ-equivalent* ($F \sim_\gamma H$) iff either γ is undefined in both F and H or there exists F', H' such that $F \doteq F'$, $H \doteq H'$ and all corrispondent nodes of γ in B.T. (F') and B.T. (H') have the same label and son number.

Definition 8. γ is a path *useful* for a set *F* iff:

i) γ is defined in all the element of *F*

ii) if γ is different from the empty path, i.e., $\gamma = \gamma' \cap < w >$ [12] then for all F, H $\in$ *F*, $F \sim_{\gamma'} H$

iii) there are F, H $\in$ *F* such that $F \not\sim_\gamma H$.

For an useful path γ we can define, in the usual way, the quotient set $F/{\sim_\gamma}$ of *F* with respect to $\sim_\gamma$ as the set of the equivalence classes of *F* determined by γ-equivalence.

Definition 9. A proper, finite set *F* is *distinct* iff either *F* contains only one term or there exists a path γ useful for *F* such that all the equivalence classes of $F/{\sim_\gamma}$ are distinct.

In [8] it is proved the following:

Lemma 4. A finite set of terms is discriminable in respect of each of its elements iff it is distinct.

Moreover, in [8], an algorithm is given to discriminate a finite, proper and distinct set *F* of terms, which uses the partition of *F* into equivalence classes induced by γ-equivalence.

---

(12) w is an integer and $\cap$ denotes the operation of concatenation of lists.

Definition 10. Given a proper set $\mathcal{F}$ and a term $F \in \mathcal{F}$, a set $\mathcal{F}' \in \mathcal{A}(\mathcal{F})$ is *fit* for F iff, for every $G \in \mathcal{F}$ such that $G \neq_{D_\infty} F$, there is $F' \in \mathcal{A}(F)$ and $G' \in \mathcal{A}(G)$ such that $F' \neq_{D_\infty} G'$ and $F', G' \in \mathcal{F}'$.

Theorem 4. A proper set $\mathcal{F}$ is discriminable in respect of $F \in \mathcal{F}$, if there exists a finite and distinct set $\mathcal{F}' \in \mathcal{A}(\mathcal{F})$ which is fit for F.

Proof. Since $\mathcal{F}'$ is finite and it is fit for F, it must be of the shape: $\{F', G_1', \ldots, G_p'\}$, i.e., for every $G \in \mathcal{F}$ such that $G \neq_{D_\infty} F$, there is an integer m $(1 \leqslant m \leqslant p)$ such that $G_m \in \mathcal{A}(G)$. By Lemma 4. $\mathcal{F}'$ is discriminable in respect of each of its elements, then in particular it is discriminable in respect of F'. I.e., there is a term δ such that:

$$\delta\, G_m' \doteq \begin{cases} \mathbf{K} & \text{if } G_m' =_{\beta} F' \\ \mathbf{O} & \text{otherwise} \qquad (1 \leqslant m \leqslant p) \end{cases}$$

but, by Lemma 3, δ is also a discriminator for $\mathcal{F}$, in respect of F. ∎

Let consider, as example, the set N of standard number. The set:

$$N' = \{\underline{0}, \lambda xy.x\Omega\}$$

belongs to $\mathcal{A}(N)$, and it is fit for $\underline{0} \doteq \lambda xy.y$.

A discriminator for N' in respect of $\underline{0}$ is the term $\underline{\delta_0} : \langle U_3^3, \mathbf{K} \rangle$, and it is easy to verify that $\underline{\delta_0}$ is a discriminator, in respect of $\underline{0}$, for N too. [13]

Note that the condition for discriminability of a set given in Theorem 4 is sufficient but not necessary: in fact we are able to show a proper set completely discriminable for which such a condition does not hold.

Let define:

$$\mathcal{L}_n = \{\lambda x_1 \ldots x_n \,.\, L \mid L \text{ is a string built on the alphabet } \{x_1, \ldots, x_n\}\}.$$

Obviously $\mathcal{L}_n$ is a proper set.

Let $\mathcal{F}_n$ be an infinite sub-set of $\mathcal{L}_n$ and let $F \doteq \lambda x_1 \ldots x_n \,.\, x_{i_1} \ldots x_{i_m} \in \mathcal{F}_n$; the only subset of $\mathcal{A}(\mathcal{F}_n)$ which is fit for F is $\mathcal{F}_n$ it-self, and then the condition of Theorem 4 is not satisfied, since $\mathcal{F}_n$ is not distinct. Nevertheless, the following theorem holds:

---

(13) In [3] the truth values are $T \doteq \mathbf{I}$ and $F \doteq \mathbf{O}$: then a discriminator of N in respect of $\underline{0}$ is $\langle \mathbf{O} \rangle$.

Theorem 5. For every n, each proper subset of L_n is (completely) discriminable.

Proof. Let $F_n = \{F_i \mid i \geqslant 0\}$ be a subset of L_n; clearly F_i $(i \geqslant 0)$ is of the shape:

$$\lambda x_1 \ldots x_n \,.\, x_{t_1} \ldots x_{t_m}$$

where $t_1 \ \ldots \ t_m$ is a string built on the alphabet $\{1, 2, \ldots, n\}$. Let $\underline{p_1}, \underline{p_2}, \ldots \underline{p_n}$ be the first n prime numbers in the standard numeral system, and let us apply each element of F_n to $\underline{p_1}, \underline{p_2} \ldots, \underline{p_n}$: we obtain a set $F'_n = \{F'_i \mid i \geqslant 0\}$, where:

$$F'_i \doteq \underline{F_i}\ \underline{p_1}\ \underline{p_2} \cdots \underline{p_n}.$$

We will prove that F'_n is a proper subset of the set of standard numbers. In fact, the standard numbers have the property that $\underline{p}\,\underline{q} \doteq \underline{q^p}$, and then,

$$F'_j \doteq \underline{p_{t_1}} \cdots \underline{p_{t_m}} \doteq \underline{p_{t_m}^{p_{t_{m-1}}^{\cdot^{\cdot^{p_{t_1}}}}}}$$

But, for F_h $(h \neq j)$, we have:

$$F'_h \doteq \underline{p_{r_1}} \cdots \underline{p_{r_{m'}}} \doteq \underline{p_{r_m}^{p_{r_{m'-1}}^{\cdot^{\cdot^{p_{r_1}}}}}}$$

where $r_1 \ \ldots \ r_{m'} \not\equiv t_1 \ \ldots \ t_m$ (since F_n is proper); then, by the unicity of the factorizaction of an integer, $F'_h \neq F'_j$. We are built in this way an injective mapping between F_n and the standard numerals: then, by property 4, F_n is (completely) discriminable. ⊠

ACKNOWLEDGMENT

The author thanks Prof. M. Coppo for his useful suggestions, and the referee for his sharp and valuable comments.

REFERENCES

[1] H. Barendregt, The λ-calculus, its syntax and semantic, North-Holland, Amsterdam, to appear.

[2] C. Böhm, Alcune proprietà delle forme β-η-normali del λ-k-calcolo, I.A.C., 696, Roma, (1968).

[3] C. Böhm, The CUCH as a Formal and Description Language, in: Formal Language Description Languages, T.B. Steel, ed. North-Holland, Amsterdam (1966), 179-197.

[4] C. Böhm, Strutture Informative e loro trasformazioni, Atti del III Seminario sul trattamento automatico delle Informazioni, Ist. Matematico "U. Dini", Firenze, (1972).

[5] C. Böhm, M. Dezani-Ciancaglini, Combinatori-al Problems, Combinator equations and Normal Forms, Lecture Notes in Computer Science, 14, (1974), 170-184.

[6] C. Böhm, M. Dezani-Ciancaglini, P. Peretti, S. Ronchi Della Rocca, A discrimination Algorithm inside λ-η-calculus, Theoretical Computer Science, 8, (1979), 271-291.

[7] A. Church, The calculi of λ-conversion, Annals of Mathematics, 6, Princenton, New Jersey, (1941).

[8] M. Coppo, M. Dezani-Ciancaglini, S. Ronchi Della Rocca, (Semi)-separability of finite sets of terms in Scott's D_∞-models of the λ-calculus, Lecture notes in Computer Science, 62, (1978), 142-164.

[9] H.B. Curry, R. Feys, W. Craig, Combinatory logic, vol. I, North Holland, Amsterdam, (1968).

[10] H.B. Curry, J.R. Hindley, S.P. Seldin, Combinatory logic, vol. II, North Holland, Amsterdam, (1972).

[11] M. Dezani, Characterization of normal forms possessing inverse in the λ-β-η-calculus, Theoretical Computer Science, 2, (1976), 323-337.

[12] R. Hindley, The discrimination Theorem holds for combinatory weak reduction (Note), Theoretical Computer Science, 8, 3, (1979), 393-394.

[13] C.P. Wadsworth, The Relation between Computational and Denotational Properties for Scott's D_∞-models of the Lambda-calculus, SIAM Journal Comput., 5,3, (1976), 488-521.

[14] C.P. Wadsworth, Some unusual λ-calculus numeral Systems, to H.B. Curry: Essays on Combinatory Logic, Lambda Calculus and Formalism, ed. J.P. Seldin and J.R. Hindley, (1980).

Vol. 77: G. V. Bochmann, Architecture of Distributed Computer Systems. VIII, 238 pages. 1979.

Vol. 78: M. Gordon, R. Milner and C. Wadsworth, Edinburgh LCF. VIII, 159 pages. 1979.

Vol. 79: Language Design and Programming Methodology. Proceedings, 1979. Edited by J. Tobias. IX, 255 pages. 1980.

Vol. 80: Pictorial Information Systems. Edited by S. K. Chang and K. S. Fu. IX, 445 pages. 1980.

Vol. 81: Data Base Techniques for Pictorial Applications. Proceedings, 1979. Edited by A. Blaser. XI, 599 pages. 1980.

Vol. 82: J. G. Sanderson, A Relational Theory of Computing. VI, 147 pages. 1980.

Vol. 83: International Symposium Programming. Proceedings, 1980. Edited by B. Robinet. VII, 341 pages. 1980.

Vol. 84: Net Theory and Applications. Proceedings, 1979. Edited by W. Brauer. XIII, 537 Seiten. 1980.

Vol. 85: Automata, Languages and Programming. Proceedings, 1980. Edited by J. de Bakker and J. van Leeuwen. VIII, 671 pages. 1980.

Vol. 86: Abstract Software Specifications. Proceedings, 1979. Edited by D. Bjørner. XIII, 567 pages. 1980

Vol. 87: 5th Conference on Automated Deduction. Proceedings, 1980. Edited by W. Bibel and R. Kowalski. VII, 385 pages. 1980.

Vol. 88: Mathematical Foundations of Computer Science 1980. Proceedings, 1980. Edited by P. Dembiński. VIII, 723 pages. 1980.

Vol. 89: Computer Aided Design - Modelling, Systems Engineering, CAD-Systems. Proceedings, 1980. Edited by J. Encarnacao. XIV, 461 pages. 1980.

Vol. 90: D. M. Sandford, Using Sophisticated Models in Resolution Theorem Proving.
XI, 239 pages. 1980

Vol. 91: D. Wood, Grammar and L Forms: An Introduction. IX, 314 pages. 1980.

Vol. 92: R. Milner, A Calculus of Communication Systems. VI, 171 pages. 1980.

Vol. 93: A. Nijholt, Context-Free Grammars: Covers, Normal Forms, and Parsing. VII, 253 pages. 1980.

Vol. 94: Semantics-Directed Compiler Generation. Proceedings, 1980. Edited by N. D. Jones. V, 489 pages. 1980.

Vol. 95: Ch. D. Marlin, Coroutines. XII, 246 pages. 1980.

Vol. 96: J. L. Peterson, Computer Programs for Spelling Correction: VI, 213 pages. 1980.

Vol. 97: S. Osaki and T. Nishio, Reliability Evaluation of Some Fault-Tolerant Computer Architectures. VI, 129 pages. 1980.

Vol. 98: Towards a Formal Description of Ada. Edited by D. Bjørner and O. N. Oest. XIV, 630 pages. 1980.

Vol. 99: I. Guessarian, Algebraic Semantics. XI, 158 pages. 1981.

Vol. 100: Graphtheoretic Concepts in Computer Science. Edited by H. Noltemeier. X, 403 pages. 1981.

Vol. 101: A. Thayse, Boolean Calculus of Differences. VII, 144 pages. 1981.

Vol. 102: J. H. Davenport, On the Integration of Algebraic Functions. 1-197 pages. 1981.

Vol. 103: H. Ledgard, A. Singer, J. Whiteside, Directions in Human Factors of Interactive Systems. VI, 190 pages. 1981.

Vol. 104: Theoretical Computer Science. Ed. by P. Deussen. VII, 261 pages. 1981.

Vol. 105: B. W. Lampson, M. Paul, H. J. Siegert, Distributed Systems – Architecture and Implementation. XIII, 510 pages. 1981.

Vol. 106: The Programming Language Ada. Reference Manual. X, 243 pages. 1981.

Vol. 107: International Colloquium on Formalization of Programming Concepts. Proceedings. Edited by J. Diaz and I. Ramos. VII, 478 pages. 1981.

Vol. 108: Graph Theory and Algorithms. Edited by N. Saito and T. Nishizeki. VI, 216 pages. 1981.

Vol. 109: Digital Image Processing Systems. Edited by L. Bolc and Zenon Kulpa. V, 353 pages. 1981.

Vol. 110: W. Dehning, H. Essig, S. Maass, The Adaptation of Virtual Man-Computer Interfaces to User Requirements in Dialogs. X, 142 pages. 1981.

Vol. 111: CONPAR 81. Edited by W. Händler. XI, 508 pages. 1981.

Vol. 112: CAAP '81. Proceedings. Edited by G. Astesiano and C. Böhm. VI, 364 pages. 1981.